BASIC ENGINEERING MECHANICS

Other titles of interest to engineers

An Introduction to Engineering Fluid Mechanics
J. A. Fox

Principles of Engineering Thermodynamics
E. M. Goodger

Analysis and Presentation of Experimental Results
R. H. Leaver and T. R. Thomas

A Background to Engineering Design
Peter Polak

Mechanics of Machines
G. H. Ryder and M. D. Bennett

Essential Solid Mechanics – Theory, worked examples and problems
B. W. Young

BASIC ENGINEERING MECHANICS

J. H. HUGHES

K. F. MARTIN

Department of Mechanical Engineering and Engineering Production, UWIST

M

First edition 1977
Reprinted 1978

Published by
THE MACMILLAN PRESS LTD
London and Basingstoke
Associated companies in Delhi Dublin
Hong Kong Johannesburg Lagos Melbourne
New York Singapore and Tokyo

ISBN 0 333 17721 5

Printed in Hong Kong

Contents

Preface

The aim of this text is to present a systematic development of elementary applied mechanics assuming a mathematical background of elementary algebra, geometry and calculus. A knowledge of the principles of Newtonian mechanics is fundamental to the solution of many engineering problems and is a prerequisite for the study of more advanced texts; it is hoped that the treatment here will provide sufficient material on which this knowledge can be based.

Our experience leads us to believe that a return to basic principles is often desirable, and is particularly necessary for new engineering undergraduates who, in spite of having been exposed to previous instruction, frequently display a lack of familiarity with the meaning and significance of certain basic ideas. It is to this category that the book is mainly addressed, namely first-year students studying for an engineering degree or equivalent, who have some acquaintance with the subject matter but who require a more thorough grounding. The book is designed to cover the applied-mechanics content of the first year of an engineering degree course but will, in many cases, also provide a significant contribution to the second-year syllabus in mechanics of machines.

The traditional subdivision into statics and dynamics has been retained in order that familiarity may initially be gained with forces and their manipulation. We do not wish to appear pedantic in this matter since we recognise that there may be an equally good case for placing the initial emphasis on the concept of mass. Unfortunately it is difficult at this elementary level to treat one without the other, and a choice has to be made.

Having expressed Newton's laws in terms of particle behaviour the development of both statics and dynamics proceeds by way of particle systems to rigid bodies and systems of rigid bodies, incorporating in some cases a limited number of elastic elements.

In order to keep the discussion within reasonable bounds certain topics have been purposely omitted in the knowledge that detailed treatments are available elsewhere. For example, a more comprehensive study of the methods of structural analysis is better left to those books devoted to this topic. More important is the limitation of the discussion in the main to topics in two dimensions. This is in accord with the elementary nature of the book, but the decision was also based on the belief that the extension to three dimensions is facilitated by a thorough grounding in the basic principles that we hope this book will provide.

The inertia-force method of solution is given prominence for dealing with problems involving acceleration. Objection is raised by many, whose views we respect, on the grounds that the method is of historical interest only, and that it is used to assist in solving problems whose solution using dynamical principles requires no such assistance. We do not think these objections are completely valid; we maintain that the method has its merits and does nothing to relieve the student of the necessity of fully understanding the dynamics of the problem.

Our justification for writing yet another book on applied mechanics is that we felt there was a gap between those of very elementary nature and the more comprehensive texts that give rather more coverage than we considered necessary at this level, and that this gap could be filled by a book based on elementary mathematical knowledge but in which the argument was carefully developed. Also, by imposing the limitations already mentioned, we hope that the book will be within the financial reach of the student we wish to address. With these considerations in mind we claim no originality for the subject matter, the principles of which have long been recognised. However, we hope that the presentation, including the worked examples, chapter summaries and hints to problems, will be found of direct assistance to the student.

The book is separated into thirteen chapters with sections and subsections identified by a decimal notation; for example, 2.3.4 refers to subsection 2.3.4 in chapter 2. Equations relating to the text also carry a decimal notation and are numbered consecutively through each chapter; for example, the fifth equation in chapter 2 is numbered 2.5. Figure numbers again use the decimal notation and are numbered consecutively through each chapter including those relating to worked examples and problems. Answers to problems are assembled at the back of the book in the hope that students will exercise some restraint before consulting these.

We wish to acknowledge the encouragement of several colleagues in making comments on certain chapters, also the unstinting help of our typists, in particular Mrs Wilma Scott.

1 Introduction

Theoretical mechanics is that aspect of applied mathematics that is relevant to the physical world and in particular to the interactions and motion of matter. The study has regard to observable phenomena and is developed from basic concepts, definitions and postulates relating to those phenomena, the development of the theory being essentially mathematical. In the study of *applied mechanics* or *engineering mechanics* we emphasise the application of the theory of mechanics to solutions of practical problems and to predictions of the behaviour of mechanical systems. In setting up the relevant theory we endeavour to follow a logical development similar to that of theoretical mechanics.

1.1 Concept: Postulate: Law

The physical *concepts* we have referred to and on which our study will be based are those measurable attributes of the material world that are simply recognised in experience and that cannot be further described or explained in terms of simpler concepts. For our present purpose they are three in number, namely extension in space, duration in time and action on matter, giving rise to corresponding *entities* or *dimensions* that can be observed and measured, namely *length, time* and *force.* On seeking relations involving these entities, other entities derived from them are immediately encountered or are found to be required in our descriptions of the behaviour of mechanical systems. These are called *derived entities* or *derived dimensions*, some examples of which are area and volume, velocity and acceleration, work and energy. Others are specially formulated to assist and simplify our descriptions. A *definition* involves the recognition or formulation of such new entities, their naming and their subsequent identification. These definitions will be introduced as they are required. It will be found that careful use of names in the sense in which they have been allocated in the definitions will enable many pitfalls to be avoided.

A *postulate* is a statement setting out, in as precise a form as possible, a relation between entities that it is thought corresponds to the behaviour of the physical world. Such postulates, which are in effect plausible assertions, form the starting point for the development of the theory, by way of deductions and predictions from those postulates.

A *law* is a statement of this kind that has been found to correspond closely with experimental observations and is justifiably accepted as being valid in the light of past and present experience. Subsequent deductions and predictions are then soundly based. The theory that we shall develop in this book is based primarily on Newton's three laws of motion, which have been found to be unassailable as far as engineering applications are concerned.

At certain points important statements are made that have been derived from the basic laws and are in a sense one stage removed from them. These merit particular attention since once they have been established the results they embody can be used to advantage without reference to the laws from which they were derived. We shall refer to a statement of this kind as a *principle*.

In applying the deductions we make to actual problems of engineering interest, it will soon become apparent that the physical objects we have to deal with do not match the assumptions made in the theory. Forces do not act on bodies at geometrical points; bodies do not make contact at points, along lines of contact or by exact matching of surfaces; surfaces are never perfectly smooth. Furthermore, rarely can all the physical factors entering into a problem be dealt with simultaneously. It is therefore necessary to idealise the problem by first simplifying the geometrical form of the objects concerned and then introducing only those physical factors that are of immediate importance. Part of the discipline of problem-solving is developed in the simplifying of the problem, its statement in mechanics forms and in recognising the significance of the simplifying assumptions made. The degree to which the problem is simplified will necessitate realistic decisions on the accuracy to which answers to problems should be stated.

1.2 Dimensions and Units

The word *dimension* is used when referring to the nature of the physical entities that are encountered, either directly or as a consequence of definition. If so-called *primary dimensions* are chosen, such as length, time and force, the dimensions of other entities are derived from these. Thus we say that velocity v has the dimensions length/time, and this is written symbolically in the form $[v] \equiv L/T$, which is read as 'the dimensions of velocity are those of length divided by time'.

The magnitude of a given quantity of some physical entity requires for its specification a *unit* and a *measure*. The unit is that amount of the same kind of entity that by common usage or legal sanction is taken as a standard reference. The measure is a number that expresses the amount of the entity as compared with that of the unit. For example, if the unit of area is the square metre (m^2) a given area may have magnitude 5 square metres (5 m^2), where the number 5 is the measure. The units that we adopt will be noted as they are required.

It follows that equations involving physical quantities are more than relations between symbols representing numbers only. Each symbol now represents the product of a measure and a unit and it further follows that if a physical equation is to have meaning each term must have the same dimensions and also the same units. The student should develop the habit of continually checking both algebraic and numerical equations to ensure correctness of dimensions and units. In itself such a check does not guarantee the validity of the equations. Thus the well known physical equation $s = ut + \frac{1}{2}at^2$ with metre, second units is dimensionally correct, since

$$[s] \equiv \text{L}$$

$$[ut] \equiv (\text{L/T}) \times \text{T} \equiv \text{L}$$

and

$$[at^2] \equiv (\text{L/T}^2) \times \text{T}^2 \equiv \text{L}$$

On the other hand $s = ut + at^2$ is dimensionally correct, but physically incorrect. In stating numerical answers to problems it is essential to state both measures and units, otherwise the answers are meaningless.

1.3 Statics and Dynamics

It is convenient to subdivide the study of mechanics into *statics* and *dynamics*, the former dealing with bodies at rest and the latter with bodies in motion. The distinction, though artificial, is traditional and is adopted for convenience. Thus the civil engineer is very much concerned with structures that are intended to remain at rest, while the mechanical engineer's concern is with equipment and processes involving the motion of bodies. It will appear later that statics can be described more fundamentally as the study of bodies in *equilibrium.* The state of equilibrium is of fundamental importance and will be referred to continually in what follows.

Statics will be studied first since this will familiarise the student with forces, their representation and their manipulation, before embarking on the study of the dynamical relations between those forces and the motion of the bodies on which they act.

At this point a further concept, that of *mass* needs to be mentioned. This concept, although of the greatest importance in dynamics, also enters into the discussion of statics, since it is necessarily involved in the description of gravitional forces, which are always present in earth-based engineering problems. A more detailed appreciation of the significance of mass will be taken up at a later stage. For the present we appeal to everyday experience of bodies set in motion by forces. Experience indicates that bodies differ in their response to a force of given magnitude, in the sense that bodies differ in the time required to attain a given speed. We say that for a given force different bodies have different accelerations. The property that distinguishes one body from another, in so far as resistance to being accelerated (or *inertia*) is concerned, is called the mass of the body. By comparing the body with a standard body, the mass of which is adopted as a unit, a measurable mass can be assigned to the body. This is a fundamental property that, as far as the engineer is concerned, is constant.

It becomes convenient to select mass as a primary dimension, together with length and time. If this is done then we are led to treat force as a derived dimension. However, in the study of statics this consideration does not arise. Force will be treated as a primary dimension, and mass will only enter the discussion when the magnitudes of gravitional forces are called for.

1.4 Mathematical Considerations

We have already noted that physical equations are symbolic statements about physical quantities. The manipulation of the symbols is in accordance with acknowledged mathematical techniques. Mechanics affords ample scope for the use of special techniques and mathematical forms, many of which will be at this stage unfamiliar to the student. For the chapters that follow a knowledge of elementary algebra, geometry and calculus is the only necessary prerequisite. An acquaintance with vector analysis will enable the characteristics of vectors to be more readily grasped and their manipulation to be more confidently undertaken; however, sufficient information is given in chapter 2 to enable the student to deal with vector quantities to the extent required in the succeeding chapters.

2 Fundamentals

In the chapters that follow the study will be directed towards describing the action of sets of forces on material bodies. Certain terms will be introduced, all of which have definite meanings that need to be borne in mind whenever they are used. This chapter deals with some of these, by setting out their working definitions.

2.1 Particle: Rigid Body: System

Definition

A *particle* is a material body whose linear dimensions are small enough to be considered irrelevant in the context of the problem in hand.

A particle can also be thought of as a quantity of matter concentrated at a point, or in the context of finite bodies, as an elementary portion of such a body.

Definition

A *rigid body* is an assembly of particles, the distance between any two of which remains fixed or can be considered to be so for the purpose of analysis.

The notion of a particle provides a convenient starting-point for our studies since the results of the analysis of particle behaviour can then be extended to a finite rigid body.

When discussing the action of forces on assemblies of particles it is essential to identify the particular assembly under consideration and the forces acting on that assembly. We therefore distinguish clearly between the so-called *system* and its *surroundings.*

Definition

A *system* is an identifiable quantity of matter or assembly of particles bounded by a geometrical surface, the *boundary*. A system may therefore comprise one or many particles, whose distances apart are not necessarily constant. The matter and space outside the boundary constitute the surroundings.

In the case of a rigid body the boundary can clearly be made to coincide with the visible physical surface of the body.

If we now recognise the surroundings as the source of the forces acting on the system we can think of the system in isolation and refer to it as a *free body*. A diagram drawn to illustrate a selected free body together with the forces acting on it is referred to as a *free-body diagram.* We shall always emphasise the importance of drawing free-body diagrams in the solution of problems.

2.2 Scalars and Vectors

When discussing physical quantities we find that some require only a statement of magnitude (measure and unit) for their complete description or specification. Thus the temperature of a water bath could be quoted as 20 degrees Celsius without any further qualification. Other quantities require in addition a statement of direction. Thus the position of a flying object would require not only the distance from the observation point to be known but also the orientation of the line of sight. These two types of quantities are called respectively *scalar* and *vector* quantities.

Definition

A *scalar* quantity is one that is specified by a statement of its magnitude.

Scalar quantities or scalars are symbolised by an algebraic symbol, for example a, representing the product of a measure and a unit. Positive and negative scalars are encountered. The sum of two or more scalars can be found by algebraic addition, provided the scalars have the same dimensions, otherwise the sum has no meaning. The addition can also be carried out by successively marking off segments of a given straight line to an appropriate scale and the sum is then given to scale by the distance between the starting and terminal points.

Definition

A *vector* quantity is one that is specified by a statement of its magnitude and its direction.

Vector quantities are symbolised in printed texts by a bold algebraic symbol, for example $\boldsymbol{a}$, implying that it is a directed quantity. In written work an algebraic symbol with a wavy underline can be used. The magnitude of the vector quantity is symbolised by the corresponding scalar symbol such as a, or if the absolute (non-negative) magnitude is intended, by $|a|$. In the figures of this book, directed arrows are used to indicate the directions of vectors when these are referred to in general terms, and a vector symbol such as $\boldsymbol{a}$ at the side of the arrow expresses the quantity referred to. However, if the directions are otherwise fixed then a scalar symbol is adequate to express the magnitude of the vector referred to.

2.3 Addition of Vectors

The manipulation of scalars according to the rules of elementary algebra and calculus should present no difficulty. The corresponding operations with vector quantities require an extension of the familiar rules to those of elementary vector algebra and calculus, which can nevertheless be reduced to operations with scalars. Initially, however, we avoid the symbolic approach by utilising the geometric properties of vectors.

A segment AB of a straight line has magnitude and direction in passing from A to B. A vector quantity can therefore be represented by such a directed straight-line segment (drawn to some arbitrary scale) since it too has the same properties. Such a representation is also given the name vector and can be referred to as the vector $\overrightarrow{AB}$, the ordering of the letters being made to correspond with the direction intended. Two vector quantities having the same magnitude and direction are said to be equal, and if their representations are respectively AB and PQ (to the same scale) we write $\overrightarrow{AB} = \overrightarrow{PQ}$.

This method of representing a vector enables us to define the addition of two vectors.

Definition

If $\overrightarrow{OA}$, $\overrightarrow{OB}$ are vectors then their addition is symbolised as $\overrightarrow{OA} + \overrightarrow{OB}$ and is defined to be the vector $\overrightarrow{OC}$, where $\overrightarrow{OC}$ is the diagonal of the parallelogram OACB (figure 2.1a). We then write $\overrightarrow{OA} + \overrightarrow{OB} = \overrightarrow{OC}$.

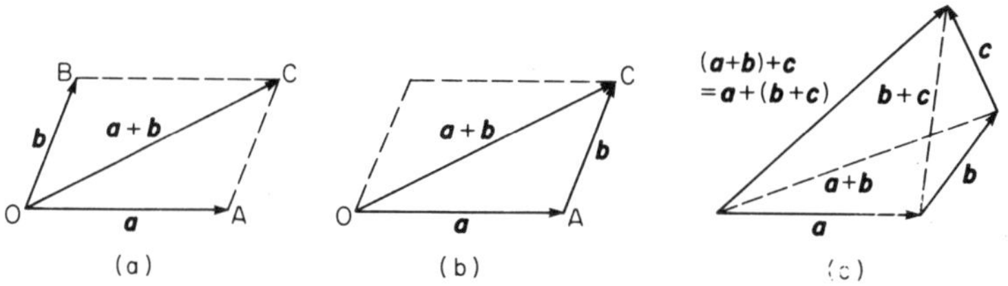

Figure 2.1

This definition conforms to our experience with most physical vector quantities of interest in so far as the vector $\overrightarrow{OC}$ does indeed represent the addition of vector quantities represented by $\overrightarrow{OA}$ and $\overrightarrow{OB}$. This definition is accordingly referred to as the *parallelogram law of addition.*

There are, however, certain quantities that do not combine in this way even though they are vector quantities having magnitude and direction. The term vector is therefore reserved for those vector quantities that satisfy the more precise definition that follows.

Definition

The quantity $\boldsymbol{a}$ is said to be a *vector* if it satisfies both of the following criteria

(1) $\boldsymbol{a}$ has magnitude and direction

(2) the sum of quantity $\boldsymbol{a}$ and a similar quantity $\boldsymbol{b}$ is the vector $\boldsymbol{a} + \boldsymbol{b}$ given by the parallelogram law (figure 2.1a).

A more useful form of the parallelogram law is the triangle law. In figure 2.1a since AC = OB we have $\boldsymbol{a} + \boldsymbol{b} = \overrightarrow{OC} = \overrightarrow{OA} + \overrightarrow{OB} = \overrightarrow{OA} + \overrightarrow{AC}$. Therefore if the two vectors $\boldsymbol{a}$, $\boldsymbol{b}$ are placed tail to head as in figure 2.1b then their sum is the vector $\overrightarrow{OC}$, the directed line from O, the starting point, to C, the terminal point.

It follows that a third vector $\boldsymbol{c}$ can be added to $(\boldsymbol{a}+\boldsymbol{b})$ and we obtain the sum $(\boldsymbol{a}+\boldsymbol{b})+\boldsymbol{c}$ (figure 2.1c). From the geometry of the figure it follows that $(\boldsymbol{a}+\boldsymbol{b})+\boldsymbol{c}=\boldsymbol{a}+(\boldsymbol{b}+\boldsymbol{c})$. Further consideration shows that other combinations are possible, such as $(\boldsymbol{c}+\boldsymbol{a})+\boldsymbol{b}$, all of which are equal. The brackets can therefore be omitted, and we conclude that any number of vectors can be added by placing them tail to head in any order, their sum being the vector extending from the starting point to the terminal point.

If $\boldsymbol{b}$ is a vector with magnitude equal to that of $\boldsymbol{a}$ but is oppositely directed, then we write $\boldsymbol{b}=-\boldsymbol{a}$. This leads us to write $\boldsymbol{b}+\boldsymbol{a}=\boldsymbol{a}+\boldsymbol{b}=0$, a result that is confirmed by the triangle law. Similarly the vector that is equal in magnitude to the vector $\boldsymbol{c}$ but oppositely directed is $-\boldsymbol{c}$. If we write $\boldsymbol{a}+(-\boldsymbol{c})=\boldsymbol{a}-\boldsymbol{c}$, this is interpreted as the subtraction of $\boldsymbol{c}$ from $\boldsymbol{a}$, an operation that is effected by adding to $\boldsymbol{a}$ a vector equal in magnitude to $\boldsymbol{c}$ but reversed in direction (figure 2.2).

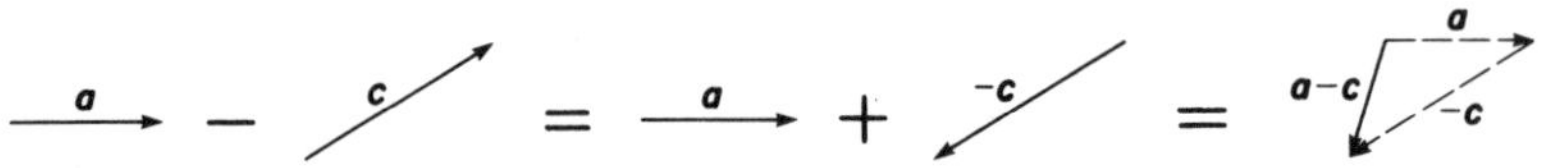

Figure 2.2

The addition and subtraction of vectors using graphical representations is a straightforward procedure if all vectors lie in one plane, and in many types of problem is the appropriate method to use. However, the manipulation of vectors in general and the establishment of significant vector relations require an extension of our notation and the use of the ideas of component and unit vector.

2.4 Components and Unit Vectors

In figure 2.3a OX, OY, OZ are three arbitrary directions and $\overrightarrow{OC}$ is the representation of a vector $\boldsymbol{a}$. With OC as diagonal, a parallelepiped is drawn with edges parallel to OX, OY, OZ. Then $\overrightarrow{OC}=\overrightarrow{OB}+\overrightarrow{BC}=\overrightarrow{OA}+\overrightarrow{AB}+\overrightarrow{BC}$, and it is evident that a vector $\boldsymbol{a}$ can be expressed in an infinite number of ways as the sum of three vectors having arbitrarily chosen directions. $\overrightarrow{OA}$, $\overrightarrow{AB}$, $\overrightarrow{BC}$ are called *vector components* of OC. It is convenient and usual to choose directions OX, OY, OZ that are mutually perpendicular and this is so in all future work. In figure 2.3b, for the rectangular parallelepiped shown, OM = LB and ON = BC and we can write $\overrightarrow{OC}=\overrightarrow{OL}+\overrightarrow{OM}+\overrightarrow{ON}$. The lengths OL, OM, ON, which are the projections of OC on to the mutually perpendicular axes OX, OY, OZ respectively, represent magnitudes that we signify as a_x, a_y, a_z. These three scalar quantities are called the *components* of the vector $\boldsymbol{a}$.

It follows immediately from the triangle law that if $\boldsymbol{a}+\boldsymbol{b}=\boldsymbol{c}$ then the components of $\boldsymbol{c}$ are respectively (a_x+b_x), (a_y+b_y), (a_z+b_z), and in general the components of the sum of two or more vectors are respectively the algebraic sums of the corresponding components of the vectors. The addition of vectors can be reduced in this way to algebraic addition of (scalar) components.

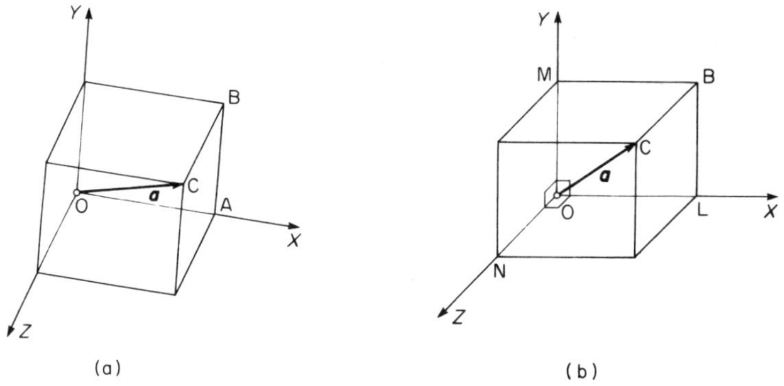

Figure 2.3

We can go further and adopt a notation that separates the magnitude and direction aspects of a vector. The magnitude of a vector $\boldsymbol{a}$ is written a as already noted, and the direction is written $\hat{\boldsymbol{a}}$, which stands for a unit vector in the direction of $\boldsymbol{a}$. Then $\boldsymbol{a} = a\hat{\boldsymbol{a}}$, in which it is understood that a stands for the scalar magnitude (having a measure and a unit). The unit vector $\hat{\boldsymbol{a}}$ can be interpreted as that vector having no dimensions, which when multiplied by the scalar a gives the vector $\boldsymbol{a}$.

If three unit vectors $\boldsymbol{i}, \boldsymbol{j}, \boldsymbol{k}$ are conventionally chosen to lie in the respective directions OX, OY, OZ then we can express the vector components of $\boldsymbol{a}$ in the respective directions as $a_x\boldsymbol{i}, a_y\boldsymbol{j}, a_z\boldsymbol{k}$, and it follows that

$$\boldsymbol{a} = a_x\boldsymbol{i} + a_y\boldsymbol{j} + a_z\boldsymbol{k}$$

and

$$\boldsymbol{a} + \boldsymbol{b} = (a_x + b_x)\boldsymbol{i} + (a_y + b_y)\boldsymbol{j} + (a_z + b_z)\boldsymbol{k}$$
$$= c_x\boldsymbol{i} + c_y\boldsymbol{j} + c_z\boldsymbol{k}$$

This provides the basis of an alternative method for defining scalars and vectors, a scalar being a quantity requiring one magnitude for its specification and a vector being a quantity requiring three magnitudes for its specification. The three magnitudes are the components, which when set down in order in the form $[a_x, a_y, a_z]$ define the vector $\boldsymbol{a}$ and serve as an alternative notation.

Although not referred to in this introductory course, the student will soon encounter quantities that involve the statement of a magnitude and two directions. Such quantities are dealt with in the manipulation of *tensors*, which can be shown to require the statement of nine scalar components that are subject again to the rules of ordinary algebra.

The directions shown for the axes OX, OY and OZ should be carefully noted and adhered to. These are described as a right-handed set of axes and are so arranged that the sense of the rotation required to bring the axis OX to coincidence with the axis OY is that of a right-handed screw advancing along the axis OZ. With the axis OZ pointing directly towards the observer, a line rotating in the XOY plane in the *corresponding* anticlockwise sense makes increasing angles with the axis OX. This anticlockwise sense is consistent with the usual convention for the positive sense of angle indication.

2.5 Units

In the study of statics the only primary dimensions involved in the discussion are length and force. The dimensions time and mass enter the discussion when dynamics is studied. The selection of units for physical quantities involving these dimensions, although having a long historical background, is now based on Newton's laws of motion, and the units are so chosen as to form sets of interrelated units. For engineering purposes two sets or unit systems are in use – the British Engineering System and the International System. The units of the latter are referred to as SI units, SI being an abbreviation of *Système International*.

The basic SI units for the quantities used in mechanics are those for length, time and mass, these being respectively the metre, the second and the kilogram. The unit for force is derived from these and is named the newton. The method by which the newton is derived is taken up in chapter 9.

On the other hand in the British Engineering System the basic units are those for length, time and force and are respectively the foot, the second and the pound-force. The unit of mass is derived from these and is named the slug. The pound-force is the gravitational pull on a body having mass of one pound at a location on the Earth's surface where the body would fall freely with acceleration 32.1416 ft/s^2.

The British Engineering System is gradually being superseded by the SI System and will be referred to very rarely in the following chapters. However, since the units are still in wide every-day use the following table is included showing the basic units of both systems and their comparative magnitudes.

	British Engineering	International (SI)
length	foot (ft) 1 ft = 0.3048 m*	metre (m) 1 m = 3.281 ft
time	second (s)	second (s)
mass	slug (32.1416 lbm*) 1 lbm = 0.454 kg	kilogram (kg) 1 kg = 2.205 lbm
force	pound-force (lbf) 1 lbf = 4.448 N	newton (N) 1 N = 0.225 lbf

The values marked with an asterisk are exact.

2.6 Solution of Problems

In the context of engineering the purpose of the study of mechanics is the analysis and solution of engineering problems, that is, given an engineering system subject to various forces and constraints, to deduce relevant information about the behaviour of the system from our knowledge of the laws appertaining to its behaviour. In the solution of every problem the following stages should be recognisable.

(1) Statement of the problem in real physical terms. In many cases the system is immediately recognisable using terms in common use. Frequently a diagram or dimensioned drawing is required to clarify the subject of the problem.
(2) Recognition of applied mechanics concepts and entities and the statement of the problem in these terms. At this stage a free-body diagram should always be drawn since this will embody the immediately relevant concepts in diagrammatic form.
(3) Statement of the applicable laws and principles, or alternatively theorems that have been derived from them, in the form of (a) physical equations or (b) graphical constructions. Usually it is possible in numerical problems to set out the work algebraically and substitute numerical values when required. This approach has advantages, the chief one being the opportunity it affords of checking the mathematical development. Whether algebraic or numerical methods are used it should become a habit to maintain a continual check on the dimensions and units of all equations.
(4) Solution of equations or interpretation of graphical constructions for the desired information.
(5) Checking of solutions. The validity of the basic laws is not open to question. However, they can be incorrectly applied and mistakes can be made in calculation. The solution should be a reasonable one as far as can be judged: for example, the order of magnitude of a numerical answer should be in accord with that of the information given in the problem. Sometimes another method of solution is possible. Usually the order in which arithmetic operations are carried out can be varied.
(6) Statement of the solution in the terms required by the problem. If numerical solutions are called for then units must be stated. The number of significant figures must not be greater than that warranted by the information given and the nature of the assumptions made. In addition the direction and sense of vector quantities must be clearly described. A succinct symbolism used for expressing the magnitude, direction and sense of a vector quantity is (by way of example) $20.75\ \text{m/s} \angle 69.5^\circ$, the angle indicated being that made with a reference direction and increasing anticlockwise.

Problem-solving is an essential part of the study of engineering mechanics since it affords practice in the construction of mathematical models and in the development of logical methods of analysis. The basic ideas are relatively few in number and the memorising of formulae is rarely required. Instead, the student should aim for precise statement of principles, concise expression, and systematic setting out of solutions.

Problems

2.1 From your experience of the following, state whether they are scalar or vector quantities.

(a) The distance between two points
(b) The height of a mountain
(c) The bearing of a ship
(d) The speed of a train
(e) The kinetic energy of a bullet
(f) The intensity of a sound
(g) The weight of a body
(h) The drag on an aeroplane

2.2 Obtain graphically the sum of the vectors $\boldsymbol{a}$ and $\boldsymbol{b}$ in the following cases.

(a) $\boldsymbol{a} = 5 \angle 60°, \boldsymbol{b} = 3 \angle 90°$
(b) $\boldsymbol{a} = 10 \angle 30°, \boldsymbol{b} = 20 \angle -30°$
(c) $\boldsymbol{a} = 200 \angle 45°, \boldsymbol{b} = -400 \angle 30°$

2.3 Obtain graphically the difference $\boldsymbol{a} - \boldsymbol{b}$ for the vectors of problem 2.2.

2.4 Obtain graphically the sum of the vectors $\boldsymbol{a} = 4 \angle 0°, \boldsymbol{b} = 5 \angle 90°, \boldsymbol{c} = 3 \angle 150°$, $\boldsymbol{d} = 6 \angle 230°$.

2.5 The following vectors all lie in the $x - y$ plane. Obtain the x- and y-componets in each case.

(a) $\boldsymbol{a} = 15 \angle 50°$
(b) $\boldsymbol{b} = 20 \angle 150°$
(c) $\boldsymbol{c} = -10 \angle -30°$

2.6 A vector $\boldsymbol{a}$ has components $a_x = 5.3, a_y = -7$. Obtain $\boldsymbol{a}$ in terms of its magnitude and direction.

2.7 Obtain the magnitude of the vector $\boldsymbol{a} = [5, 6, 3]$ and determine its inclination to (a) the $x - y$ plane and (b) the $z - x$ plane.

2.8 Obtain the magnitude of the sum of the vectors $\boldsymbol{a} = [5, 6, 3]$ and $\boldsymbol{b} = [2, 4, -7]$ and its inclination to the $x - y$ plane.

2.9 If $\boldsymbol{a} + \boldsymbol{b} = 10 \angle 60°$ and $\boldsymbol{a} - \boldsymbol{b} = 20 \angle 30°$ determine $\boldsymbol{a}$ and $\boldsymbol{b}$. (*Hint*: use the vector equations to solve for $\boldsymbol{a}$ and $\boldsymbol{b}$.)

3 Statics of a Particle

If the surroundings act on a system in such a way as to tend to change the motion of the system then we say that a force exists. The force is the action of the surroundings, but nevertheless it is usual to state that the force itself is acting on the system. The forces encountered in practice are usually found to be applied to the system as a whole or over finite areas of the system boundary. Initially, however, we consider only the forces acting on a single particle. Any one such force can immediately be characterised by its magnitude and its direction, and in addition, for a single particle, by its line of action, since this passes through the particle. In the next chapter we extend the discussion to particular assemblies of particles, namely rigid bodies.

3.1 Resultant: Components

Experiment indicates that two forces acting together on a single particle can be replaced by a single force with its line of action also passing through the particle and whose magnitude and direction are given by the triangle law. A force is therefore a vector. The single force is called the *resultant* of the two forces. The two forces have thus been summed, and if they are symbolised as $\boldsymbol{F}_1$ and $\boldsymbol{F}_2$ we write $\boldsymbol{F}_1 + \boldsymbol{F}_2 = \boldsymbol{R}$ where $\boldsymbol{R}$ is the resultant.

In accordance with the results of the preceding chapter, three or more forces on a particle can be added by placing their representations tail to head in any order, the resultant then being represented by the line joining the starting point to the terminal point. For coplanar forces the triangle law is therefore developed into the polygon law.

Definition

The *resultant* of any number of coplanar forces (a set of forces) acting simultaneously on a particle is the single force that is equivalent to the force set and that is obtained by placing their representations tail to head in any order. The resultant is then represented by the line joining the initial point to the terminal point.

The use of the word *equivalent* should be carefully noted, since it implies that the resultant has the same effect as the individual forces taken together, and can replace them. Two forces can be equal in magnitude but have different effects if, for example, their directions or lines of action differ.

A given force on a particle can be resolved into any number of vector components having specified directions. In particular, if the directions are mutually perpendicular in space the components are three in number and are referred to as rectangular vector components. If OX, OY are two chosen perpendicular directions that are coplanar with the force then the components are two in number. As with vectors in general, only the magnitudes of the vector components need to be stated. In future the word *component* will be used to stand for *rectangular scalar component* on the understanding that the directions are specified or otherwise implied.

If the line of action of a force $\boldsymbol{F}$ acting on a particle lies in the XOY plane and is inclined at an angle θ to the OX-axis, then the x- and y-components, F_x and F_y respectively are given by $F_x = F\cos\theta$, $F_y = F\sin\theta$ (figure 3.1a) and we can write

$$\begin{aligned} \boldsymbol{F} &= F_x\boldsymbol{i} + F_y\boldsymbol{j} \\ &= (F\cos\theta)\boldsymbol{i} + (F\sin\theta)\boldsymbol{j} \end{aligned} \tag{3.1}$$

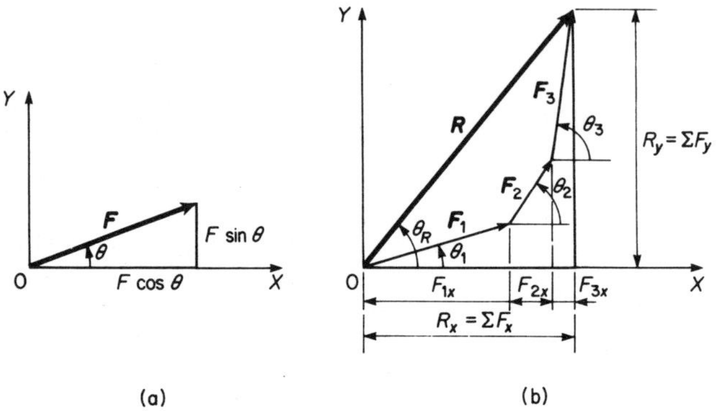

Figure 3.1

We shall be concerned mainly with sets of forces $\boldsymbol{F}_1, \boldsymbol{F}_2, \ldots$ having lines of action lying in one plane such as XOY, and referred to as *coplanar* forces. From figure 3.1b and the polygon law it follows that the x-component of the resultant $\boldsymbol{R}$ of any number of such forces is the sum of the x-components of the individual forces; and similarly for the y-components. Thus

$$\begin{aligned} R_x &= F_1\cos\theta_1 + F_2\cos\theta_2 + F_3\cos\theta_3 \\ &= F_{1x} + F_{2x} + F_{3x} \\ &= \Sigma F_x \end{aligned} \tag{3.2a}$$

and similarly

$$R_y = \Sigma F_y \tag{3.2b}$$

The symbol Σ will be used extensively to indicate the summation of quantities typified by the quantity indicated.

Conversely, if the sums of the x-components and of the y-components of a set of forces are known the resultant can be determined in magnitude and direction, since the magnitude

$$R = \sqrt{(R_x^2 + R_y^2)} \tag{3.3a}$$

and

$$\tan \theta_R = \frac{R_y}{R_x} \tag{3.3b}$$

where the magnitude of θ_R and consequently the sense of R can be determined by inspection of the signs of R_x and R_y.

3.2 Equilibrium: Newton's First Law

Definition

If the resultant force on a particle is zero then the particle is said to be in *equilibrium.*

A particle is therefore in a state of equilibrium, by definition, by reason of the resultant of the forces acting on it being zero.

The physical significance of equilibrium is contained in the first of the three fundamental laws of classical mechanics enunciated by Sir Isaac Newton (1642 - 1727). After rephrasing, the first law can be stated as follows.

The First Law

If the resultant force on a particle is zero, then the particle remains in a state of rest or constant speed in a straight line.

It follows that a particle in equilibrium is one which is either at rest or moving with constant speed in a straight line. The converse is also true, that if a particle is known to be at rest or moving at constant speed in a straight line, then it is in equilibrium and the resultant force acting on it is zero.

3.3 Conditions for Equilibrium

In practice we encounter particles that are known or are seen to be in equilibrium under the action of a set of forces, having magnitude, direction and sense, some of which may not be known initially. If the resultant of the force set is zero then the forces and their characteristics must be related in some way and must satisfy certain conditions. In order to be able to determine the unknown characteristics we have to set out the conditions that must be satisfied in a form that will permit of solution for the desired information. This can be done in two ways.

(1) If the forces acting on the particle are summed by the polygon law, then if the resultant is zero the polygon must close. We have therefore the graphical condition: for a particle to be in equilibrium the force polygon must close.
(2) Since the magnitude of the resultant $|\boldsymbol{R}| = \sqrt{[(\Sigma F_x)^2 + (\Sigma F_y)^2]}$ then for $R = 0$ the summations ΣF_x and ΣF_y must both be zero. We have therefore the analytical condition: for a particle to be in equilibrium $\Sigma F_x = 0$ and $\Sigma F_y = 0$.

3.4 Applications

In applying the graphical condition, the force polygon is built up using the known forces first and the closure of the polygon then reveals the desired unknown characteristics. It will become immediately apparent that the number of unknown characteristics that can be determined is limited to two, such as for example, the magnitude and direction of a single unknown force, or the magnitudes of two forces whose directions are initially known. The use of the graphical condition does not imply that the force polygon must necessarily be drawn to scale – a neat sketch can be used as a basis for calculation using the geometry of the polygon.

The two equations of the analytical condition will again enable two characteristics to be determined, namely two unknown components. A judicious choice of x- and y-directions will simplify the equations and enable a solution to be more readily obtained.

In the problems at the end of this chapter the solutions are to be obtained on the basis of certain assumptions, namely (1) the body is a particle; (2) if a force is applied by means of a massless cord then the direction of the force and its line of action coincide with the cord; (3) if a cord passes over a smooth pulley the two forces exerted by the cord on the pulley are equal in magnitude, this magnitude being referred to as the *tension* in the cord, (4) if a body is in contact with a smooth surface the force of the surface on the body is in a direction normal to the surface. All these assumptions need to be justified on the basis of arguments to be developed at later stages.

In certain problems the forces on the particle are acting in a vertical plane. One of those forces is invariably a gravitational force arising from the attraction of the Earth. The nature of such forces will be discussed in more detail in sections 9.1 and 10.6.1 when the dynamics of a particle is being studied. At this stage sufficient information is given to enable the gravitional force to be included with its correct magnitude.

The gravitational force on a particle is referred to as the *weight* W of the particle. Although by this terminology weight is made to appear to be a property of the particle, it should be remembered that weight is a force that depends for its existence on the presence of the earth.

When we come to consider the dynamics of a particle we shall find that if a particle mass m kg is being accelerated under the action of a force $\boldsymbol{F}$ N, the acceleration $\boldsymbol{a}$ m/s^2 is given by the equation $\boldsymbol{F} = m\boldsymbol{a}$. Now we know that a particle falling freely under the action of its weight alone descends with acceleration magnitude g, which is the same for all particles in the same locality. This was first demonstrated by Galileo (1564 - 1642). We can therefore write for any particle

$$W = mg \tag{3.4}$$

The value of g may vary slightly from one locality to another, but for our purposes we can adopt the value 9.81 m/s²; W is then in newtons if m is in kg.

Worked Example 3.1

The ends of a cord length 3.5 m are attached to points A and B as shown in figure 3.2a. A small smooth pulley carrying a body, mass m = 10 kg is placed on the cord and allowed to reach a point of equilibrium at C. Find the horizontal distance of this point from A and the tension in the cord.

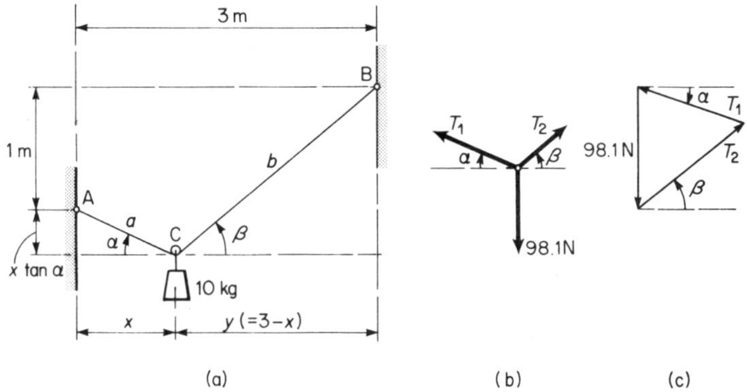

Figure 3.2

Solution

Since the distance AB = $\sqrt{10}$ = 3.16 m the cord hangs below the line AB and a diagrammatic view of the situation for the equilibrium condition is given in figure 3.2a. We now consider the equilibrium of the pulley, which can be regarded as a particle in this example. The forces acting on it are

(1) the weight of the body, $mg = 10 \times 9.81 = 98.1$ N downwards
(2) the force in the cord segment AC; this force is along CA and since the cord can only be in tension the force, denoted symbolically as T_1, is in the direction given in figure 3.2b
(3) the force in the cord segment CB; by the same reasoning this force, denoted symbolically as T_2, is in the direction again given in figure 3.2b.

Figure 3.2b is the free-body diagram for the pulley showing all the external forces acting on the pulley.

Since the forces on the pulley are in equilibrium the force triangle must close. The latter is shown drawn diagrammatically in figure 3.2c.

Making use of assumptions already stated, we can write $T_1 = T_2$ and it follows from the force triangle that $\alpha = \beta$.

From similar triangles in figure 3.2a

$$\frac{x}{a} = \frac{y}{b}$$

therefore

$$\frac{x}{a} = \frac{x + y}{a + b} = \frac{3}{3.5} = \cos\alpha$$

hence $\tan\alpha = \sqrt{13}/6$; also from figure 3.2a

$$x \tan\alpha + 1 = (3 - x)\tan\beta = (3 - x)\tan\alpha$$

giving $x = 0.667$ m.

Let $T = T_1 = T_2$ be the tension in the cord. Then from the force triangle

$$2T\sin\alpha = 98.1$$

and

$$T = 95 \text{ N}$$

3.5 Summary

(1) The resultant $\boldsymbol{R}$ of a set of forces $\boldsymbol{F}_1, \boldsymbol{F}_2, \boldsymbol{F}_3, \ldots$ acting on a particle is the single force $\boldsymbol{R} = \boldsymbol{F}_1 + \boldsymbol{F}_2 + \boldsymbol{F}_3 + \ldots$, and for coplanar forces is given by the polygon law of addition.

(2) The resultant $\boldsymbol{R}$ is equivalent to the force set.

(3) A force $\boldsymbol{F}$ can be resolved into any number of vector components. The rectangular scalar components of $\boldsymbol{F}$ are signified by F_x, F_y. For the resultant $\boldsymbol{R}$

$$R_x = \Sigma F_x \qquad (3.2a)$$

$$R_y = \Sigma F_y \qquad (3.2b)$$

The magnitude and direction of the resultant are given by

$$R = \sqrt{(R_x^2 + R_y{}^2)} \qquad (3.3a)$$

$$\theta = \tan^{-1}(R_y/R_x) \qquad (3.3b)$$

(4) A particle is in equilibrium if $\boldsymbol{R} = 0$.

(5) A particle in equilibrium is either at rest or moving with constant speed in a straight line (Newton's first law).

(6) The conditions for equilibrium of a particle are *either* (a) the graphical condition: force polygon must close; *or* (b) the analytical condition: $\Sigma F_x = 0$, $\Sigma F_y = 0$.

(7) The weight W of a particle having mass m is given by

$$W = mg \tag{3.4}$$

Problems

3.1 A particle is acted on in the horizontal plane by three forces having magnitudes 20 N, 30 N and 40 N. If the particle is in equilibrium under these forces find the directions of the 20 N and 30 N forces relative to the 40 N force. (A graphical solution is suggested.)

3.2 A particle, mass 0.2 kg, is held at rest on an inclined plane of slope 30° to the horizontal by the application of a force $\boldsymbol{P}$, which is at 60° to the horizontal. Assuming the reaction N of the plane on the particle is normal to the plane find the magnitudes of $\boldsymbol{P}$ and N. (*Hint*: Draw a free-body diagram of the three forces acting on the particle.)

3.3 Two cords, lengths 3 m and 4 m respectively, are each attached at one end to two points 5 m apart horizontally. The free ends of the cords are joined together to one end of another cord of length 2 m. To the free end of the latter is attached a small body, mass 2 kg, which is being pulled with a horizontal force 15 N to the same side as that of the 3 m cord. Determine the inclination of the 2 m cord to the vertical and the forces in the 3 m and 4 m cords. (*Hint*: Consider first the three forces acting on the body (draw a free-body diagram) and hence find the force and inclination of the 2 m cord; draw a free-body diagram for the junction of the three cords and solve for the other unknowns.)

3.4 Four forces 10 N $\angle$ 0°, 15 N $\angle$ 60°, 20 N $\angle$ 120°, and 30 N $\angle$ 180° act in the horizontal plane on a particle.

(a) Determine from a force polygon the resultant of these forces.
(b) What force is required to maintain the particle in equilibrium?

3.5 An airship, mass 10 000 kg, is tethered by four ropes equally disposed around it and each making an angle of 15° with the vertical (the four ropes would thus be on the face of a cone with vertex angle 30°). What is the lift force on the ship if the force in each rope is 20 000 N? (*Hint*: The resultant vertical force is zero.)

3.6 A particle is moving with constant velocity in a straight line on a smooth horizontal surface under the action of the following forces, which all act in the horizontal plane: 100 N $\angle$ 0°; $\boldsymbol{Q}$ $\angle$ 70°, $\boldsymbol{P}$ $\angle$ 150°; 70 N $\angle$ 225° and 80 N $\angle$ 300°. Use a polygon of forces to determine the magnitudes of $\boldsymbol{P}$ and $\boldsymbol{Q}$.

3.7 If in problem 3.6 $\boldsymbol{P}$ and $\boldsymbol{Q}$ are allowed to take up any direction as long as they are mutually perpendicular, find their magnitudes and directions if the magnitude of $\boldsymbol{P}$ is to be twice that of $\boldsymbol{Q}$.

3.8 A piece of cord of length 2.5 m is connected between two points 2 m apart horizontally. A massless pulley is placed on the cord and a force 20 N $\angle -80^\circ$ is applied to it.

(a) Show that when the pulley reaches a state of rest the angles that the two portions of the cord make with the horizontal differ by 20°.
(b) Verify that the relevant equations are satisfied when the smaller angle is 28° and hence determine the tension in the cord.

(*Hint*: Cord tension is constant. For (a) use the force triangle, for (b) criteria to be satisfied involve geometrical relations of cord length and distance between supports, relationship from (a) between the cord angles; note that only two useful equations can be derived from the force triangle – any others will be redundant.)

3.9 A particle at the point O is subject to three coplanar forces which can be represented by the vectors $\overrightarrow{OA}$, $\overrightarrow{OB}$, $\overrightarrow{OC}$. If the intersection of the medians of the triangle ABC is at G, show that the resultant of the three forces is represented by the vector $3\overrightarrow{OG}$. (*Hint*: Write $\overrightarrow{OA}$ as a summation of other vector quantities one of which is OG, and similarly for $\overrightarrow{OB}$ and $\overrightarrow{OC}$.)

3.10 A particle located at the point O of a polygon OABCO is subject to four coplanar forces, which are represented by the vectors $\overrightarrow{AO}$, $\overrightarrow{AB}$, $\overrightarrow{CO}$ and $\overrightarrow{CB}$. If P, Q are the midpoints of AC and OB respectively, show that the resultant force on the particle is represented by $4\overrightarrow{PQ}$. (*Hint*: See problem 3.9, but involve vector $\overrightarrow{PQ}$.)

4 Statics of Rigid Bodies

A rigid body, as already defined, is an assembly of particles whose relative positions remain unchanged. Actual bodies all deform to some degree when subjected to forces, but the rigid bodies we shall consider are such that changes of shape can be disregarded. The shape of a rigid body, that is, the configuration of its particles, is maintained because of forces between the particles, the forces being such that all particles are individually in equilibrium if the whole body is in equilibrium.

The forces acting on the particles of a rigid body can therefore be classified as

(1) external forces: forces whose sources are outside the body
(2) internal forces: forces between the particles of the body.

Note particularly that forces that are internal for a particular assembly of particles can become external forces for a part of the assembly. More precisely, therefore, we say: having defined the boundary of the system, those forces that have their origin in the surroundings are external forces.

On the basis of this distinction between external and internal forces we now extend the discussion of the statics of single particles to that of particle assemblies and rigid bodies in particular, with the aim of finding resultant forces and setting out the corresponding conditions which must be satisfied if equilibrium is to be achieved.

4.1 Newton's Third Law

The analysis of the forces acting on a particle system is greatly simplified when the third of Newton's laws is taken into account. This can be stated in the following form.

The Third Law

The force exerted by one particle on another is always accompanied by an equal and opposite force exerted by the second particle on the first particle, both forces acting along the line joining the particles.

The set of internal forces for a rigid body is therefore made up of pairs of equal and opposite collinear forces that cancel each other. It follows that, for the body as a whole, the external forces are the only forces that influence the resultant and the conditions for equilibrium for a rigid body.

The external forces act on different particles of the body and in general their lines of action do not meet at a point. In turn, the resultant will not only have a magnitude and a direction but also a line of action that must be found. Fortunately the results for a single particle are still applicable by virtue of a principle that we now introduce.

4.2 Principle of Transmissibility

In figure 4.1a a rigid body of arbitrary shape is indicated with a single force acting at particle A. Suppose now some other particle B is selected in the line of action of $\boldsymbol{F}$ and two forces $\boldsymbol{F}'$ and $\boldsymbol{F}''$ are applied at B as in figure 4.1b, the forces being

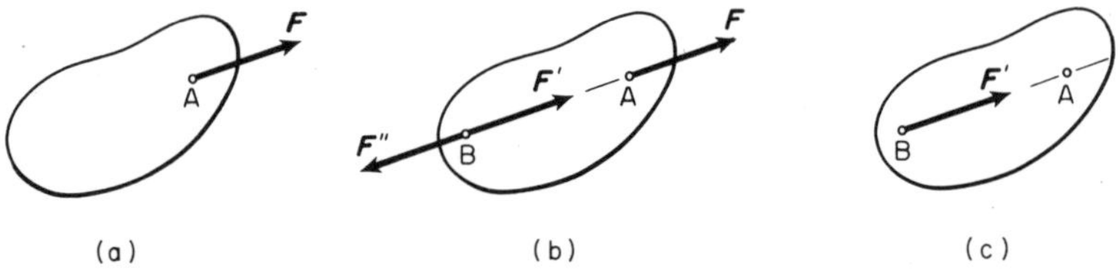

Figure 4.1

such that $\boldsymbol{F}' = -\boldsymbol{F}''$ and $\boldsymbol{F}' = \boldsymbol{F}$. This set of three forces is equivalent to $\boldsymbol{F}$ at particle A since the resultant of $\boldsymbol{F}'$ and $\boldsymbol{F}''$ is zero. However, it is evident that for the body as a whole forces $\boldsymbol{F}$ and $\boldsymbol{F}''$ have no resultant, therefore the set of forces is also equivalent to $\boldsymbol{F}'$ at particle B as indicated in figure 4.1c. Since this equivalence exists for each external force on the body we can express the result as a principle in the following form.

Principle of Transmissibility

If a set of forces acts on a rigid body then the resultant of the set or the state of equilibrium of the body is unchanged if any force $\boldsymbol{F}$ of the set acting on a particle of the body is replaced by a force $\boldsymbol{F}'$ having the same magnitude and direction acting on a different particle, provided the line of action is unchanged. The force sets are all equivalent.

Note particularly that this principle only relates to resultants and states of equilibrium. The internal forces, although their resultant is still zero, are modified if the points of application of the external forces are changed.

4.3 Resultant: Parallel Forces: Couples

The resultant of a set of forces acting on a rigid body, that is, the single force equivalent to the set, can be found by repeated application of the principle of transmissibility. The case of parallel forces will be taken up first using this method. This will enable certain new entities to be introduced, which will enable more expeditious and significant methods to be adopted for force sets in general.

(1) Two Like Parallel Forces

In figure 4.2a two parallel forces ***P*** and ***Q*** having the same sense are shown acting at the boundary of a rigid body. Choose any straight line in the body intersecting the lines of action of ***P*** and ***Q*** at points A and B. If ***P*** and ***Q*** are now moved along their lines of action to A and B respectively we obtain an equivalent force set.

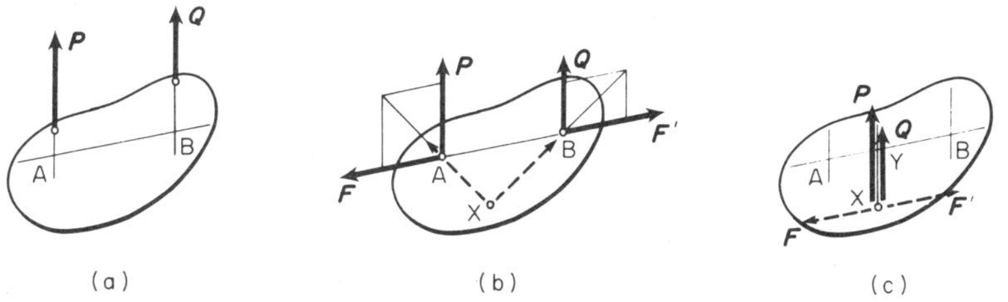

Figure 4.2

Two equal and opposite collinear forces ***F*** and ***F′*** are now introduced at A and B and combined with the forces ***P*** and ***Q*** by the triangle law to produce two forces whose lines of action meet at a point X. The point X, if it happens to be outside the body, can be taken to be an isolated particle belonging to the rigid body (figure 4.2b). At X these two forces can again be resolved into the same vector components, and after removing the two equal and opposite forces ***F*** and ***F′*** we are left with a single force (***P*** + ***Q***) whose line of action passes through the point Y of the line AB (figure 4.2c) and which is equivalent to the original force set. From the geometry of figure 4.2c and that of the force triangles we have P/F = XY/AY and Q/F' = XY/YB, from which it follows that AY/YB = Q/P. The resultant is therefore the single force ***R*** with magnitude ($P + Q$) whose line of action passes through the point Y on an arbitrary line AB and which divides it in the inverse ratio of the magnitudes of the forces.

(2) Two Unlike Parallel Forces

The same procedure applied to forces ***P*** and ***Q*** having opposite senses will show that a single force can again be found, which is equivalent to the original set but whose magnitude is the difference between the magnitudes of ***P*** and ***Q***. The point Y is again given by the relation AY/YB = Q/P but Y now lies outside AB. If $P > Q$ then the point Y lies on BA produced, that is, beyond the line of action of the larger force; the magnitude of the resultant is ($P - Q$) and its sense is that of ***P*** the larger force. Figure 4.3 illustrates two cases.

(3) Two Equal Unlike Parallel Forces

If the same procedure is attempted for forces ***P*** and ***Q*** having opposite senses but for which $P = Q$ it will become apparent that no single force can be found and that the procedure merely produces further pairs of equal unlike parallel forces, each pair being equivalent to the original set. Such pairs of forces therefore have no resultant. A force set of this kind is termed a *couple*.

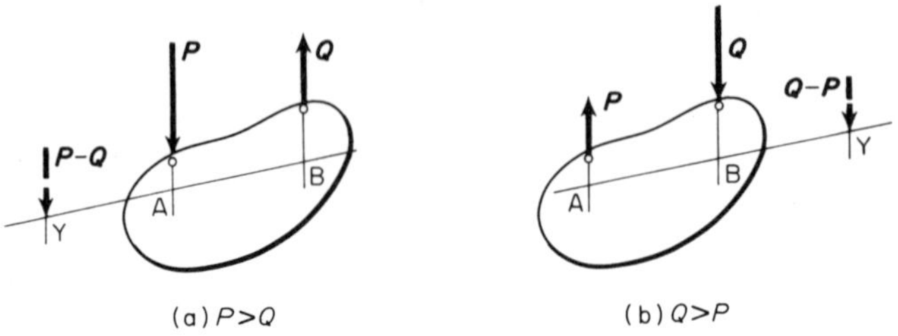

Figure 4..3

(4) Any Number of Parallel Forces

By repeated application of the above results, taking forces in pairs at each stage, we can obtain the resultant, if it exists, of any number of parallel forces. If there is no resultant then the set is reducible either to zero or to a couple as the case may be.

4.4 Centre of Parallel Forces

Consider a rigid body subject to a set of three like parallel forces $\boldsymbol{P}$, $\boldsymbol{S}$ and $\boldsymbol{Q}$ applied to particles A, B and C (figure 4.4a). The ıesultant of $\boldsymbol{P}$ and $\boldsymbol{S}$ is a force $(\boldsymbol{P} + \boldsymbol{S})$ passing through the point Y such that AY/YB = S/P. The resultant of $(\boldsymbol{P} + \boldsymbol{S})$ at Y and $\boldsymbol{Q}$ at C is a force $(\boldsymbol{P} + \boldsymbol{Q} + \boldsymbol{S})$ passing through the point Z on YC such that YZ/ZC = $Q/(P + S)$.

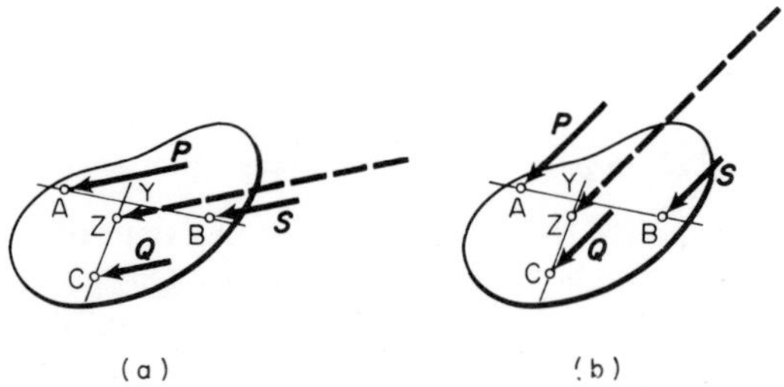

Figure 4.4

If now the lines of action of $\boldsymbol{P}$, $\boldsymbol{Q}$ and $\boldsymbol{S}$ are all rotated through the same angle, their magnitudes remaining unchanged, we obtain another set of like parallel forces, and it is evident that the line of action of their resultant passes through the same point Z as before. The lines of action of the two force sets therefore pass through a point Z, which is fixed in relation to A, B and C, that is, to the rigid body. This point is termed the *centre of parallel forces.* It follows that if the directions of $\boldsymbol{P}$, $\boldsymbol{Q}$ and $\boldsymbol{S}$ are maintained and the body is rotated instead, the resultant will always pass through the point Z in the body.

If now the set of parallel forces arises from the Earth's gravitational attraction on all particles of the body, the resultant gravitational force, the *weight* of the body, again acts through a point fixed in the body, now termed the *centre of gravity.*

It should be noted that although in practice the centre of gravity is, by the preceding argument, a point fixed in the body, this is not strictly the case since gravitational forces are not truly parallel. Thus in the case of the gravitational pull of the sun on the Earth the centre of the sun's gravitational forces varies with the orientation of the Earth. However, for the bodies of engineering applications the centre of gravity for the Earth's gravitational forces can be taken as fixed in the body with quite negligible error.

More important is the consideration that the centre of gravity is associated with gravitational *forces*. A more fundamental point can be defined that is fixed and independent of these gravitational forces. This is the mass-centre, which will be discussed in chapter 7.

4.5 Moment of a Force

A force $\boldsymbol{F}$ acting on a particle of a rigid body can be replaced by an equal force $\boldsymbol{F}'$ acting on another particle provided the line of action is not changed. However, an equal force $\boldsymbol{F}''$, acting on another particle not in the same line of action, although equal, is not equivalent; that is, $\boldsymbol{F}''$ cannot replace $\boldsymbol{F}$. The effects of $\boldsymbol{F}$ and $\boldsymbol{F}''$ on the body, in particular as far as a tendency for rotation about an axis is concerned, are different. In figure 4.5, drawn in the plane of the forces, the

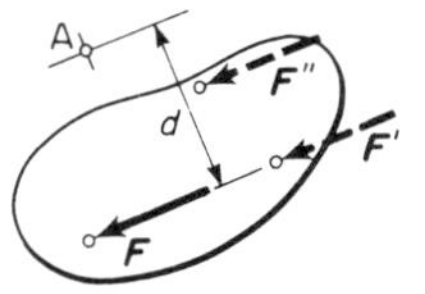

Figure 4.5

axis is at A and is perpendicular to the plane. We define the *moment* M_A of the force F (or its equivalent F') about the axis at A as the product $F \times d$, where d is the perpendicular distance from A to the line of action of the force. If the force tends to produce anticlockwise rotation then the turning effect will be considered to have a positive sense and the associated moment will also be positive.

Although the moment is defined in relation to an axis at A, for coplanar forces it is usual to refer to moments about the point A, in which case the point A can be termed the *moment-centre.*

The moment of a force has a magnitude and in the general three-dimensional case a direction would be, by definition, associated with it. In the case we have considered, the direction of an anticlockwise moment would be defined as that of the axis at A, pointing upwards from the diagram. Moments of forces can also be summed by the triangle law and are, therefore, in the general three-dimensional case, true vectors.

If arbitrary OX and OY directions are chosen in the plane of the diagram then as we have seen, F_x and F_y serve to define the magnitude and direction of the force $\boldsymbol{F}$; M_A now serves to define the line of action in relation to the axis at A, since the perpendicular distance to the line of action is given by $d = M_A/F$.

The calculation of moments is often simplified by the use of a theorem named after Varignon, which states that the moment of a force about an axis is equal to the sum of the moments of its vector components about the same axis. The result can be demonstrated with the aid of figure 4.6 in which rectangular components

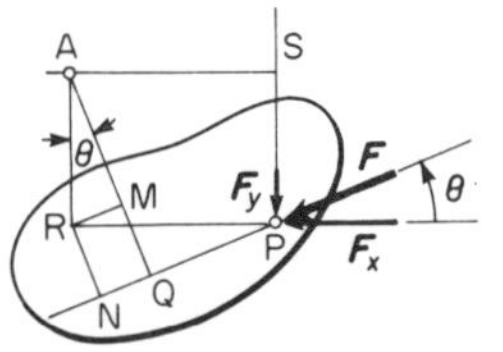

Figure 4.6

have been used for simplicity. A particle P is chosen lying in the line of action of $\boldsymbol{F}$. The chosen *vector* components of $\boldsymbol{F}$ at P are shown as $\boldsymbol{F}_x$ and $\boldsymbol{F}_y$. From the axis at A the perpendiculars AQ, AR and AS are dropped on to the lines of action of the force and its components. If RM is now made perpendicular to AQ and RN is made parallel to AQ, then

$$\begin{aligned} M_A &= F \times \text{AQ} \\ &= F(\text{AM} + \text{MQ}) = F(\text{AM} + \text{RN}) \\ &= F(\text{AR} \cos\theta + \text{PR} \sin\theta) \\ &= F_x \times \text{AR} + F_y \times \text{RP} \\ &= F_x \times \text{AR} + F_y \times \text{AS} \end{aligned}$$

thus demonstrating the theorem in this case.

4.6 Moment of a Couple

The rigid body in figure 4.7 is subject to the couple comprising two unlike

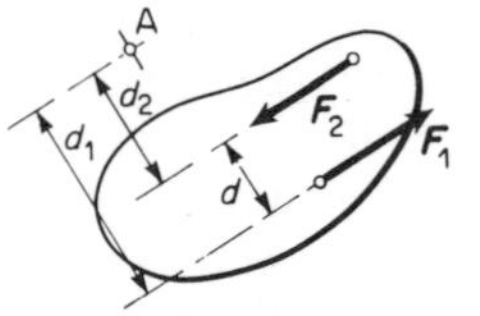

Figure 4.7

parallel forces F_1 and F_2 for which $F_1 = F_2 = F$, say. The sum of the moments of F_1 and F_2 about A is $F_1 d_1 - F_2 d_2 = F(d_1 - d_2) = Fd$. This quantity, again signified by M_A is called the *moment* of the couple. It is evident that since d is the perpendicular distance between the lines of action the moment of a couple does not depend on the position of the axis and is a property of the couple itself.

Further properties of couples follow.

(1) Two coplanar couples having the same moment and sense are equivalent (figure 4.8). It is sufficient to show that one couple can be transformed into the other;

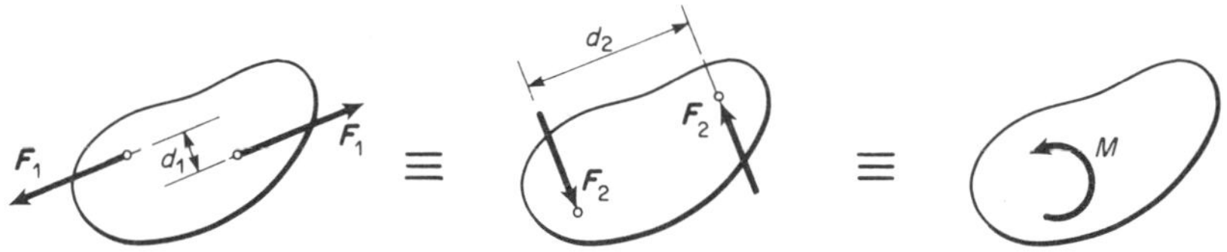

Figure 4.8

the proof, using the principle of transmissibility, is straightforward and is left as an exercise. The equivalence of couples having the same moments and sense enables us to introduce the symbol shown in the figure to represent a couple, the sense of the arrow serving to indicate the sense of the couple, and the symbol M the magnitude of the moment of the couple and its equivalents (the suffix now being omitted).

Furthermore, if the moment of a couple is known then the details of the two forces making up the couple need not be specified. A body is then said to be subject to a *torque*, a quantity that is now described by a statement of its magnitude and sense, having the same dimensions and units as those of the moment of a couple.

It is convenient in the discussion to be able to refer to a couple in general terms without particularly specifying the magnitude of its moment. For this purpose the symbol $\boldsymbol{L}$ can be used to represent the couple, the direction of $\boldsymbol{L}$ being that of the moment of the couple.

(2) Two coplanar couples are together equivalent to a single couple having a moment equal to the algebraic sum of the moments of the individual couples. The proof, utilising property 1, consists of transforming each pair of forces into corresponding equivalent pairs with the forces the same distance apart and having the same lines of action, and then summing the forces directly.

4.7 Force - Couple Sets

Although a force $\boldsymbol{F}$ acting on a rigid body can only be replaced by an equivalent single force $\boldsymbol{F}$ if the line of action is unchanged we can still obtain an equivalent force set with the force $\boldsymbol{F}$ in some other parallel line of action if it is accompanied by an appropriate couple. Thus in figure 4.9, by introducing oppositely directed

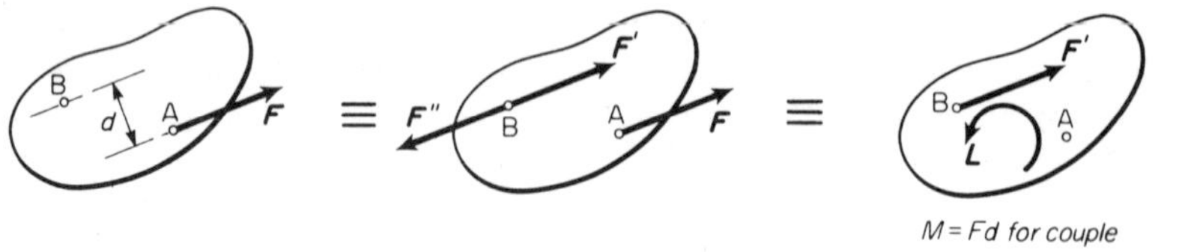

Figure 4.9

forces $\boldsymbol{F}'$ and $\boldsymbol{F}''$, for which $F' = F'' = F$, at particle B on the desired line of action, we can form a couple consisting of $\boldsymbol{F}$ and $\boldsymbol{F}''$ having moment Fd. The original force $\boldsymbol{F}$ is therefore equivalent to the single force $\boldsymbol{F}'$ acting at B, together with this couple. This couple is further equivalent to any other having the same moment. We can summarise thus: any force $\boldsymbol{F}$ acting on a particle of a rigid body may be moved to any given point (the direction being unchanged) provided a couple $\boldsymbol{L}$ is added having a moment M equal to the moment of the given force about the given point.

Conversely, a force $\boldsymbol{F}$ acting on a particle of a rigid body together with a coplanar couple $\boldsymbol{L}$ may be combined into a single force whose line of action is such that its perpendicular distance d from the particle is given by $d = M/F$, where M is the moment of the couple. To ensure that the line of action is moved in the correct direction it should be verified that the moment of the single force about the particle on the original line of action has the same sense as that of the couple originally present.

4.8 Coplanar Forces: Resultant

It has already been pointed out that the resultant of a set of forces acting on a rigid body can be found by repeated application of the principle of transmissibility and the triangle law. Any two forces are combined into a single force passing through the intersection of their lines of action. The number of forces is therefore reduced by one. The process is repeated until there remains either a single force with a defined line of action, or a couple. For particular cases the method becomes essentially graphical and will not be pursued in detail. By the use of force - couple sets more general results can be derived.

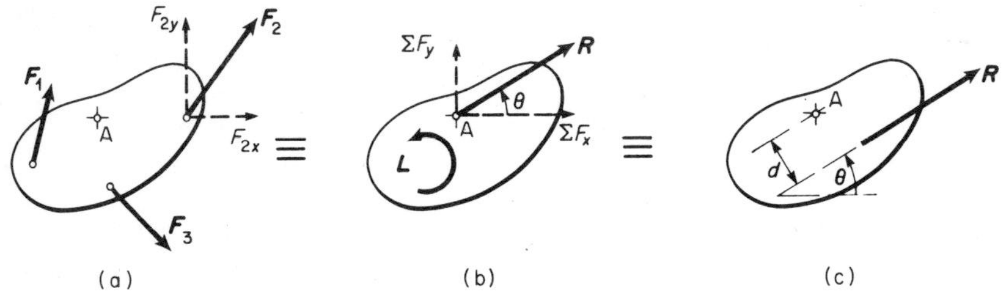

Figure 4.10

Consider the set of forces shown in figure 4.10a. Each force in turn can be replaced by an equal force at some particle A, together with a couple. The forces now acting at particle A can be combined into a single force $\boldsymbol{R}$ and the couples combined into a single couple $\boldsymbol{L}$. Thus the set is reducible to a force - couple set with the force acting at any chosen particle.

If $\boldsymbol{R}$ is not zero then from the previous section the force - couple set can in turn be replaced by a single force with a definite line of action; this force is the resultant R of the force set as shown in figure 4.10c and is given by

$$R = \sqrt{[(\Sigma F_x)^2 + (\Sigma F_y)^2]}$$

$$\tan\theta = \frac{\Sigma F_y}{\Sigma F_x}$$

$$d = \frac{\Sigma M_A}{R}$$

If $\boldsymbol{R}$ is zero then the force set has been reduced to a couple; this couple is the resultant couple $\boldsymbol{L}$ of the force set and has a moment ΣM_A.

Any set of coplanar forces acting on a rigid body can therefore be reduced either to a single force $\boldsymbol{R}$ or to a single couple $\boldsymbol{L}$ as the case may be. If in a particular case there is neither a resultant $\boldsymbol{R}$ nor a resultant couple $\boldsymbol{L}$ then the body is in equilibrium. Conversely if a body is known to be at rest then it is in equilibrium and there can be no resultant $\boldsymbol{R}$ or resultant couple $\boldsymbol{L}$.

Worked Example 4.1

In figure 4.11a the column AB has weight 2 kN, the centre of gravity being at D.

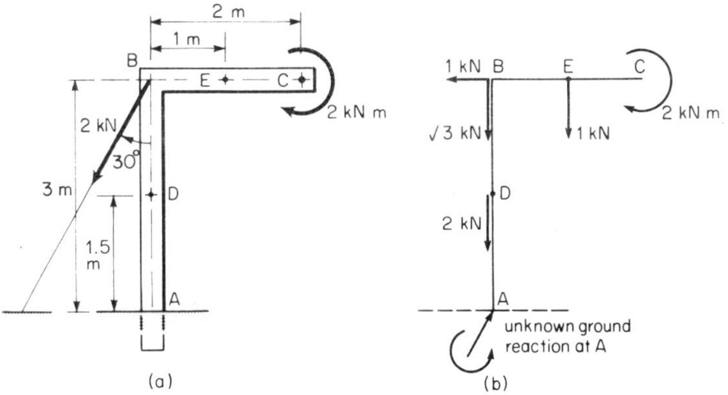

Figure 4.11

The overhanging portion, weight 1 kN with centre of gravity at E, carries a projecting pin at C to which a torque is applied as shown. The force at B is supplied by a stay wire. By transferring these forces to the foot of the column deduce the reaction of the ground on the column.

Solution

The free-body diagram is shown in figure 4.11b in which the force at B is resolved into two vector components where

$$F_{Bx} = 2 \sin 30° = 1 \text{ kN} \leftarrow$$

$$F_{By} = 2 \cos 30° = \sqrt{3} \text{ kN} \downarrow$$

The couple at C is equivalent to a couple at A having the same moment.

The 1 kN force at E is equivalent to a vertical 1 kN force at A together with a clockwise couple having moment 1 × 1 = 1 kN m.

The 2 kN force at D can be moved to A in its line of action. The 1 kN force at B is equivalent to a horizontal 1 kN force at A together with an anticlockwise couple having moment 1 × 3 = 3 kN m.

The $\sqrt{3}$ kN force at B can be moved to A in its line of action. Thus $\rightarrow\Sigma F_x$ at A = − 1 kN; $\uparrow\Sigma F_y$ at A = −1 −2 − $\sqrt{3}$ = − 4.732 kN; $\circlearrowleft\Sigma M_A$ = − 2 − 1 + 3 = 0.

For equilibrium of the body $\boldsymbol{R}$ = 0 and $\boldsymbol{L}$ = 0 and it follows that the ground reaction at A must have force components R_x = 1 kN →; R_y = 4.732 kN ↑ and that the couple exerted is found to be zero. If the applied couple at C had moment 3 kN m there would be an anticlockwise couple, moment 1 kN m, exerted by the ground on the body at A.

4.9 Conditions for Equilibrium

In practice we encounter rigid bodies that are known or are seen to be in equilibrium under the action of a set of forces. In a particular case the magnitudes, directions and lines of action of some of the forces may be known (the purpose of the rigid body being to support or transmit such forces) and it is desired to determine the characteristics of the remaining forces. To do so it is necessary to set out the conditions for equilibrium, $\boldsymbol{R}$ = 0 and $\boldsymbol{L}$ = 0, in the form of scalar equations that can be solved for the desired information. For this purpose the forces are expressed in terms of their components.

Consider again, therefore, any arbitrary set of coplanar forces acting on a rigid body. As found in section 4.8 the set must reduce either (1) to a single force $\boldsymbol{R}$ with vector components $\boldsymbol{R}_x$ and $\boldsymbol{R}_y$ where $R_x = \Sigma F_x$ and $R_y = \Sigma F_y$ in given x- and y-directions, or (2) to a couple $\boldsymbol{L}$ having moment M.

Choose three points A, B and C.

(a) If $\Sigma F_x = 0$, then $\boldsymbol{R}_x = 0$, but it is possible that the set reduces to $\boldsymbol{R}_y$ only or $\boldsymbol{L}$ only; if now $\Sigma F_y = 0$, then $\boldsymbol{R}_y = 0$, but it is possible that the set reduces to $\boldsymbol{L}$ only; if now $\Sigma M_A = 0$, then $\boldsymbol{L} = 0$. If $\boldsymbol{R}_x = 0, \boldsymbol{R}_y = 0, \boldsymbol{L} = 0$, then the body must be in equilibrium.

(b) If $\Sigma F_x = 0$, then $\boldsymbol{R}_x = 0$, but it is possible that the set reduces to $\boldsymbol{R}_y$ only or $\boldsymbol{L}$ only; if now $\Sigma M_A = 0$, then $\boldsymbol{L} = 0$, but it is possible that the set reduces to $\boldsymbol{R}_y$ only passing through A; if now $\Sigma M_B = 0$, then provided AB is not perpendicular to the x-direction, $\boldsymbol{R}_y = 0$. If $\boldsymbol{R}_x = 0, \boldsymbol{L} = 0, \boldsymbol{R}_y = 0$, then the body must be in equilibrium.

(c) If $\Sigma M_A = 0$, then $\boldsymbol{L} = 0$, but it is possible that the set reduces to $\boldsymbol{R}$ passing through A; if now $\Sigma M_B = 0$, then it is possible that the set reduces to $\boldsymbol{R}$ passing through A and B; if now $\Sigma M_C = 0$, then provided C is not on the line AB, $\boldsymbol{R} = 0$. If $\boldsymbol{L} = 0, \boldsymbol{R} = 0$ then the body must be in equilibrium.

It follows that for a rigid body to be in equilibrium, the conditions for equilibrium contained in any one of the following groups of equations must be satisfied.

$$\Sigma F_x = 0, \Sigma F_y = 0, \Sigma M_A = 0$$

where A is any point; or

$$\Sigma F_x = 0, \Sigma M_A = 0, \Sigma M_B = 0$$

where A and B are any two points not both on a line perpendicular to the x-direction; or

$$\Sigma M_A = 0, \Sigma M_B = 0, \Sigma M_C = 0$$

where A, B and C are any three points that are not collinear.

The three equations in any group can be solved for not more than three unknown quantities.

4.10 Solution of Problems

The solution of problems in statics follows the basic procedure described in the following paragraphs. The fundamental requirement for the consideration of the equilibrium of a rigid body is that all external forces are properly accounted for, therefore the importance of the free-body diagram in ensuring this cannot be overemphasised. This diagram is not a sketch to illustrate the solution but it is an integral part of the solution, and should not be omitted or drawn haphazardly.

(1) Choose the body that is known or is required to be in equilibrium. The body chosen may be a part of a larger body or one of an assembly of connected bodies that are as a whole in equilibrium. Consider the chosen body in isolation and draw a neat diagram on which the boundaries are clearly defined.

This free body or system is subject to the action of its surroundings, namely external forces that can be grouped as follows.

(a) *Applied forces*, namely all forces other than those referred to under (b) and (c).
(b) *Gravitational forces*, namely the weight of the body or its component in the plane of the other forces.
(c) *Constraint forces* or *reactions*, namely those external forces required to maintain the position or configuration of the free body and which can be regarded as being brought (simultaneously) into play as a result of the action of the applied and gravitational forces; it should be noted that there will be a constraint force or reaction at every point where the free body is in contact with any kind of support; constraint forces can therefore arise as forces exerted by other parts of a larger

body or by other connected bodies from which the chosen body has been separated.

Forces in groups (a) and (c) are also classed as *surface forces* while those of

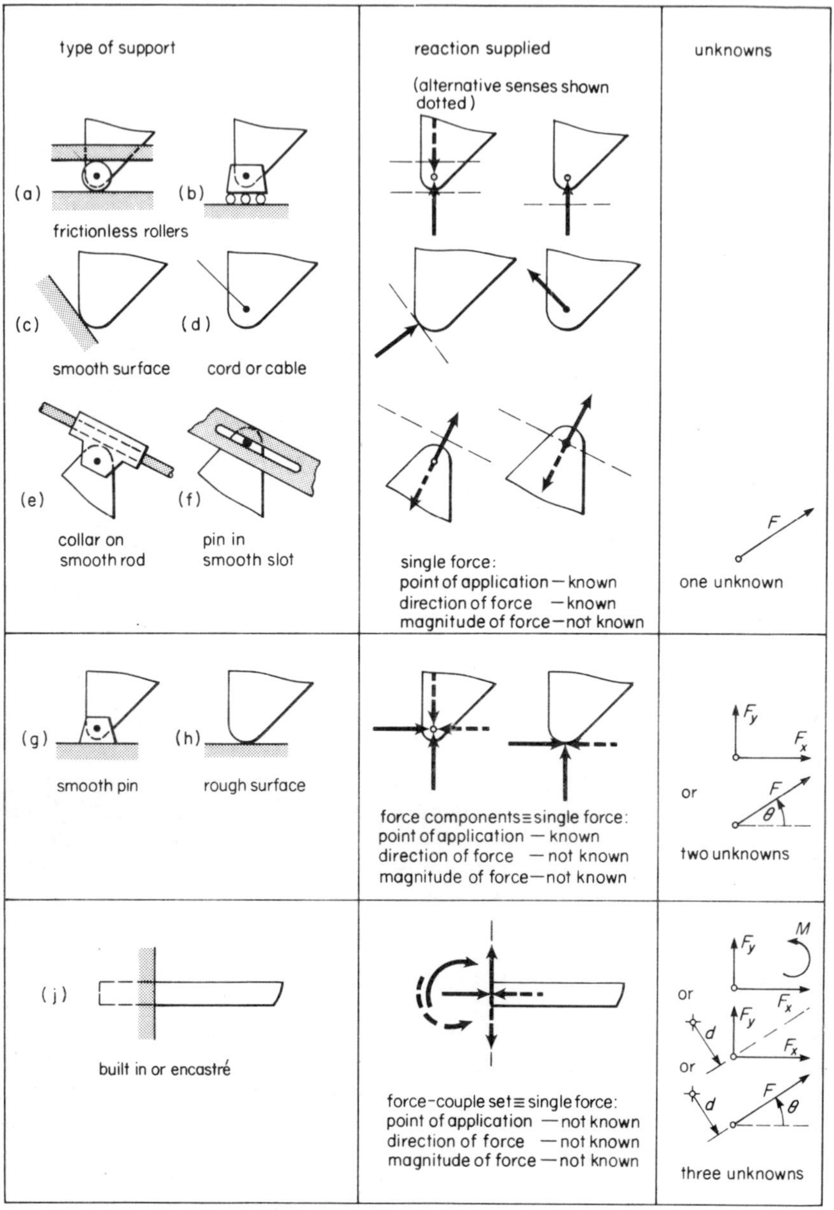

Figure 4.12

groups (b) are *body forces*. Some of these forces may not be known completely, in particular, those in group (c). However, we can note that a force in this latter category is equivalent to a force at a known point with known direction, or a force at a known point with unknown direction, or a force at an unknown point with unknown direction, the number of unknowns being respectively one, two or three in number. Figure 4.12 illustrates for reference some typical examples of constraints that are encountered in practice and the nature of the forces that can be supplied by those constraints.

(2) Show all external forces, both known and unknown, on the diagram. If any forces are unknown use appropriate symbols, preferably indicating components for ease of calculation. This diagram is the free-body diagram for the problem.

(3) Apply the conditions for equilibrium and solve the equations for the unknown quantities. This step usually calls for some thought since a suitable choice of reference directions and moment centres can often expedite a solution.

(4) Check solutions for reasonableness and arithmetical accuracy. It is always desirable to verify correctness by inserting the solution in an equation that has not already been used under paragraph (3).

4.11 Composite and Connected Bodies

Most engineering applications are concerned with assemblies of elementary rigid bodies that are connected in various ways to each other or to some fixed support. The conditions for equilibrium set out in section 4.9 are for a rigid body and obviously may be directly applied to an individual body or member, or to a group of rigidly connected members of an assembly. The conditions can also, however, be applied to a non-rigid assembly of connected members if their configuration is maintained constant by the applied forces and couples, for the assembly may then be regarded as a rigid body. Thus each individual body or member of the assembly, or group of connected members that is maintained in a fixed configuration, can be chosen as a free body and the forces required for equilibrium of that free body determined.

An essential fact to note is that by Newton's third law there are at each connection mutual forces or reactions that are equal in magnitude and oppositely directed along the same line of action. Thus in figure 4.12 a force shown as acting on the body from the support is accompanied by an equal and opposite force acting on the support from the body. However, as far as the free-body diagram of the body is concerned, it is only the force on the body that is indicated. Similarly, for two connected bodies, the free-body diagram for one body must show (as one of the external forces) only the force exerted by the other body to which it is connected.

Following from the conditions for equilibrium it is possible to write three, and only three, independent equations for each individual body or member. Thus an assembly of three connected members gives rise to nine independent equations allowing a solution for nine unknowns. However, in certain problems, depending on the number of unknowns required to be evaluated, it will not be necessary to set up all the independent equations. Free-body diagrams of connected groups do not give rise to further independent equations although they may yield more useful combinations of the same equations.

Typical of the engineering examples of connected bodies are trusses, frames, mechanisms and composite beams. The analysis of these assemblies is directed towards the determination of the internal forces in the members, but a necessary preliminary is the determination of all the external forces acting on each member separately using the principles already discussed.

In the analysis of connected bodies it is useful to recognise two types of member that are frequently encountered: these are referred to as *two-force* and *three-force* members respectively.

(1) Two-force member: if a body is in equilibrium under the action of two forces only then the two forces have equal magnitudes and opposite senses in the same line of action.
(2) Three-force member: if a body is in equilibrium under the action of three forces then their lines of action meet at a point, unless they are parallel.

If these conditions were not met then clearly there would be a resultant force in case (1) and a resulting moment about the intersection of two of the forces in case (2), and a state of equilibrium could not exist.

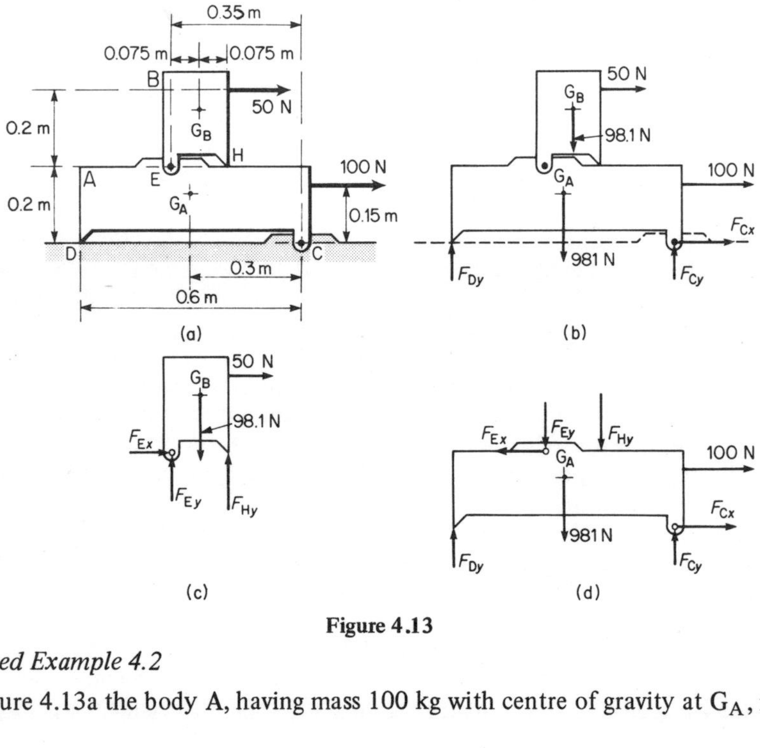

Figure 4.13

Worked Example 4.2

In figure 4.13a the body A, having mass 100 kg with centre of gravity at G_A, is hinged to a fixed point C and a leg at D rests on a smooth surface. The body B, having mass 10 kg, with centre of gravity at G_B, is hinged to A at E and is also supported at H, the contact surfaces at this point being smooth. If the system is in equilibrium under the applied 50 N force and 100 N force indicated, determine the horizontal and vertical components of (1) the reactions of the fixed surface on the body A at D and C and (2) the reactions of the body B on body A at E and H.

Solution

Following section 4.10, free-body diagrams are drawn for the assembly of body A and body B as shown in figure 4.13b, for body B alone, as in figure 4.13c, and for body A alone, as in figure 4.13d.

Using the categories of section 4.10 we insert into each free-body diagram forces as follows.

(a) The applied forces; in the case of figure 4.13b these are the 50 N and 100 N forces, in figure 4.13c it is the 50 N force only and in figure 4.13d the 100 N force only.
(b) Gravitational forces; these are the weights of A and B acting through G_A and G_B respectively in figure 4.13b and the separate weights in figures 4.13c and 4.13d.
(c) Constraint forces or reactions exerted by other parts of the system on the particular free body being considered, as follows.

(i) In the case of figure 4.13b the reactions of the fixed surface on body A at points C and D; from a study of figure 4.12 it follows that there can only be an upward force at D, called F_{Dy}, while at C the force can have both horizontal and vertical components that are unknown in sense; senses have therefore been assumed for these components, which are symbolised by F_{Cy} and F_{Cx}.
(ii) In the case of figure 4.13c the reactions of body A on body B at points E and H; F_{Hy} can only be vertically upwards but the senses of the two components F_{Ex} and F_{Ey} at E are again unknown and have to be assumed.
(iii) In the case of figure 4.13d the reactions of the fixed surface on body A at points C and D as already described in (i) and the reactions of body B on body A at points E and H; the senses of F_{Hy}, F_{Ex} and F_{Ey} in figure 4.13d must be opposite to those in figure 4.13c to conform with the third law. Note that the sense of F_{Ex} and F_{Ey} in figure 4.13d must follow logically from the senses assumed in figure 4.13c.

Each separate member gives rise to three independent equations and since all six unknowns (F_{Ex}, F_{Ey}, F_{Hy}, F_{Dy}, F_{Cx} and F_{Cy}) are to be evaluated it will be necessary to write down all six independent equations in order to provide a solution. Equilibrium conditions as set out in section 4.9 are used to write down the six equations, each equation preferably containing only one unknown. The latter can often be accomplished by judicious choice of moment-centres or reference directions.

From figure 4.13b: $\Sigma M_C = 0$

$$981 \times 0.3 + 98.1 \times 0.275 - 50 \times 0.4 - 100 \times 0.15 - 0.6 \times F_{Dy} = 0$$

therefore

$$F_{Dy} = 477 \text{ N}$$

$\Sigma F_x = 0$

$$50 + 100 + F_{Cx} = 0$$

therefore

$$F_{Cx} = -150 \text{ N}$$

$\Sigma F_y = 0$

$$F_{Dy} + F_{Cy} - 981 - 98.1 = 0$$

therefore

$$F_{Cy} = 602 \text{ N}$$

From figure 4.13c: $\Sigma M_E = 0$

$$F_{Hy} \times 0.15 - 98.1 \times 0.075 - 50 \times 0.2 = 0$$

therefore

$$F_{Hy} = 115.7 \text{ N}$$

$\Sigma F_x = 0$

$$F_{Ex} + 50 = 0$$

therefore

$$F_{Ex} = -50 \text{ N}$$

$\Sigma F_y = 0$

$$F_{Ey} + F_{Hy} - 98.1 = 0$$

therefore

$$F_{Ey} = -17.6 \text{ N}$$

Strictly figure 4.13d is unnecessary since all the required answers are available; it should be retained, however, and used to check the already calculated values. This can be done by inserting these values into the following equation, deduced from figure 4.13d, for the equilibrium of body A.

$$\Sigma M_C = F_{Ex} \times 0.2 + F_{Ey} \times 0.35 + 981 \times 0.3 + F_{Hy} \times 0.2 - 100 \times 0.15 - F_{Dy} \times 0.6$$

Inserting values the right-hand side becomes

$$(-50) \times 0.2 + (-17.6) \times 0.35 + 981 \times 0.3 + 115.7$$

$$\times\ 0.2 - 100 \times 0.15 - 477 \times 0.6 = 0$$

as required. (Note the retention of the negative signs in the values of F_{Ex} and F_{Ey}.)

The required answers are as follows.

(1) Reactions of the fixed surface on A at C and D are F_{Cx}, F_{Cy} and F_{Dy} in figure 4.13b (or d); taking account of the assumed directions and the signs obtained for the quantities the correct reaction components are

at C $\leftarrow$ 150 N, $\uparrow$ 602 N

at D $\uparrow$ 477 N

(2) Reactions of the body B on body A at E and H and are F_{Ex}, F_{Ey} and F_{Hy} in figure 4.13d; taking account of signs the correct reaction components are

at E $\rightarrow$ 50 N, $\uparrow$ 17.6 N

at H $\uparrow$ 115.7 N

4.12 Simple Trusses

A simple truss is a structure made up of straight two-force members usually of uniform cross-section; the connections are regarded as being made by smooth pinned joints, and the members are so connected that the structure as a whole is rigid under the action of a set of forces applied at the joints in the plane of the truss.

The simplest rigid assembly is the triangle made up of three members shown in figure 4.14a, the members being represented in the diagram by straight lines.

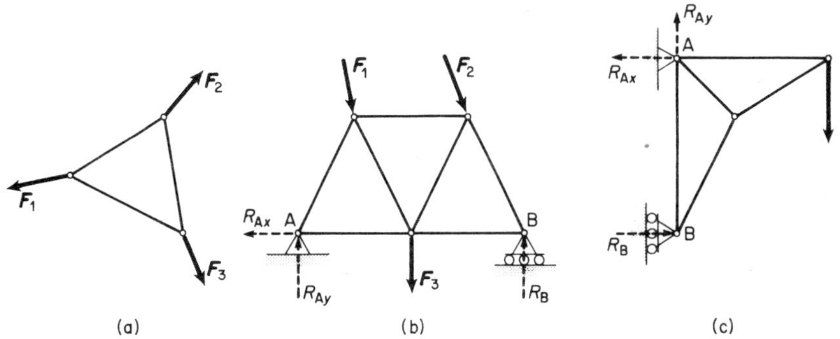

Figure 4.14

The figure shows other configurations developed by connecting further members in pairs to form additional rigid triangles, the whole assembly or truss then being connected to or being in contact with the supports at appropriate joints.

The support shown diagrammatically at A in figures 4.14b and c represents a smooth pinned joint as detailed in figure 4.12g; the support at B in both figures represents a roller support as detailed in figure 4.12b.

The truss is in equilibrium under the action of applied forces or loads, and the reactions from the supports. The conditions for equilibrium require in general that the supports supply three reaction components and conversely, the conditions will enable three such components to be determined. Figures 4.14b and c show how three components of reaction are supplied by a roller and pinned joint. A simple truss supported in this manner is said to be statically determinate. If the truss in figure 4.14b or 4.14c had been connected to the support at two pinned joints then a further reaction component would have been supplied. The determination of the four components would have required some additional condition beyond those relating to equilibrium. The truss is then said to be statically indeterminate. We shall only consider trusses and other assemblies that are statically determinate.

Each pin of the truss is in equilibrium under the action of the forces exerted by the members meeting at the joint. In addition there are applied forces and reactions at certain joints. Since by definition the members are two-force members, the forces (having equal magnitude and opposite sense) exerted *on any member* by the pins at the ends of that member have lines of action coinciding with the line joining the pins. By the third law the directions of the member forces *on the pin* at each joint are therefore known; the magnitude and sense of each member force remain to be determined. Applying the conditions for equilibrium of a particle, treating the pin as a free body, the member forces at each joint can now be determined, analytically or graphically, provided the unknown forces at the joint considered do not exceed two in number. The analysis commences at a joint where a solution for the member forces is possible and then moves from joint to joint, noting that since the members are two-force members the solutions are progressively carried over from one joint to another, care being taken to change the sense of a member force on passing to an adjacent joint. It is advisable to calculate the support reactions initially by treating the whole truss as a free body; this can afford a check on the correctness of the analysis since the pins at the support joints must be individually in equilibrium.

The complete solution yields the forces exerted on the pins. These forces, reversed in sense, are the forces on the members. If these latter forces are directed inwards along a member then it is said to be in compression, and if outwards, in tension. A member in compression can be referred to as a strut, and a member in tension as a tie.

Worked Example 4.3

The truss in figure 4.15 is pin jointed at A and B and carries applied loads, magnitudes 1000 N and 2000 N, as shown. Determine (a) the force in each member stating whether it is a strut or tie and (b) the reactions of the truss on the wall at A and B.

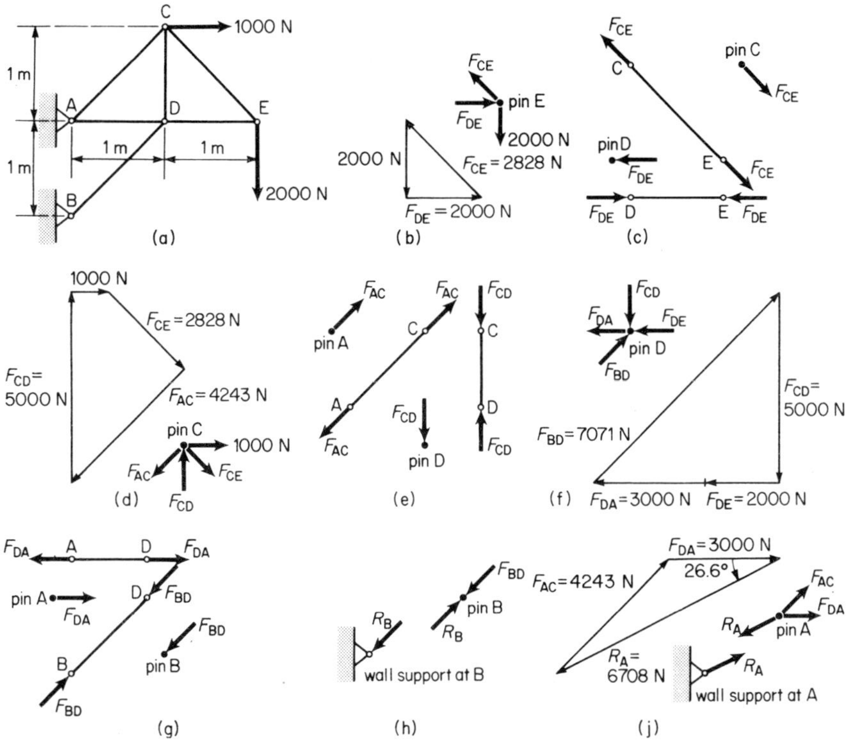

Figure 4.15

Solution

This type of problem lends itself to graphical solution since the force directions are known to be along the lines representing the members. We start at a pin where there are only two unknown forces – for example, in this particular case it is advantageous to start by considering the equilibrium of pin E.

The forces acting at E are the applied 2000 N force and the forces in the members CE and DE, the directions of which lie along these members. A triangle of forces using the three directions and the sense of the 2000 N force is drawn to close in figure 4.15b and the senses and values of F_{CE} and F_{DE} are deduced from this.

The force exerted by pin E on link CE is opposite to that in figure 4.15b and it follows that the forces exerted on CE are as indicated on figure 4.15c, therefore CE is a tie. By the same arguments the forces exerted on DE are as indicated, showing it to be a strut. From the force acting at end C of CE the force exerted by CE on pin C is deduced and shown in figure 4.15c. The force exerted by DE on pin D is also indicated.

We may now consider the equilibrium of pin C, which has four forces acting upon it: the 1000 N applied force, the known force F_{CE} exerted on C in the direction indicated in figure 4.15c and the forces in the members AC and CD. A polygon of forces is drawn (figure 4.15d) and from this the values and senses of F_{CD} and F_{AC} are deduced. The forces exerted on AC and CD are thus as given in figure 4.15e, indicating AC as a tie and CD as a strut. The force exerted by AC on pin A is deduced and also that exerted by CD on pin D.

There are now only two unknown forces acting at pin D and the relevant force polygon, figure 4.15f, is drawn to give the values and senses of F_{DA} and F_{BD}.

The forces acting on AD and BD (figure 4.15g) show AD as a tie and BD as a strut. The diagram also shows the forces exerted by AD on pin A and BD on pin B. From the latter, the forces on pin B are shown in figure 4.15h and the action of the truss on the wall at this point is 7071 N $\angle$ 225°.

Of the three forces acting on pin A, two are known and one is unknown in magnitude and direction. The triangle of forces in figure 4.15j gives the solution for $\boldsymbol{R}_A$ (of the wall on the pin) as 6708 N $\angle$ 206.6°; it follows that the action of the truss on the wall is 6708 N $\angle$ 26.6°.

The force $\boldsymbol{R}_A$ (wall on truss) could be determined from a polygon of forces for the complete truss; this would include the applied 1000 N and 2000 N forces, the force $\boldsymbol{R}_B$ exerted by the wall on the truss at B, and will be closed by $\boldsymbol{R}_A$. $\boldsymbol{R}_B$ could have been calculated, since its direction must be that of BD, from the free-body diagram for the whole truss, by taking moments about the point A.

The required answers are given below.

Member	Force (N)	Strut (S) or Tie (T)
AC	4243	T
AD	3000	T
BD	7071	S
CD	5000	S
CE	2828	T
DE	2000	S

4.13 Simple Frames

A simple frame is an assembly of connected bodies, some or all of which are multi-force members. For our present purpose the connections are regarded as being made by smooth pinned joints. The frame will not necessarily need to be rigid when independent of the support, and the applied forces can act at points other than the connections.

Typical configurations are shown in figure 4.16. Figure 4.16b illustrates a simple case in which the frame is no longer rigid if detached from its supports. There are now four reaction components at the supports (and two internal equal and opposite pairs of components at the joint C) but these can all be determined using equilibrium conditions only since six independent equations can be written (see section 4.11). The vertical wall between joints A and B can, if necessary, be

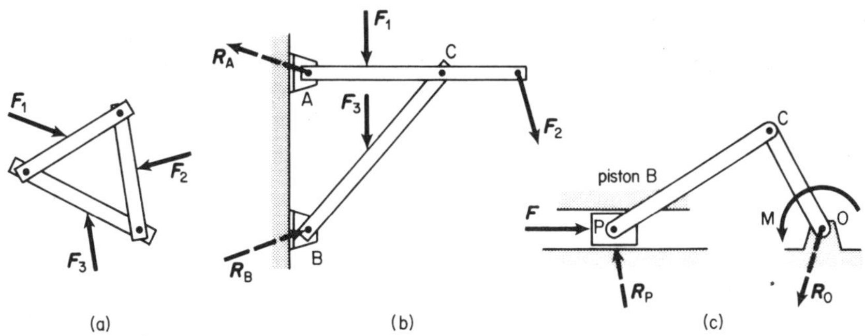

Figure 4.16

regarded as a member ensuring the rigidity of the frame. The frames in figures 4.16a and b are both fixed in position and configuration and are therefore being maintained in a state of equilibrium.

Figure 4.16c is strictly that of a *mechanism*; this is discussed further in chapter 11 but is included here with frames since the force analysis is the same for both. A mechanism can be regarded as a frame in which one of the support constraints has been removed; this implies that the configuration of the mechanism can change, but it is the purpose of a mechanism to transform movement at one point to a corresponding movement at another. If there are forces acting on the mechanism then it is called a *machine* in which an applied force or torque at one point is required to balance a force or torque applied at another point. We may thus regard a particular configuration of the machine as being fixed under the applied forces. From section 4.11 it follows that any member of the machine, or any group of connected members, can be selected as a free body that is in equilibrium. For example, the free body consisting of the whole engine mechanism of figure 4.16c is in equilibrium under the external forces $\boldsymbol{F}$, $\boldsymbol{R}_\mathrm{P}$, $\boldsymbol{R}_\mathrm{O}$ and the torque M; the free body consisting of the connecting rod CP and piston B is in equilibrium under the external force set consisting of $\boldsymbol{F}$, $\boldsymbol{R}_\mathrm{P}$ and the action of OC on CP through the pin at C. Mechanisms are one example where it is sometimes advantageous to consider a connected group of members as a free body rather than each individual member.

Whereas for a truss the analysis involved the equilibrium conditions at the joints, for a frame the analysis proceeds by isolating each member, or group of connected members as discussed above, as a free body in equilibrium under the action of the applied forces and the action of those other members that are connected to it at pinned joints. The pin at a joint can be imagined to be an integral part of one of the members meeting at the joint. At each pin there exists an opposing pair of internal reaction components. The worked example 4.4 indicates a suitable method of indicating the pin reaction components, by which each component in a pair of opposed components is assigned one symbol; the senses in which the reaction components are assumed to be positive for the member under consideration are then entered on the free-body diagram of that member; at the same time the opposing components are entered on the free-body diagrams of the adjacent members.

By repeated application of the conditions of equilibrium the forces on each isolated member can be determined. It is here that care should be taken to avoid a multiplicity of simultaneous equations by judicious selection of reference directions and moment-centres. It should again be noted (see section 4.11) that the number of independent equations available is three times the number of separate members although this number of equations may not need to be set up in a particular problem. The unknown should be determined and inserted in successive equations as the solution proceeds.

Worked Example 4.4

Determine the forces acting on each member of the frame shown in figure 4.17a when carrying the 4 kN load at the point F.

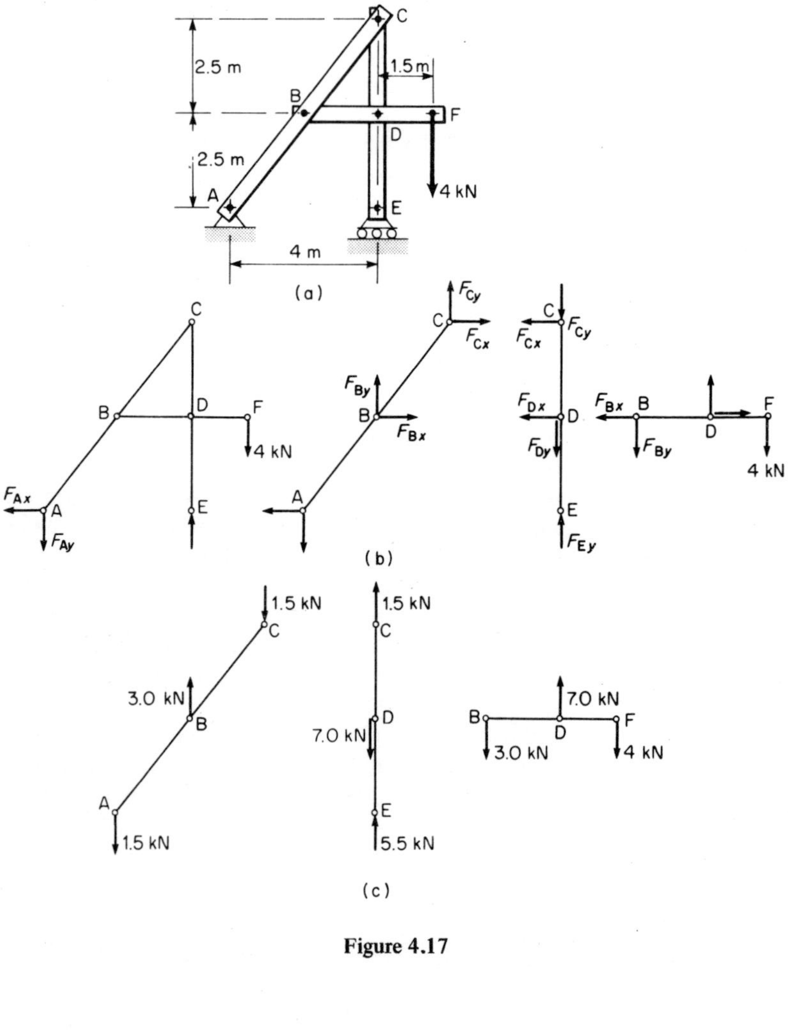

Figure 4.17

Solution

Since the roller at E and the pin at A together supply three reaction components they can be determined without difficulty. Free-body diagrams of the frame and of the individual members are drawn (figure 4.17b). The pin reaction components at each joint are entered in opposing pairs with one symbol assigned to each pair, as illustrated at joint C. There is clearly no need to letter all components since the symbol assigned has the same form at each joint. The direction of the opposing pairs is initially assumed but it is essential that the sense in which a component is shown be maintained throughout the analysis until all the magnitudes have been determined. Following the arguments at the end of the previous section only nine independent equations can be written and since from the free-body diagrams there are nine unknowns, all nine equations must be utilised. Equilibrium conditions are thus used to write down nine equations, preferably those equations that contain only one unknown.

For the frame

$\Sigma M_A = 0$

$$F_{Ey} \times 4 - 4 \times 5.5 = 0 \qquad F_{Ey} = 5.5 \text{ kN}$$

$\Sigma F_y = 0$

$$-F_{Ay} + 5.5 - 4.0 = 0 \qquad F_{Ay} = 1.5 \text{ kN}$$

$\Sigma F_x = 0$

$$-F_{Ax} = 0 \qquad F_{Ax} = 0$$

Member BF

$\Sigma M_B = 0$

$$-4 \times 3.5 + F_{Dy} \times 2 = 0 \qquad F_{Dy} = 7.0 \text{ kN}$$

$\Sigma F_y = 0$

$$-F_{By} + 7.0 - 4.0 = 0 \qquad F_{By} = 3.0 \text{ kN}$$

Member CE

$\Sigma M_C = 0$

$$-F_{Dx} \times 2.5 = 0 \qquad F_{Dx} = 0$$

$\Sigma F_x = 0$

$$-F_{Cx} - 0 = 0 \qquad F_{Cx} = 0$$

$\Sigma F_y = 0$

$$-F_{Cy} - 7.0 + 5.5 = 0 \qquad F_{Cy} = -1.5 \text{ kN}$$

Member BF

$\Sigma F_x = 0$

$$-F_{Bx} + 0 = 0 \qquad F_{Bx} = 0$$

As a check verify equilibrium of member AC

$$\Sigma M_C = F_{Ay} \times 4 - F_{Ax} \times 5 - F_{By} \times 2 + F_{Bx} \times 2.5$$
$$= 1.5 \times 4 - 3.0 \times 2$$
$$= 0$$

The forces are now shown with correct senses on the free-body diagrams of the members (figure 4.17c).

Worked Example 4.5

For the frame shown in figure 4.18a determine the magnitude of the forces on the three pins at A, B and C.

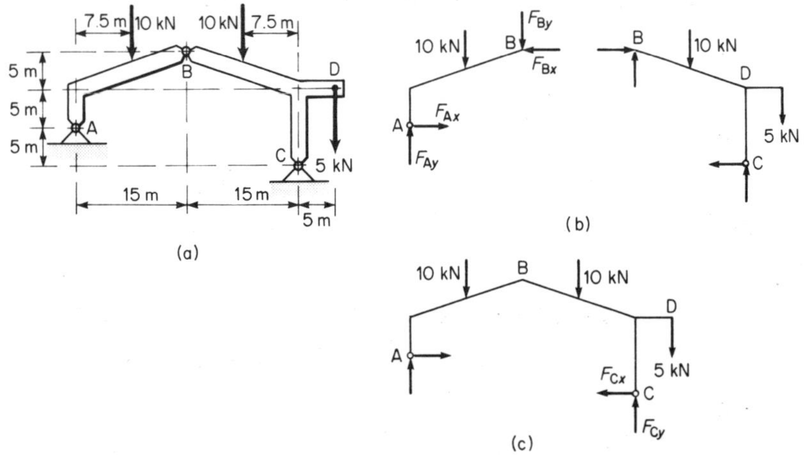

Figure 4.18

Solution

Since the pins at A and C supply four reaction components they cannot be determined completely from the equilibrium conditions for the frame. Free-body diagrams are drawn as in figure 4.18b for the two parts of the frame and in figure 4.18c for the frame as a whole.

Member AB

$\Sigma M_A = 0$

$$-F_{By} \times 15 + F_{Bx} \times 10 - 10 \times 7.5 = 0$$

Member BC

$\Sigma M_C = 0$

$$-F_{By} \times 15 - F_{Bx} \times 15 + 10 \times 7.5 - 5 \times 5 = 0$$

Solving these two equations, $F_{Bx} = 5$ kN and $F_{By} = -5/3$ kN.

Member AB

$\Sigma F_x = 0$

$$F_{Ax} - 5 = 0 \qquad F_{Ax} = 5 \text{ kN}$$

$\Sigma F_y = 0$

$$F_{Ay} - 10 - (-5/3) = 0 \qquad F_{Ay} = 25/3 \text{ kN}$$

Member BC

$\Sigma F_x = 0$

$$-F_{Cx} + 5 = 0 \qquad F_{Cx} = 5 \text{ kN}$$

$\Sigma F_y = 0$

$$F_{Cy} + (-5/3) - 10 - 5 = 0 \qquad F_{Cy} = 50/3 \text{ kN}$$

Check for the frame

$$\Sigma M_A = -10 \times 15/2 - 10 \times 45/2 - 5 \times 35 - 5 \times 5 + 50/3 \times 30 = 0$$

$$F_A = \sqrt{(5^2 + 8.33^2)} = 9.72 \text{ kN}$$

$$F_B = \sqrt{(5^2 + 1.67^2)} = 5.27 \text{ kN}$$

$$F_C = \sqrt{(5^2 + 16.67^2)} = 17.38 \text{ kN}$$

Worked Example 4.6

The mechanism of figure 4.19a has massless links and is in equilibrium under the applied force F acting on link BC and an unknown torque Q acting on a shaft at D which is rigidly fixed to the link CD. Determine this torque and the reactions of the supports on the mechanism at A and D.

Solution

Free-body diagrams are drawn, see figures 4.19b, c, d and e for the whole mechanism and the separate links. The reactions shown in the diagrams have been drawn in an arbitrary fashion when with experience many of them could be drawn in the correct direction (if not in sense); these directions can be decided in the following manner.

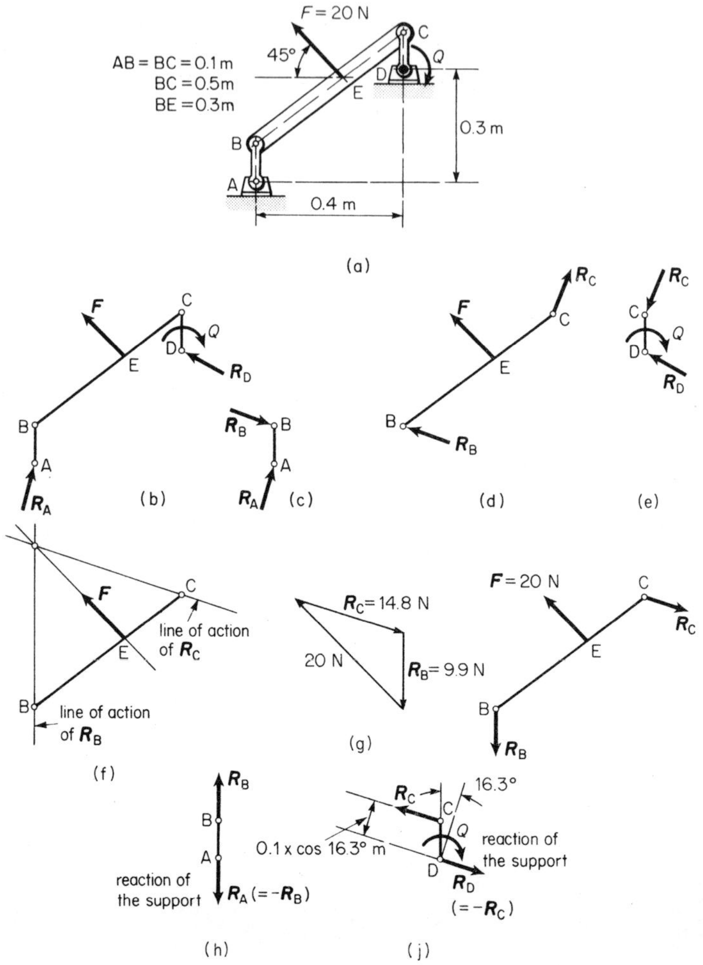

Figure 4.19

(1) Forces R_B and R_A acting on AB must lie along the line joining the hinged ends since it is a two-force member.
(2) The three forces acting on link BC are in equilibrium and therefore pass through one point. Figure 4.19f is drawn utilising this fact and the known direction of R_B to find the direction (but not the sense) of R_C.

With their directions known a triangle of forces, figure 4.19g, is drawn of the three forces acting on BC, the sense of the forces following from that of F.

The forces on the link AB are as indicated in figure 4.19h (showing it to be a tie) and it follows that the reaction of the support on the mechanism at A is 9.9 N$\angle -90°$.

The force $\boldsymbol{R}_D$ acting on link CD was indicated randomly in figure 4.19e, but it follows from the equilibrium of the link that $\boldsymbol{R}_D$ must be equal and opposite to $\boldsymbol{R}_C$; the diagram is drawn correctly in figure 4.19j, the direction of $\boldsymbol{R}_C$ in this being opposite, of course, to that in figure 4.19g. The external torque Q must be clockwise for equilibrium and its value can be determined by moments about any point; for example

$\Sigma M_D = 0$

$$R_C \times (0.1 \cos 16.3°) - Q = 0$$

therefore

$$Q = 1.42 \text{ N m}$$

The reaction of the support on the mechanism at D is, as indicated in figure 4.19j, 14.8 N $\angle -16.3°$.

4.14 Summary

(1) By Newton's third law the set of internal forces reduce to zero, therefore the external forces are the only forces that influence the resultant and the conditions for equilibrium of a rigid body.
(2) By the principle of transmissibility a force on a rigid body can be moved to any point on the line of action without affecting the resultant or the conditions for equilibrium.
(3) Two unequal parallel forces have a resultant, but two forces having equal magnitude but opposing senses have no resultant and are referred to as a couple.
(4) A set of parallel forces acting on specific particles of a body pass through a point whose position in the body does not depend on the orientation of the body.
(5) The magnitude M of the moment of a force magnitude F about a point is defined to be Fd, where d is the perpendicular distance from the point to the line of action of the force. For coplanar forces moments having anticlockwise sense are taken as being positive.
(6) The magnitude M of the moment of a couple $\boldsymbol{L}$ is Fd where d is the perpendicular distance between the two unlike parallel forces, each having magnitude F, constituting the couple. The moment is a property of the couple and does not depend on the choice of a moment centre.
(7) Coplanar couples having the same moment and sense are equivalent. A body can then be said to be subject to a torque of specific magnitude. Coplanar torques are added algebraically.
(8) A force acting on a rigid body may be replaced by an equal force having a parallel line of action, provided a couple having the correct moment and sense is introduced.
(9) For a set of coplanar forces acting on a rigid body, the magnitude R of the resultant, the direction θ relative to the x-direction and the distance d of the line of action from a point A are given by

$$R = \sqrt{[(\Sigma F_x)^2 + (\Sigma F_y)^2]}$$

$$\tan\theta = \frac{\Sigma F_y}{\Sigma F_x}$$

$$d = \frac{\Sigma M_A}{R}$$

(10) Any set of coplanar forces acting on a rigid body can be reduced either to a single resultant force $\boldsymbol{R}$, or if there is no resultant force, to a single couple $\boldsymbol{L}$. If there is neither a single resultant nor a single couple then $\boldsymbol{R} = 0$ and $\boldsymbol{L} = 0$ and the body is in equilibrium.
(11) If a body is in equilibrium then $\boldsymbol{R} = 0$ and $\boldsymbol{L} = 0$, and any one of the following groups of equations must be satisfied.

(a) $\Sigma F_x = 0$, $\Sigma F_y = 0$, $\Sigma M_A = 0$, where A is any point.
(b) $\Sigma F_x = 0$, $\Sigma M_A = 0$, $\Sigma M_B = 0$, where A and B are any two points not both on a line perpendicular to the x-direction.
(c) $\Sigma M_A = 0$, $\Sigma M_B = 0$, $\Sigma M_C = 0$, where A, B and C are any three points that are not collinear.

Note that for a rigid body only three independent scalar equations are supplied by the equilibrium conditions.
(12) In the solution of problems a free-body diagram should always be drawn on which all external forces must be indicated.
(13) A free-body diagram may be of any single member, or of any assembly or group of connected members whose configuration is fixed under the action of the external forces and couples.
(14) The external forces to be inserted in a free-body diagram are

(a) applied forces
(b) gravitational forces (weight)
(c) reactions of all other bodies and supports (note that there will be a reaction at every point of contact or support)

(15) For a two-force member in equilibrium the forces must have the same line of action; thus if the forces are applied at the ends of the member their lines of action lie along the line joining the ends.
(16) For a three-force member in equilibrium the lines of action of the three forces intersect at one point.
(17) For a system of n connected members there are $3n$ independent equations of equilibrium available; any further equations of equilibrium beyond this number will be redundant.
(18) A truss is an assembly of pin-jointed two-force members and is analysed by applying the conditions for equilibrium of a particle to each pin.
(19) A frame is an assembly of pin-jointed multi-force members and is analysed by applying the conditions for equilibrium to each member, a group of connected members, or the whole frame. A frame may be non-rigid when a support constraint is removed. When used to transmit forces and couples it is then classed as

a mechanism or a machine. If at any instant in a particular fixed configuration it is in equilibrium under the applied forces and couples.
(20) Graphical methods sometimes facilitate a solution especially if the system consists of two- and three-force members.

Problems

4.1 Reduce the force-couple set in figure 4.20 to (a) a single force through A plus a couple and (b) a single force. State the magnitude, direction and position where necessary.

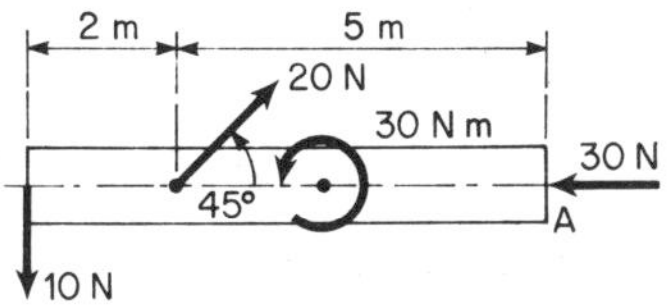

Figure 4.20

4.2 The massless rectangular plate ABCD in figure 4.21 is maintained in equilibrium by the forces shown. Determine the magnitude of P and the reaction of the plate on the hinge. (*Hint*: See conditions for equilibrium; draw a free-body diagram.)

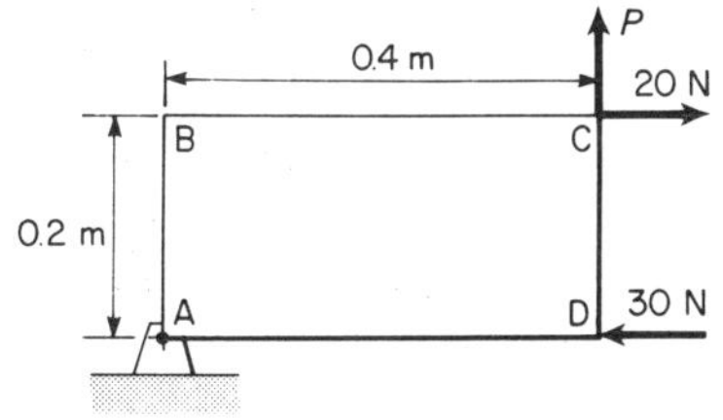

Figure 4.21

4.3 The rectangular plate in figure 4.22 is lying in the horizontal plane and is in equilibrium under two known forces (1000 N and 600 N) and three forces R_1, R_2, R_3 known in direction (lying along the full lines) but not in magnitude.

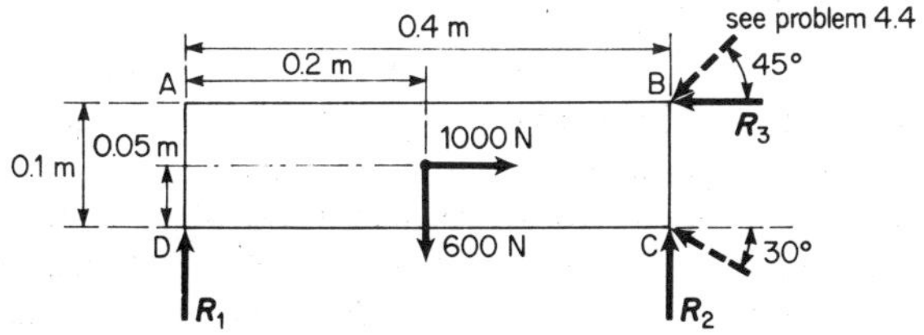

Figure 4.22

(a) Can the magnitudes of R_1, R_2 and R_3 be determined?
(b) If so, calculate these magnitudes.

(*Hint*: How many equations are available? A judicious choice of moment-centre simplifies the calculation.)

4.4 As for problem 4.3 but R_2 and R_3 are now inclined along the broken lines.

4.5 A rigid rod ABC, length 1 m, carries a vertical load 500 N at 0.2 m from A. The rod is supported in a horizontal position by three frictionless roller supports, one at A, another at B, 0.4 m from A, and the other at C.

(a) Can the vertical reactions at A, B and C be evaluated?
(b) If it assumed that $R_B = 0.5\ R_C$ calculate their values.
(c) How is the system in (a) described?

(*Hint*: Draw a free-body diagram and see conditions for equilibrium.)

4.6 Determine the reaction of the wall on the rod in figure 4.23 if this is in equilibrium under the applied forces 10 N and 20 N and has weight 5 N. (The centre of gravity is at G.) (*Hint*: Draw a free-body diagram assuming force components and a couple at the fixed end.)

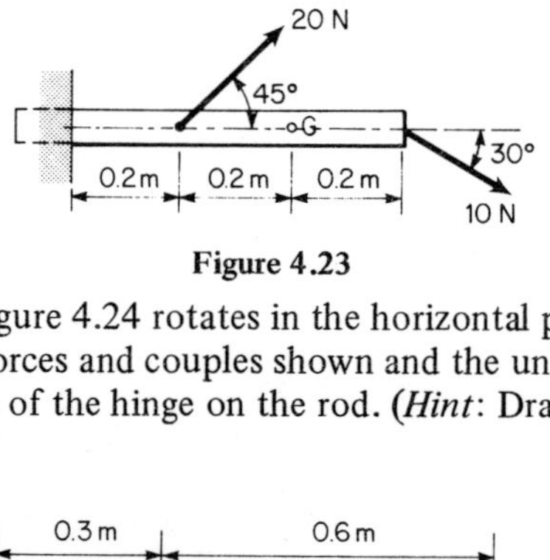

Figure 4.23

4.7 The hinged rod in figure 4.24 rotates in the horizontal plane but is held in equilibrium under the forces and couples shown and the unknown force F. Determine F and the reaction of the hinge on the rod. (*Hint*: Draw a free-body diagram.)

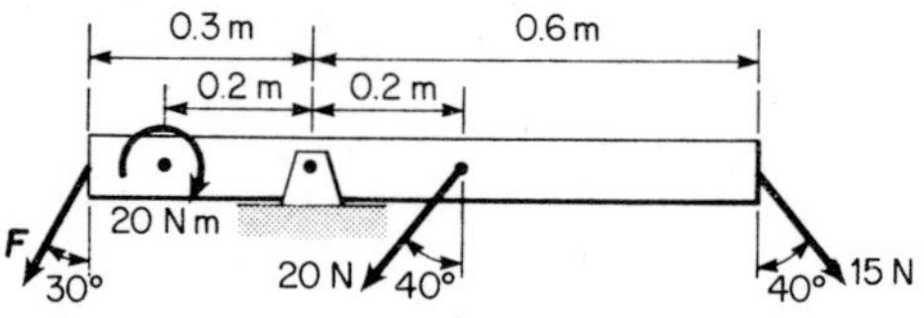

Figure 4.24

4.8 A rod AB, length 1 m, is hinged at its top end A so that it can rotate in a vertical plane. A force, magnitude P, at $\angle\ 0^\circ$ is applied at B in this vertical plane such that AB makes an angle 30° to the vertical. If the mass of the rod is 20 kg and its centre of gravity is 0.6 m from A, determine P and the reaction of the hinge on the body without taking moments. (*Hint*: Only three forces act on the rod.)

4.9 A massless square lamina ABCD with 1 m side is lying in a vertical plane and has cords attached to two top corners A and B and one bottom corner C. The cords from the two top corners are passed over frictionless pulleys and a body, mass 20 kg, is attached to that from A and a body, mass 40 kg, to the other. If a body, mass 50 kg, is attached to the third cord and the lamina rests in equilibrium find (a) the angles the upper cords make with the vertical and (b) the angle the top edge of the lamina makes with the horizontal. A graphical solution is acceptable. (*Hint*: Use the triangle of forces and the fact that only three forces act on the lamina; the tensions in each cord are uniform.)

4.10 An elastic cord of constant 20 N/m is fixed, just taut, between two points that are 1 m apart horizontally. If a particle of mass 2 kg is slowly dropped on to the cord show that when a state of equilibrium is reached all the relevant conditions are satisfied when the cord tension is 12.46 N and each portion of the cord makes an angle of 38° to the vertical. (*Hint*: Make use of the relationship of tension and extension; use conditions for equilibrium, also geometrical relationships.)

4.11 (a) A uniform beam of mass *m* is placed against a smooth vertical wall and stands on a smooth horizontal floor; can it ever be in equilibrium?

(b) A uniform ladder of length 10 m and mass 100 kg is placed to rest in a vertical plane with each end resting on a smooth surface, the surfaces being inclined towards each other and each making an angle of 45° with the horizontal. If a mass of 60 kg is placed at 3 m from one end find the angle of inclination of the ladder when it reaches a state of equilibrium. Is the 60 kg mass above or below the centre of gravity of the ladder? (*Hint*: Consider equilibrium of the free-body diagrams; for (b) note directions of reactions of surfaces and use conditions for equilibrium.)

4.12 The assembly of two connected bodies in figure 4.25 is in equilibrium in the situation shown. Find the applied force ***P***, the reactions of body A on body B at C and D, and the reactions of the surface on A. (*Hint*: Draw free-body diagrams

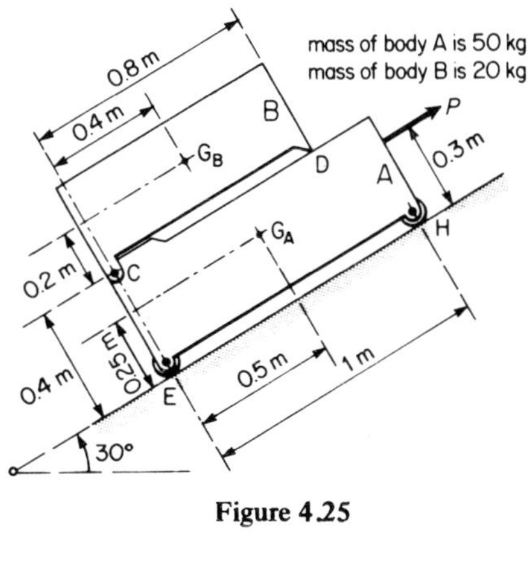

Figure 4.25

for each body and the assembly; make use of those directly applicable and make reference directions suitable for the problem. It is simpler for moment and force summations to resolve all forces parallel to and perpendicular to the chosen reference directions.)

4.13 The assembly in figure 4.26 is in equilibrium with the applied force $\boldsymbol{P}$ such that the reaction at C is just zero. Determine $\boldsymbol{P}$ and the reactions of body B on A at D and E. (*Hint*: See problem 4.12.)

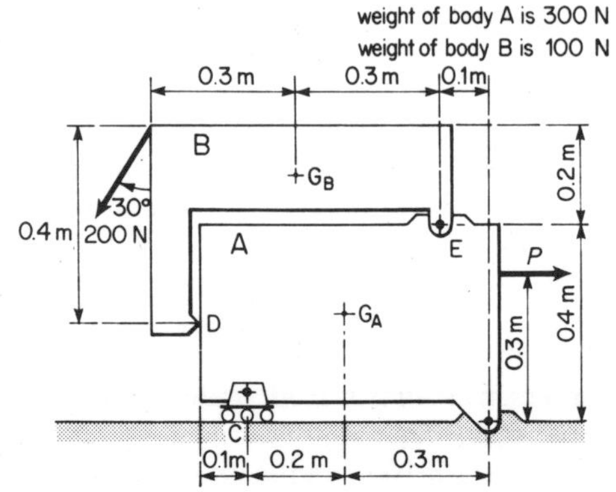

Figure 4.26

4.14 Determine, for the truss in figure 4.27 the forces in each member, stating whether each is a strut or a tie. Find also the reactions at A and E of the supports on the frame. (*Hint*: Consider the free body of the whole truss to find the reaction at E.)

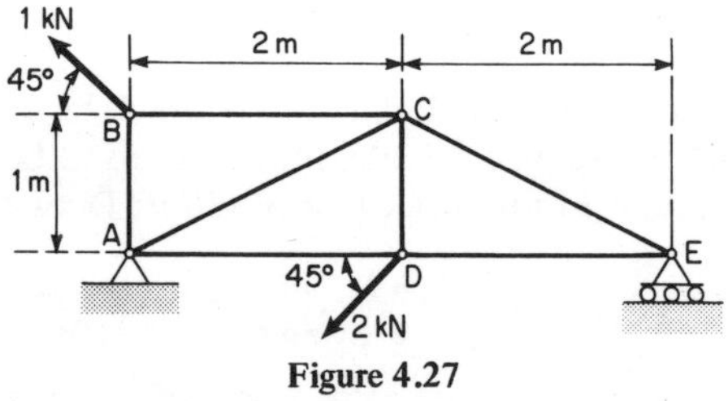

Figure 4.27

4.15 Determine the forces (stating whether struts or ties) in each member of the truss in figure 4.28. Would the force in AB be reduced if a member were used to connect AD rather than BE?

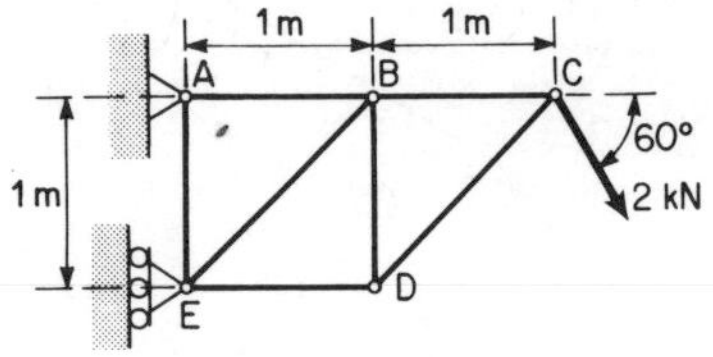

Figure 4.28

4.16 Determine the forces in each member of the truss in figure 4.29.

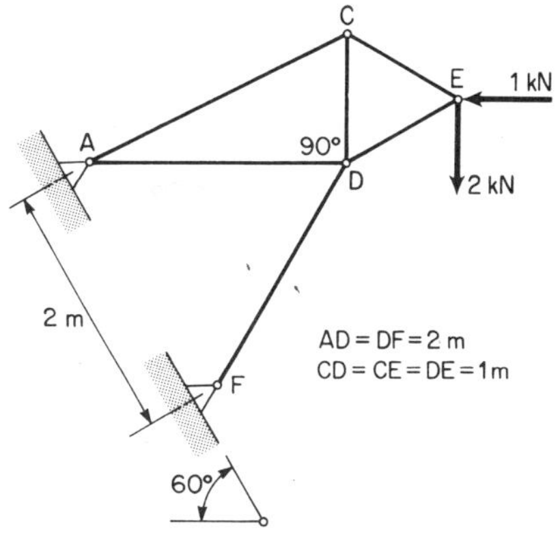

Figure 4.29

4.17 The mechanism in figure 4.30 is in equilibrium under the applied force of 20 N and a torque Q applied to the link AB through a shaft at A. Determine this applied torque and the reactions of the supports on the mechanism at A and D. Ignore the mass of the links. (*Hint*: See worked example 4.6.)

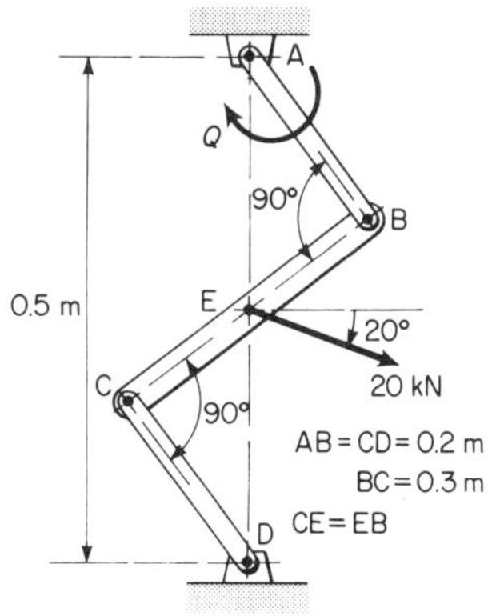

Figure 4.30

4.18 Determine the values of the reactions of the supports at A and C on the frame, which lies in the horizontal plane, of figure 4.31. What is the reaction of AB on BC at B? (*Hint*: Draw separate free-body diagrams; an analytical method is probably most suitable.)

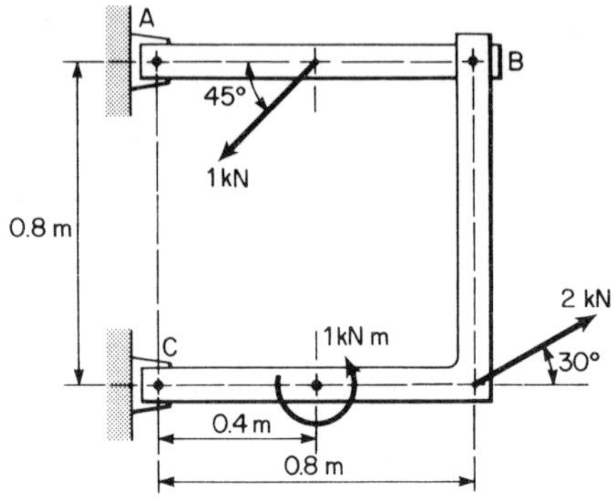

Figure 4.31

4.19 In the mechanism shown in figure 4.32 the small roller D moves in a frictionless fixed horizontal slide and roller A in a frictionless guide in link FC. Determine the external torque required on AE to maintain equilibrium when a force of 10 kN is applied at D, and also the reactions of the mechanism on the supports at D, E and F. Ignore the mass of the links. (*Hint*: Draw free-body diagrams; use knowledge of two-force and three-force members; consider a graphical solution.)

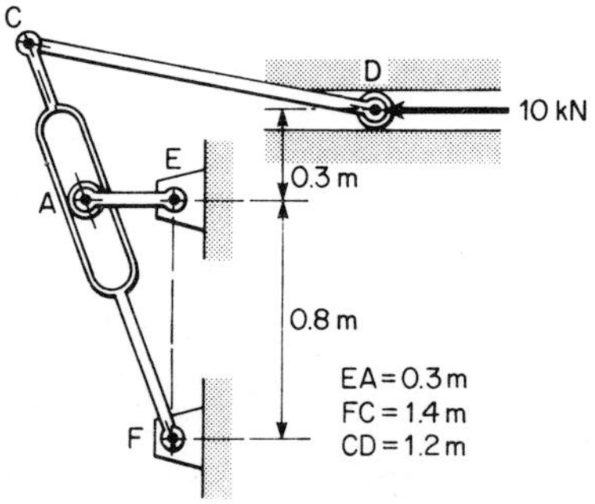

Figure 4.32

4.20 The mechanism in figure 4.33 lies in the vertical plane. The only external forces acting are the applied force of 20 kN on the roller E, which moves in a vertical frictionless guide, and the weight of AC, which is also 20 kN and acts through G. Determine the torque required on OA to maintain the mechanism in equilibrium. What are the reactions of the mechanism on its supports at O, D and E? A partly graphical solution is suggested. (*Hint*: Draw a scale diagram to determine angles and distances, consider the free-body diagrams of roller E and links CD, BE, AC and OA.)

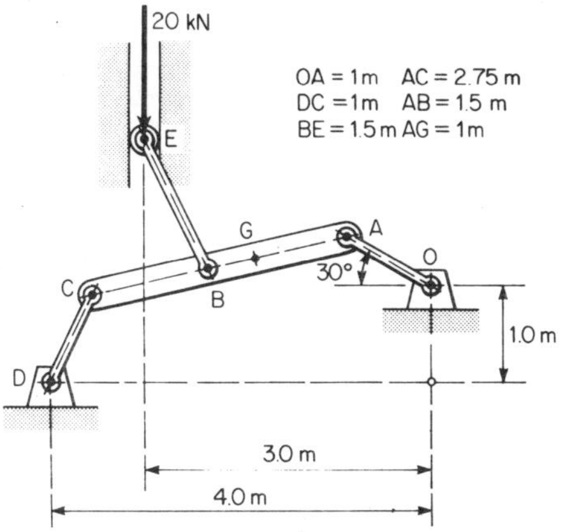

Figure 4.33

4.21 In the mechanism shown in figure 4.34 the link BC has mass 10 kg, its centre of gravity being at D. The small block E has mass 5 kg and the mass of link AB is to be neglected. Determine the external couple required on AB to maintain the mechanism in equilibrium under the weights of BC and E and the applied force of 1000 N for the position shown. (*Hint*: The use of a free-body diagram of a connected group of members will facilitate a solution.)

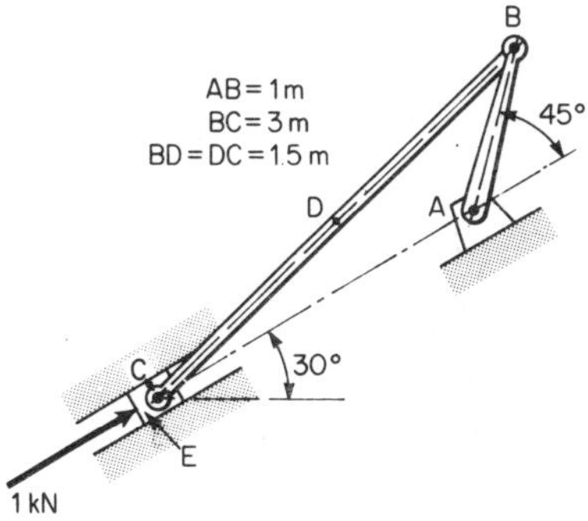

Figure 4.34

4.22 A platform is supported by two cross-frames AB and CD hinged at E as shown in figure 4.35. The frames are free to slide at B and C, AB is hinged to the ground at A and CD hinged to the platform at D. The platform and the load it carries have a total mass M and the centre of gravity of this may be assumed to be

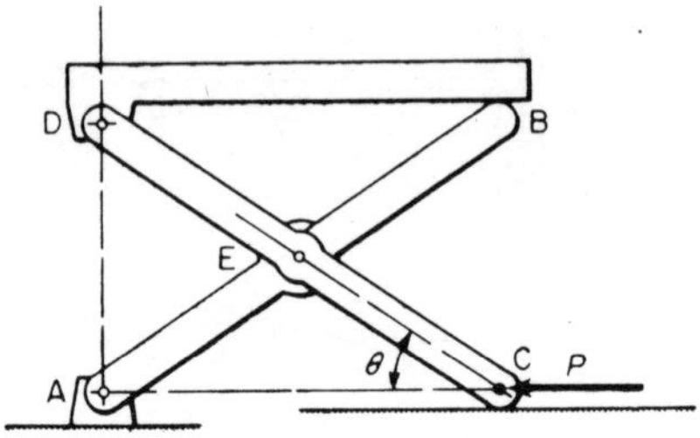

Figure 4.35

directly above E. Each frame has a mass m and its centre of gravity is at E. AB = CD = $2a$.

If friction is negligible find the value of P required to maintain the system in equilibrium when $\theta = 30°$. (*Hint*: Draw free-body diagrams for the whole frame, the platform, and DC.)

5 Friction

If two bodies are in contact and an attempt is made to slide one body relative to the other, forces tangential to the surface of contact will be developed tending to impede relative motion between the bodies. These forces are called *friction* forces and exist primarily because of the roughness of the contact surfaces; however, there are other contributory factors. If two surfaces are described in problems as being smooth then it is implied that friction forces do not exist or that they can be disregarded.

The magnitude and characteristics of the resultant friction force on the body under consideration depend on the applied forces, the nature of the materials in contact, and the nature of the contact surfaces. The presence of contaminants such as oxide films and lubricants can affect the friction force considerably; thus a film of lubricant may be sufficient to keep the surfaces apart and the friction force is then dependent only on the properties of the lubricant film.

We shall concern ourselves only with *dry friction* (also referred to as Coulomb friction), that is, the friction at surfaces not purposely contaminated or lubricated. The characteristics of dry friction are fairly consistent and as such can be applied to problems in engineering mechanics.

Friction is a complex phenomenon that has been studied for many centuries. In recent years the interdisciplinary nature of the study has been recognised and is now the subject of the wide-ranging field of *tribology*, defined as 'the science and technology of interacting surfaces in relative motion and of the practices relating thereto'.

5.1 Characteristics of Dry Friction

Suppose a block of mass m is lying on a horizontal surface and a horizontal force $\boldsymbol{P}$ is applied to it. If the magnitude of $\boldsymbol{P}$ is slowly increased from zero then the block remains stationary until, at a certain value of $\boldsymbol{P}$, the block is on the point of moving.

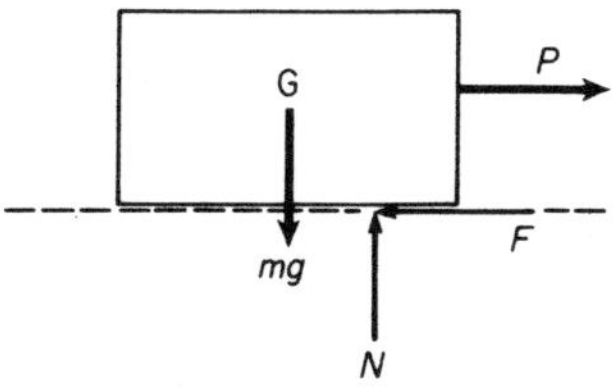

Figure 5.1

The forces acting on the block are shown in figure 5.1; these are the weight mg, the force $\boldsymbol{P}$, the normal component N of the reaction of the surface, and the friction component F, in the directions shown. The action of the block on the surface consists of a downwards force, component N, and a friction force component F to the right. If the block is at rest then $F = P$ and $N = mg$. As P is increased the magnitude of the friction component also increases until it reaches a limiting value F_L when motion is about to occur. If motion does occur it is found that P can be reduced in magnitude while maintaining the body in motion at constant speed, implying that the limiting friction component has also decreased in magnitude. Continued increase in P causes the block to move at increasing speed, the friction component remaining at its reduced magnitude or possibly decreasing further in magnitude. The variation of F with P is indicated in figure 5.2.

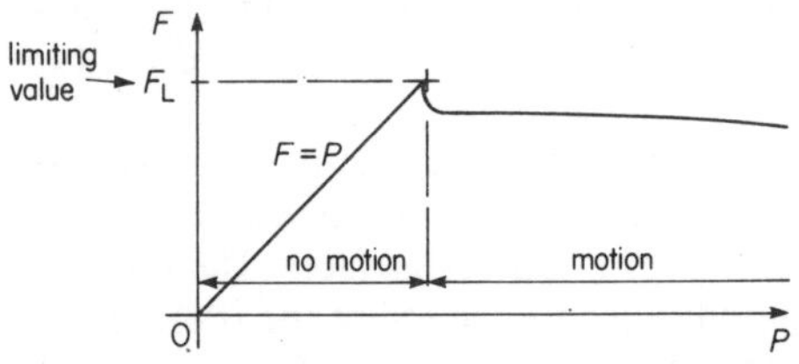

Figure 5.2

If the normal force between the surface and the block, as represented by the magnitude N, is varied by adding loads to the block, it is found that the limiting value F_L of the friction force component (as determined by observing P) is proportional to the magnitude N of the normal reaction component.

In the case discussed, N is equal to the weight of the block and the superimposed loads. However, in most cases this is not so. Thus in figure 5.3a, in

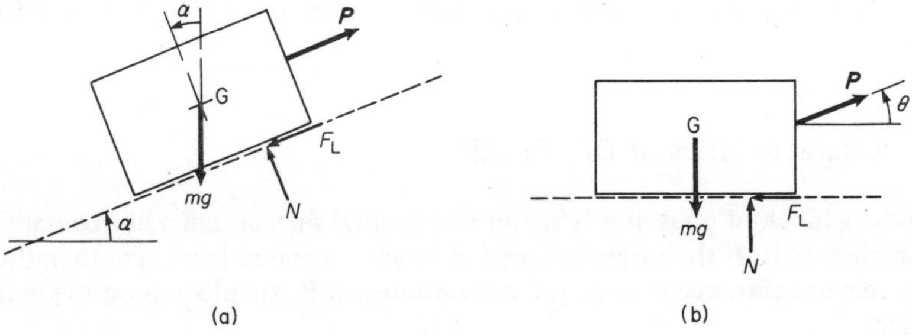

Figure 5.3

which the surface is inclined to the horizontal, $N = mg \cos \alpha$; in figure 5.3b the force P is shown inclined to the surface and $N = mg - P \sin \theta$, showing that N can also depend on the applied force.

The proportionality between the limiting value F_L of the friction force (component) F and the normal reaction (component) N has long been recognised and can be described as a *law of friction*. If the relation is written

$$F_L = \mu N \tag{5.1}$$

the proportionality constant μ is called the *coefficient of friction* between the contact surfaces. The coefficient has a *static* value μ_s corresponding to the limiting friction force at impending motion, and a lesser *kinetic* value μ_k corresponding to the friction force when motion is occurring. However, in this book we shall express the coefficient as μ, without a subscript, on the understanding that the value assigned to this is the relevant one for the particular conditions of the problem being solved. In the few problems where situations of both impending and actual motion occur we shall assume that the value given applies in both cases; in the cases where motion occurs we shall also assume that the coefficient of friction is independent of the relative velocity of sliding.

The basic facts relating to the magnitude of the friction force may be summarised as follows.

(1) If no motion takes place the magnitude of the friction force can have any value between zero and μN, depending on the applied forces, but will always be such that the body is in equilibrium.
(2) There is a limiting value to the magnitude of the friction force that can be generated; this value is μN.
(3) The limiting value of the friction force is only attained if relative motion is about to take place between the two surfaces in contact.
(4) When relative motion occurs the friction force on a body is in the opposite direction to that of the relative motion of the body.
(5) The coefficient of friction μ depends on the nature of the materials in contact and the nature of the surfaces; in this respect any quoted figures must be taken as purely nominal values – for example, touching a surface with the hand can change the characteristics of the surface.
(6) The friction force is independent of the area of contact.
(7) If motion occurs the kinetic friction force is less than the limiting static friction force.

5.1.1 Angle of Friction

The friction force $\boldsymbol{F}$ and the normal reaction $\boldsymbol{N}$ are the vector components of the total reaction $\boldsymbol{R}$ of the surface on the body – see figure 5.4a in which all other external forces have been omitted for clarity; the total reaction is seen to be inclined at an angle β to the normal. If motion is impending, F has attained the value $F_L = \mu N$, and β has attained a limiting value ϕ such that

$$\tan \phi = \frac{F_L}{N} = \mu \qquad (5.2)$$

The limiting angle ϕ is termed the *angle of friction*.

It follows from this that before motion takes place $\boldsymbol{R}$ can possibly lie in any direction within $\pm\,\phi$ from $\boldsymbol{N}$ (figure 5.4b), its actual direction depending on the other external forces. When motion is impending, $\boldsymbol{R}$ is at an angle ϕ to $\boldsymbol{N}$ in such a direction that $\boldsymbol{F}_L$ opposes the impending motion.

Suppose the body is resting on a surface inclined at a small angle α to the horizontal (as in figure 5.4a) and α is then gradually increased. Initially the body

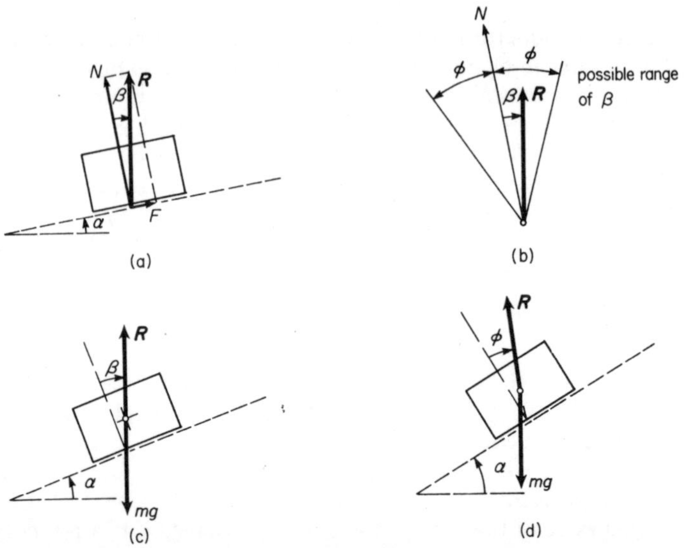

Figure 5.4

remains at rest and the reaction $\boldsymbol{R}$ of the surface is equal and opposite to the weight mg, since these are the only two forces acting (figure 5.4c). Resolving parallel and perpendicular to the plane

$$F = mg \sin \beta$$

$$N = mg \cos \beta$$

with $\beta = \alpha$. When β attains the value ϕ, the inclination of $\boldsymbol{R}$ to the normal has reached its limiting value, but $\boldsymbol{R}$ is still equal and opposite to the weight. A further increase in α cannot increase the inclination of $\boldsymbol{R}$ to the normal and there is now an unbalanced force on the body (figure 5.4d), which now moves down the plane. The value of α when motion is impending is called the *angle of repose* and is equal to the angle of friction ϕ.

Worked Example 5.1

In figure 5.5a an external force $\boldsymbol{P}$ acts on a body of mass 10 kg; the coefficient of friction $\mu = 0.4$.

(a) Find the magnitude and direction of the friction force and the value of the normal reaction for (i) $P = 10$ N (ii) $P = 30$ N (iii) $P = 56.6$ N (iv) $P = 70$ N.
(b) At what value of P is motion impending up the plane?

Solution

A free-body diagram is drawn as in figure 5.5b; the friction force $\boldsymbol{F}$ is unknown in either magnitude or sense and this is indicated in the figure.

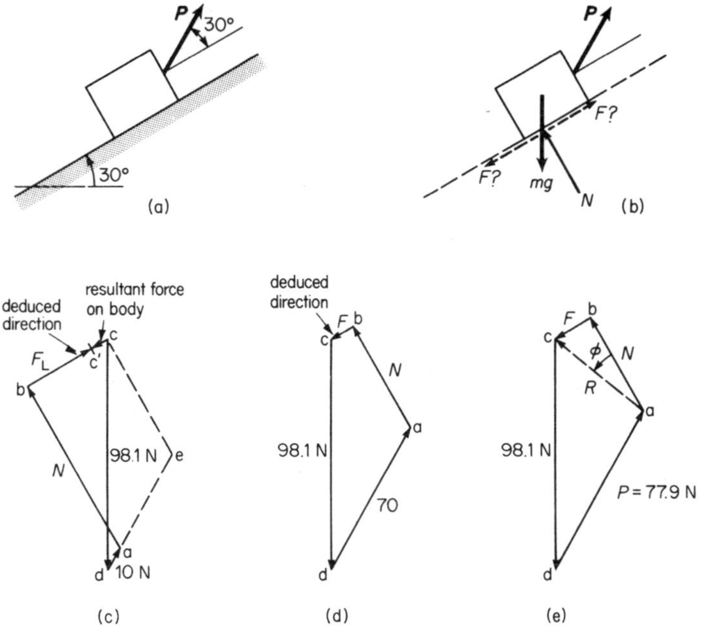

Figure 5.5

In this type of problem, in which it is not known whether motion is not impending, is impending, or is taking place, the method of solution is to examine the equilibrium of the body. The problem lends itself to a graphical solution and a polygon of forces is drawn to close with two lines parallel to the two unknowns N and F; N is not restricted and can take up the value represented in the polygon, but the value of F generated by the surface is restricted to a maximum value of $F_L = \mu N$. A comparison of the value of F required to close the polygon with the limiting value F_L will determine whether equilibrium is possible. If the polygon requirement is less than F_L it can be generated by the surface, and it follows that the body is at rest with non-limiting friction; if the polygon requirement is equal to F_L, motion is impending; if the polygon requirement is greater than F_L it cannot be generated, the actual friction force is F_L, and motion occurs due to the unbalanced resultant force on the body.

(a) (i) Figure 5.5c shows the polygon of forces for $P = 10$ N. Vectors cd and da represent the known forces $mg = 98.1$ N and $P = 10$ N and line ab is drawn parallel to N. The closing line bc parallel to F represents the polygon requirement for F to maintain equilibrium; this is directed up the plane and measures 40.4 N. Now vector ab (N) measures 80 N and thus F is limited to $F_L = 0.4 \times 80 = 32$ N and is represented by the vector bc′. With $F = F_L = 32$ N up the plane c′ is the terminal point of the force summation and there is thus a resultant force on the body of 8.4 N down the plane represented by cc′ causing motion in that direction.

(ii) The same general figure 5.5c suffices for $P = \text{da} = 30$ N. With this value $N = \text{ab} = 70$ N and bc = 23.1 N. The maximum value possible for F is F_L =

0.4 × 70 = 28 N and thus F can take the value 23.1 N. The friction force is therefore 23.1 N up the slope, the body remains in equilibrium, and motion is not imminent because $F < F_L$.

(iii) In this case $\boldsymbol{P}$ is represented by de (figure 5.5c) and N by ec = 56.6 N; F = bc = 0 and motion is not imminent.

(iv) See figure 5.5d; with da = 70 N, N = ab = 49.9 N and F (= bc) needs to be 11.6 N, directed down the plane. F_L = 0.4 × 49.9 = 20 N and F can therefore take the value 11.6 N. The friction force is 11.6 N down the slope and motion is not imminent because $F < F_L$.

(b) For impending motion up the plane $F_L/N = 0.4$ and it is more convenient to use $\boldsymbol{R}$ (= ac) in the diagram, drawn at an angle ϕ (= $\tan^{-1}$ 0.4) to the direction of N. Take care that ϕ is placed on the correct side of N; this is determined by the direction of F_L, which, in this case, because impending motion is upwards, is down the slope. A line to represent $\boldsymbol{R}$ is thus drawn, in the correct direction, through c to locate a on the vector representing $\boldsymbol{P}$. P then measures 77.9 N, N = 46 N and $F = F_L$ = 18.4 N.

Note the following from this problem.

(1) The normal reaction N can vary according to the external forces. In this case it varies from 80 N to 46 N.
(2) Because N varies the limiting value of friction is different for the two directions, being 32 N for motion down the plane and 18.4 N for motion up the plane.
(3) The value of the friction force can consequently take up any value (depending on the value of P) between 32 N down and 18.4 N up.

Worked Example 5.2

A uniform plank having mass 10 kg and length 6 m is lying at 20° to the horizontal across a channel as in figure 5.6a. The value of μ at the contact surfaces is 0.3.

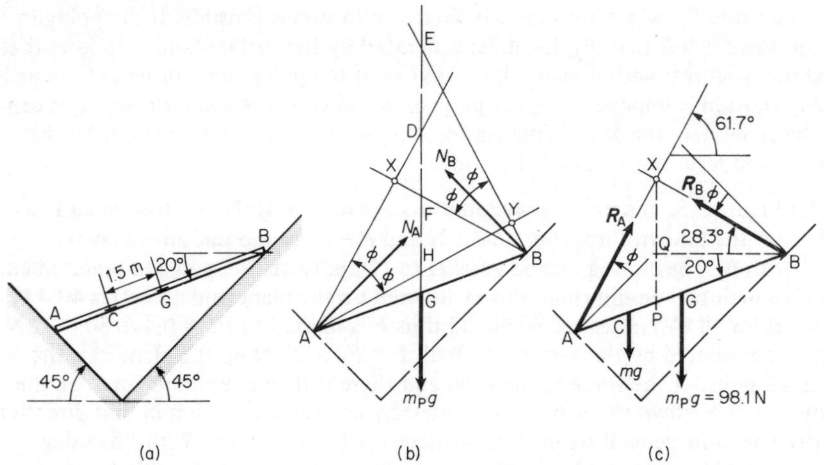

Figure 5.6

(a) Show that the plank is in equilibrium in this situation.
(b) Determine the mass m of a small body, which, if placed at the point C 1.5 m from the centre, will make the plank just slip.

Solution

(a) If the body is in equilibrium under non-limiting friction conditions it is impossible to calculate the actual end reactions since there are four unknowns (magnitudes and directions of $\boldsymbol{R}_A$ and $\boldsymbol{R}_B$) and only three equations are available from the equilibrium of the body.

The method of solution is to assume impending motion at both points of contact, being the condition for the body as a whole to be on the point of moving, and to demonstrate from the equilibrium of the body either that this is so, or not. In the latter case the body remains at rest in equilibrium.

Note that the force set acting initially on the plank consists of three forces – the total reactions at A and B, $\boldsymbol{R}_A$ and $\boldsymbol{R}_B$, and the weight $m_P g$ acting through the centre of gravity. For equilibrium these three forces must pass through one point.

The angle of friction $\phi = \tan^{-1} 0.3 = 16.7°$.

The total reaction $\boldsymbol{R}_A$ has a possible direction of $\pm\phi$ from N_A and similarly with $\boldsymbol{R}_B$; the ranges of these directions are shown in figure 5.6b. If the assumption is made that upward motion is imminent at A and thus downward motion is imminent at B then $\boldsymbol{R}_A$ takes the line AH (figure 5.6b) and $\boldsymbol{R}_B$ the line BE. This is clearly inconsistent with the assumed impending motion since this requires the body to be still in equilibrium and the lines of action of both $\boldsymbol{R}_A$ and $\boldsymbol{R}_B$ to pass through the same point on the line of action of the weight $m_P \boldsymbol{g}$. Also, for this implied clockwise impending motion, $\boldsymbol{R}_A$ and $\boldsymbol{R}_B$ pass through the point Y and moments about this point imply anticlockwise motion, which is again incompatible with the original assumption.

If the assumption is made that downward motion is imminent at A and upward motion is imminent at B then $\boldsymbol{R}_A$ takes the line AD and $\boldsymbol{R}_B$ the line BF. This is again inconsistent with the assumed impending motion since $\boldsymbol{R}_A$ and $\boldsymbol{R}_B$ do not pass through one point on the line of action of $m_P \boldsymbol{g}$. Also, for this implied anticlockwise impending motion, $\boldsymbol{R}_A$ and $\boldsymbol{R}_B$ pass through the point X and moments about this point imply clockwise motion, again incompatible with the implied anticlockwise impending motion.

The plank must therefore be in equilibrium under non-limiting friction conditions, the two reactions $\boldsymbol{R}_A$ and $\boldsymbol{R}_B$ not having their maximum inclination to their respective normals but both passing through the same point between F and D on the line of action of $m_P \boldsymbol{g}$.

It also follows that motion of the plank will be imminent if its position is changed so that the points F and D coincide, and that the plank cannot be in equilibrium if the ranges HD and FE do not overlap.

(b) In problems the statement 'to just slip' can be assumed to have the same meaning as 'imminent motion'.

The force set in this case consists of the four forces $\boldsymbol{R}_A$, $\boldsymbol{R}_B$, $m_P \boldsymbol{g}$ and the weight of the small body, $m\boldsymbol{g}$. $\boldsymbol{R}_A$ and $\boldsymbol{R}_B$ take up their limiting positions at ϕ to the normal and, assuming impending downward motion at A, their directions are known, as shown in the free-body diagram figure 5.6c; the angles in this figure can easily be verified. There are thus only three unknowns – the magnitudes of $\boldsymbol{R}_A$, $\boldsymbol{R}_B$ and $\boldsymbol{mg}$. Since the body is in a condition of imminent motion the forces are in equilibrium and thus

$\Sigma F_x = 0$

$$R_A \cos 61.7° - R_B \cos 28.3° = 0 \tag{5.3}$$

$\Sigma F_y = 0$

$$R_A \sin 61.7° + R_B \sin 28.3° - 98.1 - mg = 0 \tag{5.4}$$

$\Sigma M_B = 0$

$$6R_A \sin 41.7° - 98.1 \times 3 \cos 20° - mg \times 4.5 \cos 20° = 0 \tag{5.5}$$

The solution of these three simultaneous equations gives $m = 9.74$ kg.

A partly graphical solution is also possible: with impending downwards motion at A the reactions $\boldsymbol{R}_A$ and $\boldsymbol{R}_B$ intersect at X (figure 5.6c), which must, for equilibrium, be on the line of action of the resultant of $\boldsymbol{m_P g}$ and $\boldsymbol{mg}$. If this line intersects AB at P then $mg \times \text{CP} = m_P g \times \text{PG}$ and $m = m_P \times \text{PG/CP}$.

The location of P is relatively simple in this case because the triangle AXB has a right angle at X. Then

$$\text{BX} = \text{AB} \cos 48.3° = 3.99 \text{ m}$$
$$\text{BQ} = \text{BX} \cos 28.3° = 3.51 \text{ m}$$

and

$$\text{BP} = \frac{\text{BQ}}{\cos 20°} = 3.74 \text{ m}$$

$$\text{PG} = 0.74 \text{ m}$$

and

$$\text{CP} = 0.76 \text{ m}$$

Thus

$$m = \frac{10 \times 0.74}{0.76} = 9.74 \text{ kg}$$

Worked Example 5.3

A cylinder of radius a rests with its curved surface on a horizontal floor. A uniform straight plank of length $2L$ lies symmetrically across it, such that its centre is in contact with the cylinder and its lower end rests on the floor. The coefficients of friction at all three points of contact are equal. The system is in equilibrium with impending motion at only one of the three points of contact. Show that this point is never the point of contact of the cylinder with the ground but is the point of contact of the plank with the floor or with the cylinder depending on the inequalities

$$3a^2 \lessgtr L^2$$

Confirm that for equilibrium conditions to exist in these situations μ must be $> 1/\sqrt{3}$.

Solution

Note that both bodies are acted on by three forces. The cylinder (see figure 5.7a) has the reactions at A and B and its own weight, the plank the reactions at B and C and its own weight. Free-body diagrams of the cylinder and plank are drawn in figures 5.7b and d making use of the fact that when only three forces act on a body in equilibrium they must pass through one point.

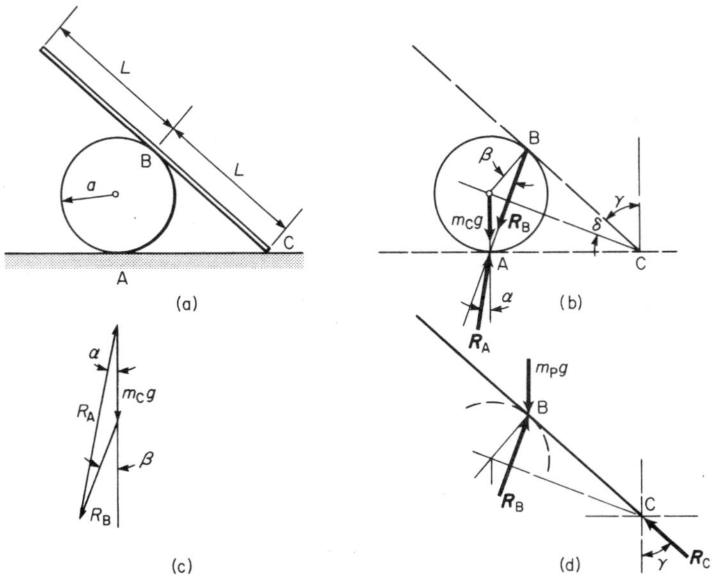

Figure 5.7

For the cylinder, $\boldsymbol{R}_B$ must thus pass through A and it follows, see figure 5.7c, that α must be less than β for the force triangle to close. For the plank, $\boldsymbol{R}_C$ must pass through B and therefore its line of action is inclined to the normal at C at the same angle as BC, namely γ.

For impending motion to occur at only one of the points A, B or C the angle between the total reaction and the normal at that point must attain the value ϕ. This point cannot be A since β is always greater than α and thus limiting friction will always occur at B before it occurs at A.

From the geometry of figure 5.7b

$$\beta = \delta = \tfrac{1}{2}\left(\frac{\pi}{2} - \gamma\right) \tag{5.6}$$

Consider first the case in which motion is impending at both B and C simultaneously and $\beta = \gamma = \phi_1$. From equation 5.6

$$\beta = \tfrac{1}{2}\left(\frac{\pi}{2} - \beta\right)$$

and

$$\beta = \gamma = \phi_1 = 30^\circ$$

also

$$\tan \delta = \frac{a}{L} = \frac{1}{\sqrt{3}} \tag{5.7}$$

in this case, since $\delta = \beta = 30^\circ$.

If now $\gamma > 30^\circ$ then (from equation 5.6) $\beta < 30^\circ$. Motion can become imminent at C first for some value of ϕ, ϕ_2 that is greater than 30° (when $\gamma = \phi_2, \beta < \phi_2$). It follows that for motion to be impending at C first δ $(= \beta) < 30^\circ$ and

$$\frac{a}{L} < \frac{1}{\sqrt{3}}$$

On the other hand if $\gamma < 30^\circ$ then $\beta > 30^\circ$ and motion can become imminent at B first for some value of ϕ, ϕ_3 that is greater than 30° (when $\beta = \phi_3, \gamma < \phi_3$). It follows that for motion to be impending at B first δ $(= \beta) > 30^\circ$ and

$$\frac{a}{L} > \frac{1}{\sqrt{3}}$$

Motion is therefore impending at one of the points C or B depending on the inequalities $a/L \lessgtr 1/\sqrt{3}$ or $3a^2 \lessgtr L^2$.

It has been demonstrated (1) for motion to become imminent at C first that $\phi = \phi_2$ must be $> 30^\circ$, (2) for motion to become imminent at B first that $\phi = \phi_3$ must be $> 30^\circ$. It follows that the system can only be in equilibrium with motion imminent at either B or C (A being excluded as already shown) if $\phi > 30^\circ$, that is $\mu > 1/\sqrt{3}$.

5.1.2 Note on the Solution of Problems

A geometrical solution, as illustrated by the preceding worked example, in which the total reaction $\boldsymbol{R}$ of the surface is used rather than its components F and N, is often preferable since the angle β between $\boldsymbol{R}$ and the normal can be seen and compared with its limiting value ϕ. However, the analytical method in which F and N are treated separately can always be used; the ratio F/N at each contact must then be compared with its limiting value μ.

For problems involving bodies at rest or in a state of impending motion the conditions for equilibrium are applied in graphical or analytical form. In all cases free-body diagrams should always be drawn.

(1) To ascertain whether a body is in a state of rest and that motion is not impending under given applied forces, demonstrate that for equilibrium the friction forces are not simultaneously limiting. Friction forces can be inserted in the free-body diagrams with arbitrary directions, subject of course to compliance with the third law.
(2) To ascertain whether motion is impending demonstrate that for equilibrium the friction forces are limiting at all points of contact.
(3) To show that motion will occur under given applied forces set all friction forces to their limiting values and demonstrate that there is then a resultant force or couple acting on the body. Friction forces must be inserted with the correct sense corresponding to the assumed direction of relative motion.
(4) Corresponding considerations arise for problems in which it is necessary to ascertain the applied forces required to ensure states of rest or impending motion.

5.2 Some Practical Applications

5.2.1 Wedges

Wedges are used for raising large loads or adjusting levels of heavy machinery by the application of relatively small forces. Wedges have the useful attribute of being self-locking.

Worked Example 5.4

If the wedge in figure 5.8a is used to lift the block, mass 1000 kg, which abuts against the vertical wall, find P if $\mu = 0.2$ at all contact surfaces. Ignore the mass of the wedge.

Solution

We assume motion is just about to take place and that limiting friction exists at all surfaces. Draw free-body diagrams for both bodies, as in figures 5.8b and c. The directions of the total reactions in these figures are determined from the directions of the respective friction forces, which themselves are determined by the relative motion of the surface; their lines of action are not known.

$$\phi = \tan^{-1} 0.2 = 11.3°$$

Triangles of forces (since the forces are in equilibrium at impending motion) are

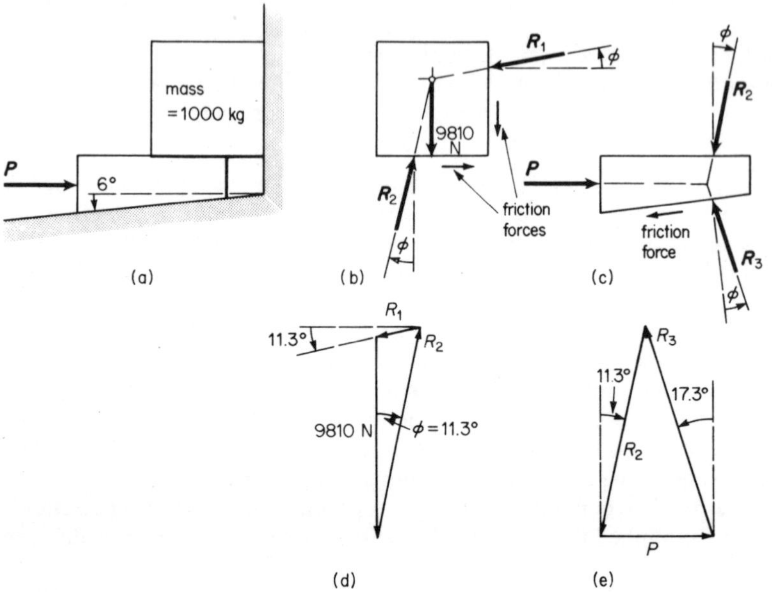

Figure 5.8

given in figures 5.8d and e for the block and the wedge respectively. Note that the reaction R_2 is common to both triangles.

From figure 5.8e, $P = 5230$ N.

5.2.2 Jamming or Self-locking

If a state of impending motion in a body or mechanism cannot be brought about by an applied force however large the force may be, then the body or mechanism is said to be jammed or self-locked. The condition is frequently associated with a force in a particular direction, and a reversal of the force will release the body or mechanism.

Consider again the case of a block mass m resting on an inclined plane but now subject to a horizontal force $\boldsymbol{P}$ (figure 5.9a). If $\alpha < \phi$ and P is zero then conditions will be as illustrated in figure 5.9b, which is the same as figure 5.4c. The block will

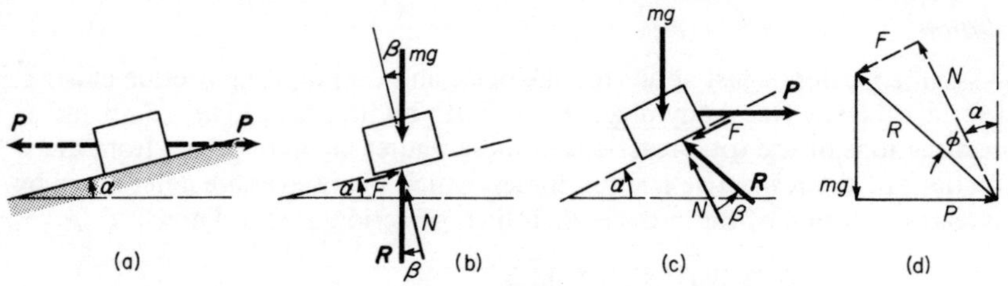

Figure 5.9

be in equilibrium and will remain so whatever additional loads are placed on the block. A force $\boldsymbol{P}$ to the left and of sufficient magnitude is now required to initiate motion down the plane.

If $\alpha > \phi$ and $\boldsymbol{P}$ is zero, F on the block is directed up the plane but is insufficient to maintain equilibrium and the block moves down the plane. If $\boldsymbol{P}$ acting to the right is gradually increased in magnitude, F first reduces to zero and is then directed down the plane as in figure 5.9c, the block being in equilibrium if $\beta < \phi$. Eventually $\boldsymbol{P}$ is such that F has reached its limiting value and the angle β has reached the value ϕ when motion is impending up the plane. From the polygon of forces at impending motion, with $\beta = \phi$ (figure 5.9d), $P = mg \tan (\alpha + \phi)$. For a given value of ϕ, if α is now increased $\boldsymbol{P}$ must also be increased for impending motion and the required value of $\boldsymbol{P}$ can become indefinitely large as α tends to the value $(\pi/2 - \phi)$; the block is then said to be jammed as regards motion up the plane. Note, however, that if $\boldsymbol{P}$ is removed, then because α is larger than ϕ, the 'angle of repose', the block will begin to move down the plane.

Deformation of the body and of the surface have not been taken into account and in practice there will clearly be limitations on the magnitude of the forces which can be applied.

Worked Example 5.5

The arm AB in figure 5.10a is attached to a shaft which can rotate in frictionless bearings. The rigid rod CD, mass 20 kg, slides in vertical guides and is in contact with the arm AB at C. If $\phi = 15°$ at each sliding contact, determine

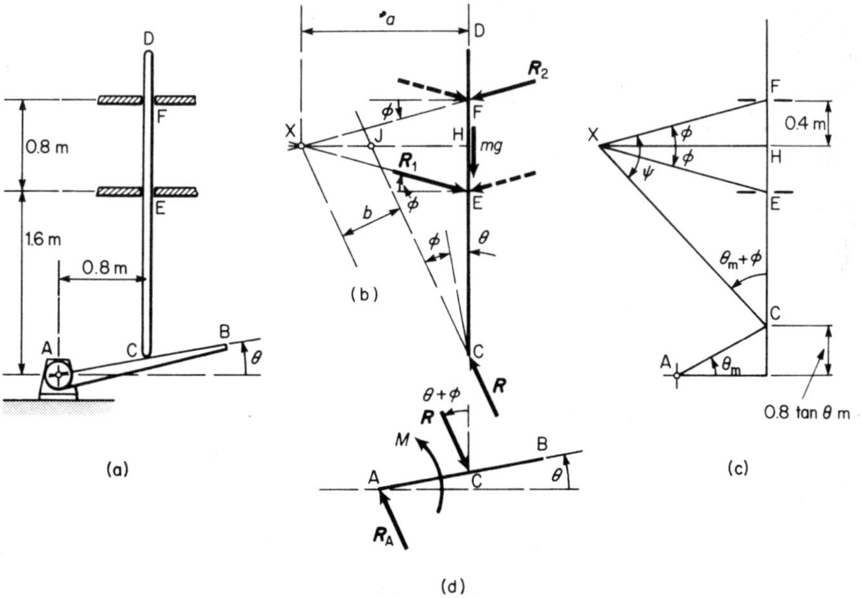

Figure 5.10

(a) the value of the angle θ at which it is impossible to lift CD by the application of an anticlockwise torque to the shaft.

(b) If $\theta = 10°$, the magnitude of the torque required to initiate upward movement of CD.

Assume that the diameter of the rod and the thickness of the arm are small.

Solution

(a) We need to consider the equilibrium of both CD and AB and it is therefore necessary to determine the directions of the total reactions at the sliding surfaces at C, E and F. The impending motion at E and F is upwards and thus the friction forces exerted by the guides at these points on CD must be downwards. For this upward impending motion, the impending motion of the contact point of CD on AB is towards B; the friction force exerted by AB on CD at C opposes this motion and is thus towards A. The free-body diagram, figure 5.10b, is now drawn with the angles between the reactions and their respective normals all equal to ϕ because motion is everywhere imminent. The reactions $\boldsymbol{R}_1$ and $\boldsymbol{R}_2$ could, without further consideration, lie along either the broken lines or full lines but moments about F and C will confirm that, for equilibrium, they must lie along the full lines.

It is now necessary to consider whether CD can be in equilibrium under the external forces and in this type of situation the moment equation for equilibrium is useful. In this case the lines of action of $\boldsymbol{R}_1$ and $\boldsymbol{R}_2$ intersect at X, a fixed point, as long as motion is impending at E and F. The moments of the other two forces $\boldsymbol{mg}$ and $\boldsymbol{R}$ must thus sum to zero about X and it follows that for equilibrium to be possible $\boldsymbol{R}$ must lie to the right of X such that $bR = amg$. As the angle θ is increased b will decrease and R $(= amg/b)$ will increase becoming infinite as $b \to 0$. Thus an infinite value of R is necessary for the rod CD to be in equilibrium when the line of action of R passes through X. This decides the maximum value of $\theta(\theta_m)$ beyond which CD may not be lifted, the mechanism then being jammed as regards upward motion.

From the geometry of figure 5.10c

$$\frac{XF}{\sin(\theta_m + \phi)} = \frac{CF}{\sin\psi} \tag{5.8}$$

$$XF = \frac{0.4}{\sin\phi} = 1.5455$$

$$CF = 2.4 - 0.8\tan\theta_m$$

$$\psi = 90 - (\phi + \theta_m) + \phi = 90 - \theta_m$$

and

$$\sin\psi = \cos\theta_m$$

also

$$\sin(\theta_m + \phi) = \sin\theta_m\cos\phi + \cos\theta_m\sin\phi$$
$$= 0.966\sin\theta_m + 0.2588\cos\theta_m$$

Equation 5.8 becomes

$$1.5455 = \left(\frac{2.4 - 0.8 \tan \theta_m}{\cos \theta_m}\right)(0.966 \sin \theta_m + 0.2588 \cos \theta_m)$$

The solution of this is

$$\theta_m = 65.4^\circ \text{ or } 28.7^\circ$$

The smaller value that satisfies the equation, namely 28.7°, is the solution. It is thus impossible to raise AB beyond $\theta = 28.7^\circ$.

(b) If $\theta = 10^\circ$ then

$$\begin{aligned} CH &= 2.0 - 0.8 \tan 10^\circ \qquad \text{(see figure 5.10b)} \\ &= 1.859 \text{ m} \\ HJ &= CH \tan (\theta + \phi) \\ &= 1.859 \tan 25^\circ = 0.867 \text{ m} \end{aligned}$$

$$a = \frac{0.4}{\tan \phi} = 1.492 \text{ m}$$

$$XJ = a - HJ = 0.626 \text{ m}$$

and

$$b = XJ \cos 25^\circ = 0.567 \text{ m}$$

Thus

$$R = \frac{amg}{b} = \frac{1.492 \times 20 \times 9.81}{0.567} = 516 \text{ N}$$

From the equilibrium of AB, whose free-body diagram is given in figure 5.10d

$\Sigma M_A = 0$

$$AC \times R \cos \phi - M = 0$$

$$AC = \frac{0.8}{\cos 10^\circ} = 0.812 \text{ m}$$

and

$$\begin{aligned} M &= 0.812 \times 516 \times \cos 15^\circ \\ &= 405 \text{ N m} \end{aligned}$$

This is the torque required to produce imminent motion upwards when $\theta = 10^\circ$.

5.2.3 Screws

We shall only consider screws with threads of square cross-section. The thread winds spirally around a cylindrical shaft and can slide in a corresponding square groove, cut spirally inside a cylindrical guide. An axial load on the shaft is transmitted to the guide through the thread and an external torque is then usually required to make the load move axially through movement of the thread in its surrounding guide, which is assumed fixed in the following analysis.

A view of a thread on a spindle or shaft and a cross-section of the guide or nut is given in figure 5.11a. Two definitions relating to the thread are

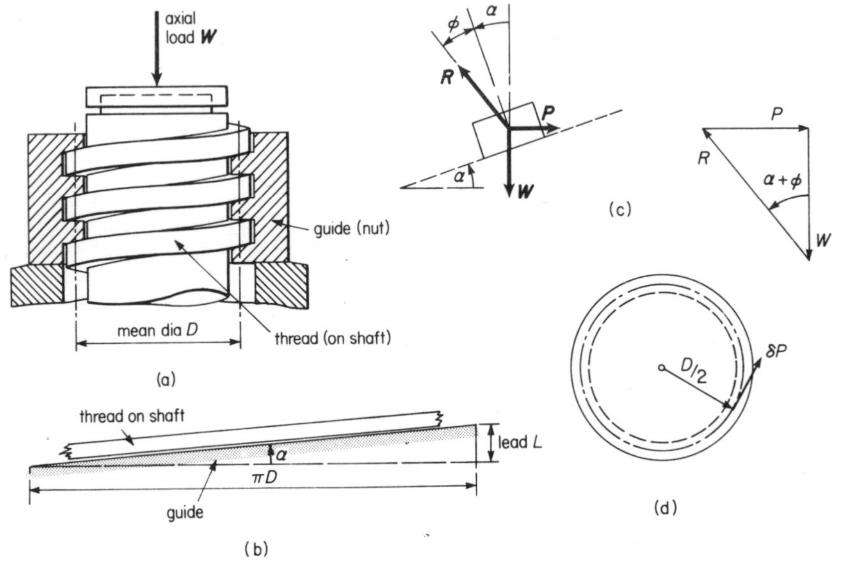

Figure 5.11

(1) lead L is the axial distance travelled by the shaft during one complete revolution of the shaft
(2) mean diameter D is the mean diameter measured midway between the top and bottom of the thread.

If one could unwind the spiral, diameter D, made by the thread and lay it down flat it would form an inclined plane, as indicated in figure 5.11b, the angle α being given by $\tan \alpha = L/\pi D$. Thus the loaded screw thread in its guide or nut is mechanically equivalent to a block carrying the same load on a plane inclined at angle α. If the external couple applied to the shaft is applied in a plane perpendicular to the axis of the shaft it follows that a force $\boldsymbol{P}$ applied to the block, equivalent to the set of elementary forces distributed around the thread, is horizontal and in the plane of the figure. The equivalent system is shown in figure 5.11c for motion up the plane and it is seen to be exactly that discussed in the previous section; from that section the horizontal force $\boldsymbol{P}$ required to move such a block up an inclined plane of slope α is $W \tan (\alpha + \phi)$ where W is now the axial load.

Reverting now to the case of the threaded shaft the force **P** is made up of a set of elementary forces that are tangential to the thread in a plane perpendicular to its axis, see figure 5.11d, and in total they constitute a torque about the axis having magnitude

$$M = P \times \frac{D}{2} = W \tan(\alpha + \phi) \times \frac{D}{2} \tag{5.9}$$

This is the axial torque required to move a spindle against an axial load W and is applied in the sense of the angular movement of the shaft. Conversely this may be thought of as the frictional torque exerted by the guide that has to be overcome in rotating the shaft to cause axial movement against the load. Its magnitude may be rewritten as

$$M = \frac{WD}{2} \tan(\alpha + \phi) = \frac{WD}{2} \frac{\tan\alpha + \tan\phi}{1 - \tan\alpha \tan\phi}$$

$$= \frac{WD}{2} \frac{L + \mu\pi D}{\pi D - \mu L}$$

since $\tan\alpha = L/\pi D$ and $\tan\phi = \mu$.

If the shaft is fixed and the guide, now carrying the axial load, is rotated to produce axial movement the relationships are unchanged.

By further reference to section 5.2.2 it is apparent that if $\alpha < \phi$ then the loaded shaft can remain in equilibrium in its guide with no external torque applied, friction then being non-limiting. However, if motion is required in the direction of the axial load then for the block, $P = W\tan(\phi - \alpha)$ in the opposite direction to that given in figure 5.11c and a corresponding torque M is required on the shaft, where

$$M = W\tan(\phi - \alpha)\frac{D}{2} \tag{5.10}$$

in the sense of the angular movement of the shaft.

If $\alpha > \phi$ then for the block, motion will occur down the plane unless a force magnitude $P = W\tan(\alpha - \phi)$ is applied in the direction shown in figure 5.11c. The external torque required to be applied to the shaft to maintain the load in position is

$$M = W\tan(\alpha - \phi)\frac{D}{2} \tag{5.11}$$

in the opposite sense to the impending angular motion. If this torque is not applied the shaft will move axially in the direction of the load.

Worked Example 5.6

A screw is to lift a vertical load of weight 2000 N. If $\mu = 0.2$ and the lead is 0.05 m

(a) find the mean diameter if, when the external couple is removed, the load just remains stationary; (b) if the mean diameter is made half the value determined in (a) find the external torque required (i) to raise the load and (ii) to lower the load; (c) repeat (b) if the mean diameter is made twice that determined in (a).

Solution

(a) The limiting value of α for the load to remain stationary is

$$\alpha = \phi = \tan^{-1} 0.2 = 11.3°$$

Also

$$\tan \alpha = \frac{L}{\pi D}$$

$$0.2 = \frac{0.05}{\pi D}$$

giving $D = 0.08$ m.

(b) (i) If D is actually $0.08/2 = 0.04$ m then

$$\tan \alpha = \frac{0.05}{\pi \times 0.04}$$

$$\alpha = 21.7° \text{ and } \alpha > \phi$$

From equation 5.9

$$M = 2000 \tan (21.7° + 11.3°)\frac{0.04}{2}$$

$$= 26.0 \text{ N m}$$

in the sense of the angular motion.

(ii) Since $\alpha > \phi$, a torque is required to stop the load descending of its own accord. From equation 5.11

$$M = 2000 \tan (21.7° - 11.3°)\frac{0.04}{2}$$

$$= 7.3 \text{ N m}$$

in the opposite sense to that of the angular motion.

(c) (i) $D = 0.16$ m

$$\tan \alpha = \frac{0.05}{\pi \times 0.16}$$

$$\alpha = 5.68° \text{ and } \alpha < \phi$$

From equation 5.9

$$M = 2000 \tan (5.68° + 11.3°)\frac{0.16}{2}$$

$$= 48.8 \text{ N m}$$

in the sense of the angular motion.

(ii) From equation 5.10

$$M = 2000 \tan (11.3° - 5.68°)\frac{0.16}{2}$$

$$= 15.7 \text{ N m}$$

in the sense of the angular motion.

5.2.4 Belt and Cable Friction

Belts are often used for the transmission of a torque from one shaft to another and have some advantages over gearing; transmission is carried out by having cylindrical pulleys fixed to each shaft and a continuous belt passing around the rims of the pulleys.

Friction is necessary between the belt and the pulleys in order that a torque can be transmitted, and the maximum torque that can be transmitted is related to the coefficient of friction between them.

Consider a belt passing over a pulley radius R, which is mounted on a shaft and is being driven by the belt in a clockwise direction as shown in figure 5.12a. The

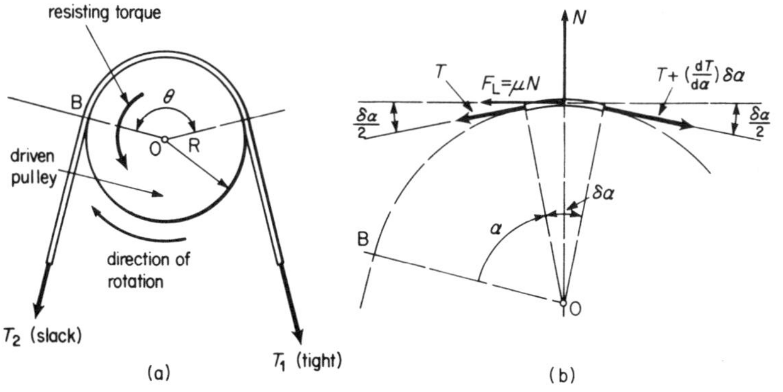

Figure 5.12

belt laps the pulley by an angle θ, T_1 is the tension in the tight side of the belt (because this side is being pulled) and T_2 ($< T_1$) is the tension in the slack side when the shaft is subject to an opposing torque and rotating at constant speed.

We require to know the relationship between T_1 and T_2 in the limiting condition when the belt is about to slip on the pulley. In this condition the forces acting on a small circumferential element of the belt, length $R\delta\alpha$, at an angle α from the tangent point on the slack side are as shown in figure 5.12b. The tension is T at angle α and has increased to $[T + (\mathrm{d}T/\mathrm{d}\alpha)\delta\alpha]$ over the angular distance $\delta\alpha$; there is the normal reaction N of the pulley on the element of the belt and the friction force opposing the imminent motion has its limiting value μN. The mass of the belt has been disregarded.

Choosing the normal and tangential directions at the centre of the element

$$\Sigma F_\mathrm{n} = 0$$

$$-T\sin\left(\frac{\delta\alpha}{2}\right) - \left[T + \left(\frac{\mathrm{d}T}{\mathrm{d}\alpha}\right)\delta\alpha\right]\sin\left(\frac{\delta\alpha}{2}\right) + N = 0$$

$$\Sigma F_\mathrm{t} = 0$$

$$-T\cos\left(\frac{\delta\alpha}{2}\right) + \left[T + \left(\frac{\mathrm{d}T}{\mathrm{d}\alpha}\right)\delta\alpha\right]\cos\left(\frac{\delta\alpha}{2}\right) - \mu N = 0$$

Since $\delta\alpha$ is small, $\sin(\delta\alpha/2) = \delta\alpha/2$, $\cos(\delta\alpha/2) = 1$ and since $(\delta\alpha)^2/2$ is second order, terms involving it can be neglected. The equations then reduce to

$$-T\delta\alpha + N = 0$$

$$\left(\frac{\mathrm{d}T}{\mathrm{d}\alpha}\right)\delta\alpha - \mu N = 0$$

Eliminating N

$$\frac{\mathrm{d}T}{\mathrm{d}\alpha} = \mu T$$

Integrating

$$\int_{T_2}^{T_1}\frac{\mathrm{d}T}{T} = \mu\int_0^\theta \mathrm{d}\alpha$$

$$\ln\frac{T_1}{T_2} = \mu\theta$$

and

$$\frac{T_1}{T_2} = \mathrm{e}^{\mu\theta} \tag{5.12}$$

It follows that

$$T_1 - T_2 = T_1\left(1 - \frac{T_2}{T_1}\right)$$

$$= T_1 (1 - e^{-\mu\theta})$$
$$= T_2 (e^{\mu\theta} - 1) \qquad (5.13)$$

These are the required relations between T_1 and T_2 when the belt is about to slip. The onset of slipping is therefore governed not by the difference in the tension but by either their ratio, or their difference in comparison with the tension in one side or the other of the belt.

The same relationships hold for a belt passing over a driving pulley except that now of course the pulley is driving the belt; thus the same equations, 5.12 and 5.13 can be applied in the limiting condition.

For the case of a fixed pulley or cylinder the same relationships are applicable and cover the case of a rope passing around a fixed bollard. The ratio of tensions can become very large in the latter case where the rope is wound around a number of times; for example for two complete turns round a bollard with $\mu = 0.2$, $T_1/T_2 = 12.3$.

Worked Example 5.7

A pulley 1.5 m diameter is mounted on a shaft. If $\mu = 0.2$ and θ is 120° what is the maximum torque that can be transmitted along the shaft if the maximum allowable tension in the belt is 2000 N?

Solution

$$\text{Torque} = (T_1 - T_2)R \qquad \text{(see figure 5.12a)}$$

$$= T_1 \left(1 - \frac{T_2}{T_1}\right) R$$

If we put $T_1/T_2 = k$ whether slipping is imminent or not, then torque = $T_1(1 - 1/k)R$ and is obviously at its maximum when both T_1 and k have their largest values. Since $T_1 > T_2$ we set T_1 equal to the maximum allowable tension, which is 2000 N. The largest value of k occurs when the belt slips and

$$k = \frac{T_1}{T_2} = \exp\left(0.2 \times \frac{120}{180}\pi\right) = 1.52$$

(note θ is in radians).
Thus the maximum torque that can be transmitted is

$$2000\left(1 - \frac{1}{1.52}\right) \times 0.75 = 513 \text{ N m}$$

5.2.5 Friction in Thrust Bearings and Clutches

Although the theoretical analysis is the same for thrust bearings and clutches there is a distinct practical difference in that we wish to reduce friction in the bearing to conserve energy, whereas we wish to increase it in the clutch in order that it may transmit more torque. Figure 5.13 shows schematic sections of a

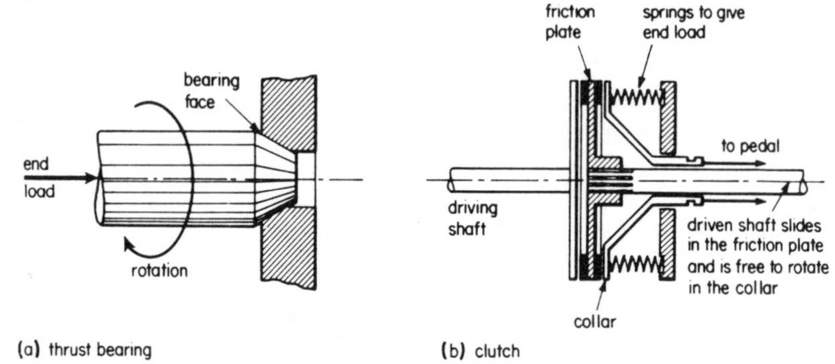

Figure 5.13

thrust bearing and a clutch. In figure 5.13a the rotating shaft carrying an axial load is supported by the stationary frustum-shaped bearing. In figure 5.13b, which shows the clutch, the springs press the face of the collar on to a renewable friction plate and torque is transmitted from the driving to the driven shaft. (The action of depressing a clutch pedal in a car is to compress the spring and draw the clutch faces apart.)

The theory for both can be covered by examining the thrust bearing as shown in figure 5.14a. Consider an elementary frustum between radii r and $r + \delta r$, shown

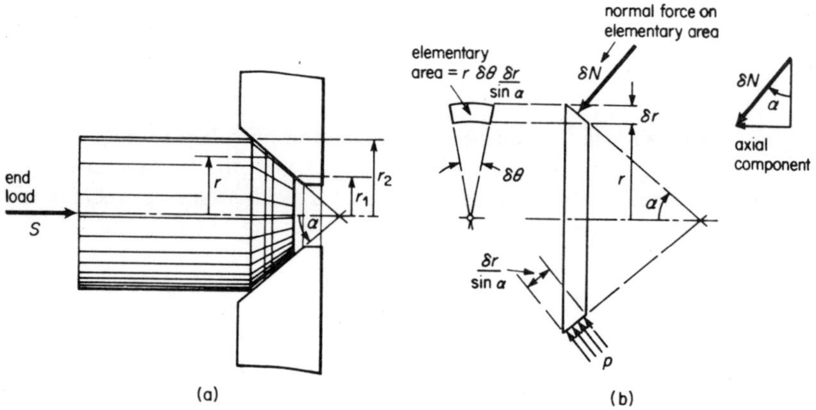

Figure 5.14

enlarged in figure 5.14b. The normal pressure, or force per unit area, between the bearing and the conical end of the shaft is denoted as p. The normal force δN on an elementary length of the frustum which subtends an angle $\delta\theta$ at the axis of the shaft is

$$\delta N = p \frac{\delta r}{\sin \alpha} r \delta\theta$$

and the axial component of this is

$$\delta N \sin \alpha = p \frac{\delta r}{\sin \alpha} r \delta\theta \sin \alpha$$

$$= p \delta r \, r \delta\theta$$

The axial force δS on the elementary frustum is thus

$$\delta S = \sum_{\theta = 0}^{2\pi} p \delta r \, r \delta\theta = p \delta r \, r 2\pi$$

and the total axial force

$$S = 2\pi \int_{r_1}^{r_2} pr \, \mathrm{d}r \qquad (5.14)$$

The tangential friction force on the elementary length of the elementary frustum is

$$\mu \delta N = \mu p \frac{\delta r}{\sin \alpha} r \delta\theta$$

and the total tangential frictional force on the elementary frustum is

$$\sum_{\theta = 0}^{2\pi} \mu p \frac{\delta r}{\sin \alpha} r \delta\theta = \mu p \frac{\delta r}{\sin \alpha} r 2\pi$$

The torque required to overcome friction on the elementary frustum is thus

$$\delta M_\mathrm{f} = \mu p \frac{\delta r}{\sin \alpha} r \, 2\pi r$$

$$= \frac{2\pi\mu}{\sin \alpha} pr^2 \delta r$$

For the whole bearing the torque M_f required to overcome friction is

$$M_\mathrm{f} = \frac{2\pi}{\sin \alpha} \int_{r_1}^{r_2} \mu pr^2 \mathrm{d}r \qquad (5.15)$$

Equations 5.14 and 5.15 cannot be evaluated until the relationships between p, μ and r are known. The usually accepted assumptions are

(1) μ is constant
(2) if the surfaces are new or fit together perfectly p is constant
(3) if the surfaces are worn the wear is assumed uniform over the bearing.

If wear is assumed to be proportional to the product of pressure and distance moved by one surface relative to the other, then since the relative distance moved is proportional to the radius r, wear is proportional to $p \times r$; then (3) also implies that, for uniformly worn surfaces, pr = constant.

For new surfaces p = constant and equations 5.14 and 5.15 become

$$S = 2\pi p \int_{r_1}^{r_2} r\,\mathrm{d}r = \pi p(r_2^2 - r_1^2) \tag{5.16}$$

$$M_f = \frac{2\pi\mu p}{\sin\alpha}\int_{r_1}^{r_2} r^2\,\mathrm{d}r = \frac{2\pi\mu p}{3\sin\alpha}(r_2^3 - r_1^3)$$

$$= \frac{2\mu S}{3\sin\alpha}\frac{(r_2^3 - r_1^3)}{(r_2^2 - r_1^2)} \tag{5.17}$$

For uniform wear pr = a constant C and the equations become

$$S = 2\pi C\int_{r_1}^{r_2}\mathrm{d}r = 2\pi C(r_2 - r_1) \tag{5.18}$$

and

$$M_f = \frac{2\pi\mu C}{\sin\alpha}\int_{r_1}^{r_2} r\,\mathrm{d}r = \frac{\pi\mu C}{\sin\alpha}(r_2^2 - r_1^2)$$

$$= \frac{\mu S}{\sin\alpha}\left(\frac{r_2 + r_1}{2}\right) \tag{5.19}$$

For clutches, M_f represents the friction torque that can be transmitted before slipping occurs and S represents the spring force acting on the clutch plates. For bearings M_f represents the torque that needs to be applied to the shaft to overcome the friction of the bearing.

In order to obtain a conservative value for M_f use the larger of the values given by equations 5.17 and 5.19 for bearings and the smaller of the values for clutches.

Worked Example 5.8

A flat clutch plate has $r_1 = 0.2$ m, $r_2 = 0.25$ m, $\mu = 0.4$. Find the spring force required in order that a torque magnitude 250 Nm may be transmitted.

Solution

Equation 5.17

$$S = \frac{3(\sin \alpha) M_f (r_2^2 - r_1^2)}{2\mu(r_2^3 - r_1^3)}$$

$$= \frac{3 \times 1 \times 250\,(0.25^2 - 0.2^2)}{2 \times 0.4\,(0.25^3 - 0.2^3)}$$

$$= 2766 \text{ N}$$

Equation 5.19

$$S = \frac{250 \times 1}{0.4}\left(\frac{2}{0.25 + 0.2}\right)$$

$$= 2778 \text{ N}$$

In order to ensure that the clutch will transmit at least the required torque, the larger value is chosen, that is, $S = 2778$ N.

5.2.6 Bearing and Axle Friction

Bearings are used to provide lateral support to rotating shafts and axles. The simplest form is the journal bearing in which the accurately turned end of the shaft or axle, the journal, rotates in a fixed plain cylindrical bearing having slightly larger internal diameter – as in figure 5.15a, in which the radial clearance

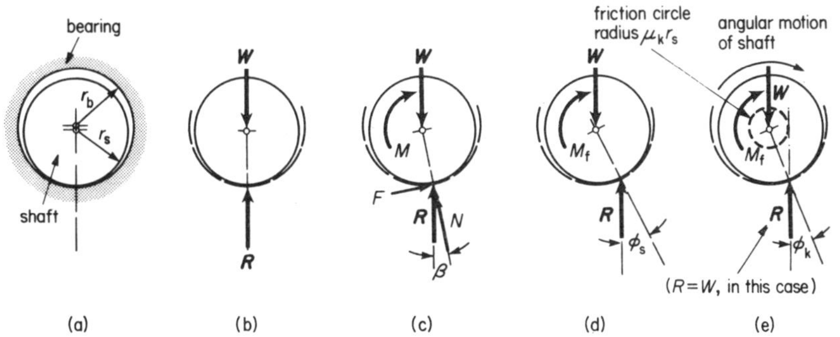

Figure 5.15

$(r_b - r_s)$ has been exaggerated. We shall only consider the case where dry friction exists, or, more realistically, where the bearing is lubricated but the amount of lubricant is insufficient to separate the surfaces at the point of contact.

Figure 5.15b shows a free-body diagram of the shaft, when at rest. If now a small couple is applied tending to rotate the shaft, the shaft rolls up the bearing surface and takes up the position shown in figure 5.15c when equilibrium is reached, with the total reaction **R** still equal in magnitude to W, the lateral load. As the couple is increased the angle β becomes equal to ϕ_s (figure 5.15d) and slipping occurs. The shaft moves back to a new equilibrium position with β equal to ϕ_k corresponding to the coefficient of kinetic friction (see figure 5.15e) and the shaft continues to rotate as long as the couple is maintained, for example by a driving motor. Note in particular that the shaft rolls up the bearing in the opposite sense to that of the angular motion of the shaft; a useful rule of thumb is that the sense of the moment of **R** about the axis must oppose that of the rotation.

When the shaft is rotating at constant speed it is in equilibrium and the point of application of the vertical reaction **R** is clearly dependent upon the value of $\phi_k = \tan^{-1} \mu_k$. If the magnitude of the continuously applied torque is M_f, then

$$M_f = Wr_s \sin \phi_k$$

In practice ϕ_k is very small and $\sin \phi_k \approx \tan \phi_k = \mu_k$ and justifiably $M_f = \mu_k Wr_s$.

M_f is the so-called friction torque, being the magnitude of the torque required to be applied to the shaft in order to maintain angular motion when the lateral load W is being supported. The line of action of **R** is always tangential to the circle radius $\mu_k r_s$ (figure 5.15e); this is called the *friction circle* and is the same for any lateral load if μ_k is constant.

Consider now a wheeled container being hauled along a horizontal track as in figure 5.16a; it has wheels at back and front but only one wheel and axle are shown.

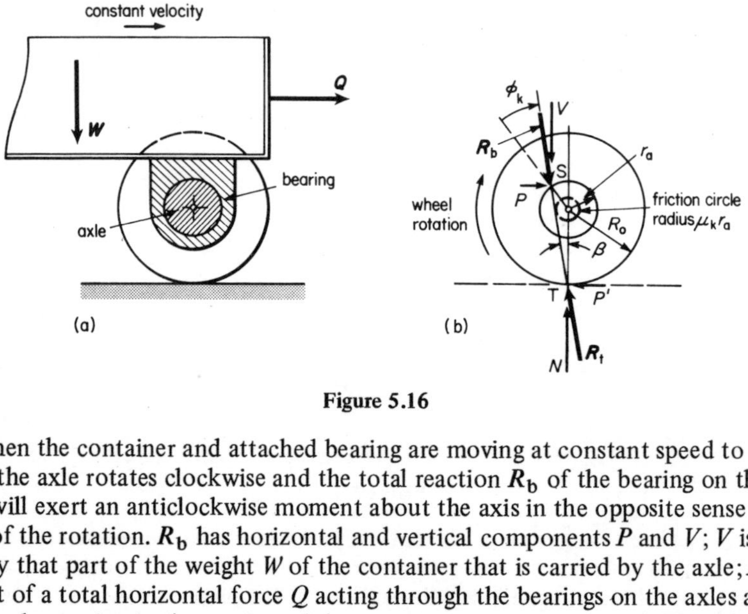

Figure 5.16

When the container and attached bearing are moving at constant speed to the right the axle rotates clockwise and the total reaction $\boldsymbol{R}_b$ of the bearing on the axle will exert an anticlockwise moment about the axis in the opposite sense to that of the rotation. $\boldsymbol{R}_b$ has horizontal and vertical components P and V; V is clearly that part of the weight W of the container that is carried by the axle; P is part of a total horizontal force Q acting through the bearings on the axles and required to maintain the motion of the container at constant speed. The other force acting on the wheel and axle is the total reaction of the track, $\boldsymbol{R}_t$, which has components N and P'. The free-body diagram of the wheel and axle is given in figure 5.16b and it follows for equilibrium that $P' = P$, $N = V$ and that the total reactions $\boldsymbol{R}_b$ and $\boldsymbol{R}_t$ must have the same line of action passing through the points S and T, inclined at angle β to the vertical. It follows that $P = V \tan \beta$. Since $\boldsymbol{R}_b$ is inclined at angle ϕ_k to the radial direction at S it follows from figure 5.16b that the common line of action is again tangential to the friction circle radius $r_a \sin \phi_k$, where r_a is the radius of the axle. Again if μ_k is small the radius of the circle is $\mu_k r_a$ as before. Now $\sin \beta = \mu_k r_a/R_o$, where R_o is the radius of the wheel, and if r_a is small compared with R_o then $\sin \beta \approx \tan \beta$, therefore

$$P = V \times \mu_k r_a/R_o$$

If the track friction force P', as given by the equation for P above, cannot be generated due to limiting friction conditions at the track (P' is limited to $\mu_{track} N$) and μ_k at the bearing is sufficiently large, the wheel and axle cannot rotate and slipping occurs at the track. On the other hand, if the bearing is well lubricated, μ_k is small and the force P to maintain the motion is correspondingly reduced.

Assuming all the wheels and axles of the container are the same size and since

$$Q = \Sigma(P \text{ on all axles})$$

$$W = \Sigma(V \text{ on all axles})$$

then

$$Q = \mu_k W r_a/R_o$$

being the force required to overcome friction at the bearings due to the vertical load W.

5.3 Summary

(1) The friction force component F of the total reaction $\boldsymbol{R}$ of a surface at a rough contact is in the opposite sense to that in which the body is tending to move relative to the adjoining surface.
(2) The friction force F due to dry friction has a limiting value $F_L = \mu N$, which is attained when motion is impending. If motion is not impending the value of F is less than μN.
(3) The angle of friction $\phi = \tan^{-1} \mu$ is the angle that the total reaction $\boldsymbol{R}$ of a surface on a body makes to the normal when friction is limiting; as such it is the limiting value of β, the angle between $\boldsymbol{R}$ and its normal component N.
(4) The values of N and F are both dependent upon the other external forces.
(5) Under non-limiting friction conditions where more than one sliding face is involved it is usually not possible, when only one body is involved, to determine the individual reactions.
(6) For a body to be in a condition of impending motion limiting friction must exist at all sliding surfaces.
(7) In solving problems make use of equilibrium conditions in graphical or analytical form; use free-body diagrams. If equilibrium exists and all the friction forces are not limiting simultaneously, the body is at rest and motion is not impending; if equilibrium exists when all the friction forces are limiting simultaneously then motion is impending; if when all the friction forces have limiting values there is a resultant force or couple, motion of the body is occurring; see section 5.1.2.
(8) When slipping *just* occurs it is assumed that the conditions for equilibrium can be applied.

(9) Wedges: examine the equilibrium of each body with limiting friction present, appropriately directed.
(10) Jamming: consider the equilibrium of each body assuming limiting friction is present and ascertain the conditions under which applied forces become indefinitely large.
(11) Screws: equivalent model is a block (carrying a load) on an inclined plane with an applied force perpendicular to the load; equations are derived on the assumption of limiting friction; note differences for motion against and with the load, particularly for the latter when $\phi > \alpha$.
(12) Belts

$$\frac{T_1}{T_2} = e^{\mu\theta} \tag{5.12}$$

(θ in radians). Note other limitations such as maximum allowable tension.
(13) Clutches and thrust bearings

$$S = 2\pi \int_{r_1}^{r_2} pr\,\mathrm{d}r \tag{5.14}$$

$$M_f = \frac{2\pi\mu}{\sin\alpha} \int_{r_1}^{r_2} pr^2\,\mathrm{d}r \tag{5.15}$$

For new surfaces assume p = constant; for uniformly worn surfaces assume pr = constant.
(14) Bearing and axle friction: the total reaction $\boldsymbol{R}$ of a bearing on an axle has a line of action that is always tangential to the friction circle and such that the moment of $\boldsymbol{R}$ about the axis is in the opposite sense to that of the angular motion of the axle; the friction circle has radius μr_s where r_s is the radius of the axle.

Problems

5.1 A bar rests on two pegs and makes an angle β with the horizontal. The coefficients of friction are μ_1 at one peg, which is distance a from G, the centre of gravity of the bar, and μ_2 at the other peg at distance b from G. Show that for an equilibrium condition to exist

$$\tan\beta < \frac{\mu_1 b + \mu_2 a}{(a + b)}$$

(*Hint*: Draw a free-body diagram inserting forces at contact points as separate normal (N) and friction forces (F); from the equilibrium condition obtain three equations.)

5.2 A circular cylinder of mass m with its axis horizontal is supported in contact with a rough vertical wall by a string wrapped partly around it and attached to a point on the wall above the cylinder.

If the angle between the string and the wall is β show that the coefficient of friction must be not less than cosec β, and that the normal force on the wall is $mg \tan(\beta/2)$. (*Hint*: As problem 5.1.)

5.3 A uniform circular hoop has a small body, with a mass equal to its own, attached to a point on its rim, and is hung over a rough horizontal peg. Show that if the angle of friction is greater than $\pi/6$ the system can rest with any point of the hoop in contact with the peg. (*Hint*: The external forces due to the weights of the attached mass and the hoop itself are vertical; the direction of the total reaction at the peg follows from this; use a free-body diagram with the peg at θ to the vertical through the hoop centre.)

5.4 The body in figure 5.17 is stationary when the force P is gradually applied.

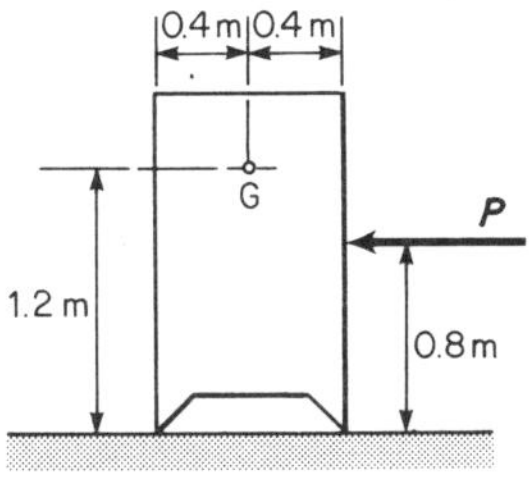

Figure 5.17

What is the critical value of μ so that the body tips before it slides? Assume accelerations are zero. (*Hint*: Use a free-body diagram inserting N and F at each contact point; use equilibrium conditions; if it tips one normal reaction must be zero.)

5.5 A uniform plank of length 5 m and mass 50 kg is standing on a horizontal floor and leaning against a vertical wall. At the floor the coefficient of friction is 0.3 and at the wall 0.2.

(a) The plank is placed at an angle $\tan^{-1} 0.6$ to the vertical. Can the magnitudes of the normal reactions and the friction forces at the wall and the floor be determined?
(b) Determine the ranges of values that are possible for these forces.
(c) Can the plank slip in the position given in (a)?
(d) Can the friction force at the floor ever be zero when the plank is in equilibrium?
(e) At what angle to the vertical must the plank be placed for slip to be about to occur?

(*Hint*: As problem 5.1; see worked example 5.2 for (a), (b) and (c); for (e) use total reactions and a geometrical approach.)

5.6 A uniform rod mass m and length L is lying on a rough horizontal table. A horizontal force P is applied to the rod perpendicular to its axis at a distance

$kL(k > \frac{1}{2})$ from one end so that it just moves. Show that the rod rotates about a point distance hL from the same end and that $P = \mu mg(1 - 2h)$, where $h = k - (k^2 - k + \frac{1}{2})^{\frac{1}{2}}$. (*Hint*: Choose an elementary length of the cylinder and thus decide the frictional force acting on it; then consider the equilibrium of the whole cylinder.)

5.7 A thin sheet of metal is to be positioned for feeding into a press by use of a simple jig consisting of two cylindrical rods, which are parallel and at the same level. The sheet is passed to the right over rod 1 and then under rod 2 and positioned so that its centre of gravity is 2 m to the left of rod 1. The whole jig is now tilted anticlockwise until the sheet slides out of the jig. Find the distance required between the two rods if the sheet is to slide when at an angle of 60° to the horizontal. For both sliding surfaces $\mu = 0.2$. Ignore the thickness of the sheet and the diameter of the rods. (*Hint*: Consider the equilibrium of the forces on the sheet under limiting friction.)

(5.8) Two rough uniform cylinders, having equal diameters but unequal masses m_1 and m_2 $(m_1 < m_2)$ rest in line contact with each other on a plane whose inclination β is $< 45°$, the axes of the cylinders being horizontal.

If μ is the same for all surfaces, show that the frictional forces at each contact surface are equal and that for equilibrium to be possible the heavier cylinder must be uppermost and that $\mu > (m_2 + m_1)/(m_2 - m_1)$. (*Hint*: Draw the separate free-body diagrams inserting (with assumed directions) the friction and normal forces at each contact point; use the equilibrium condition to answer the questions; compare the ratio of F/N with μ.)

5.9 Two cylinders lie in equilibrium on a rough inclined plane, in line contact with one another with their axes horizontal. The upper cylinder radius a, is heavy but the lower cylinder, radius b, has negligible weight. Show that if β is the inclination of the plane to the horizontal then $b > (a \tan^2 \beta)/4$, that the coefficient of friction between the heavy cylinder and the plane must be at least $1/[2 \cot \beta - \sqrt{(a/b)}]$, and that the other coefficients of friction must be at least equal to $\sqrt{(b/a)}$. (*Hint*: As for problem 5.8; all friction and normal force components can be evaluated in terms of the weight and the dimensions of the heavy cylinder (use the geometrical relationships). The condition for b arises from the fact that all normal components must be > 0.)

5.10 (a) If for the wedge and block shown in figure 5.18 $Q = 100$ N and $\theta = 40°$, find the value of P required to lift the block.
(b) If $P = 100$ N and $\theta = 40°$ find the value of Q required to make the block descend.
(c) What should be the value of θ so that the wedge is locked (i) for any value of P, (ii) any value of Q?

Disregard the masses of the wedge and the block and take μ at all surfaces as 0.3. (*Hint*: Use the total reactions and hence draw triangles of forces for each body; solve for Q and P. For (c) examine the relevant triangle to find the critical values of θ.)

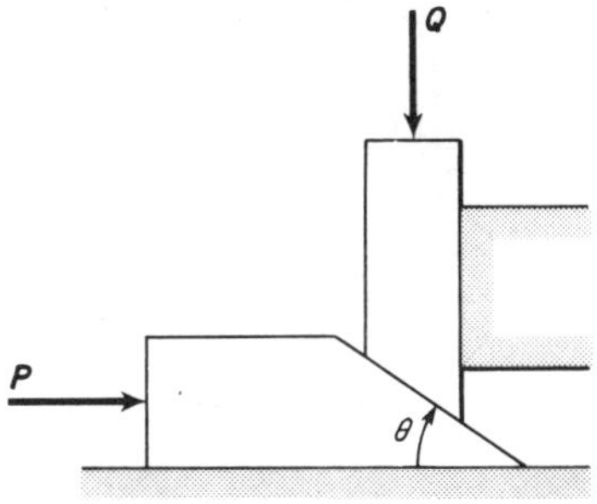

Figure 5.18

5.11 The mechanism shown in figure 5.19 lies in the horizontal plane. AB slides in guides and carries a pin E which slides in a slot on the arm CD. The latter rotates freely on an axle at D. The coefficient of friction at each of the sliding surfaces at R, S and E is 0.35.

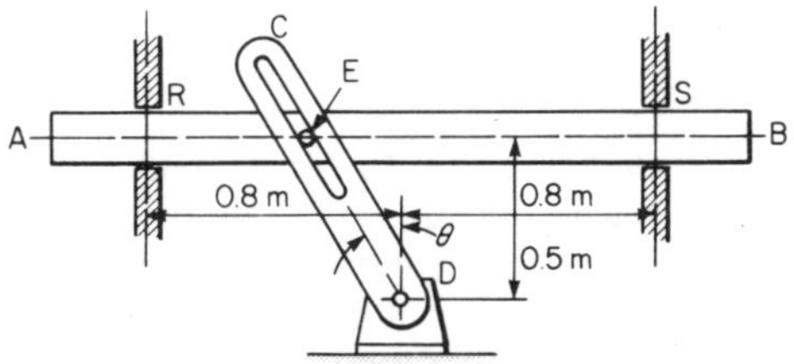

Figure 5.19

(a) Find the maximum value of θ at which a couple applied to CD can move AB to the left.
(b) Determine the value of the axial force that needs to be applied at B to move AB to the left when a clockwise couple having moment 50 N m is applied to CD and θ has the value found in (a). Ignore the thickness of AB.

(*Hint*: Use separate free-body diagrams; the directions of the total reactions are determined from the relative movement. For (a) consider equilibrium of AB, for (b) consider equilibrium of both AB and CD.)

5.12 The mechanism shown in figure 5.20 lies in the horizontal plane and consists of a rigid rod AB with cylindrical ends sliding in mutually perpendicular slots. μ for all sliding surfaces is 0.3.

(a) If $Q = 0$ determine the minimum value of θ at which P can move AB.
(b) If P is removed, can Q cause movement at the value of θ determined in (a)?
(c) If $Q = 100$ N and $\theta = 40°$ determine the magnitude of P so that motion is impending (i) in the anticlockwise direction (ii) in the clockwise direction.

(*Hint*: Use a free-body diagram with limiting friction conditions. For (a) consider the case when there is no resultant moment on AB to cause motion; similarly for (b). Solve (c) using a polygon or triangles of forces, or by moments.)

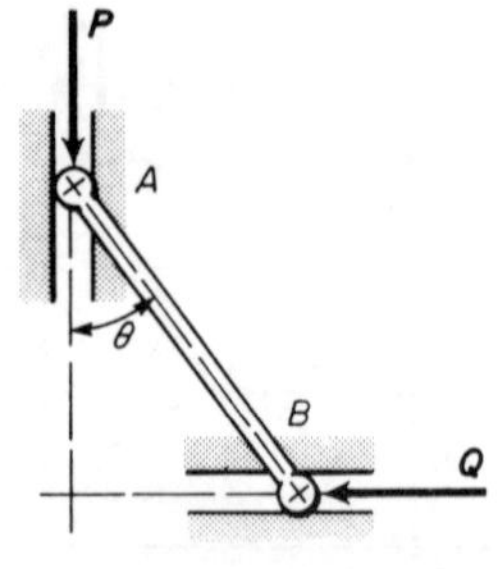

Figure 5.20

5.13 The force required to tether a ship is 10 000 N. The rope used to do this is wound four complete turns around a cylindrical post, the remaining length L lying on the ground. If the rope has mass 20 kg/m and the coefficient of friction between the rope and the floor and between the rope and the post are 0.1 and 0.2 respectively, determine L to prevent slipping. Disregard the mass of the rope except where it contacts the floor. (*Hint*: Use the ratio of tensions and the force required to move the rope in contact with the floor.)

5.14 A belt-and-pulley system consists of a driving pulley A with outside radius 0.2 m and a driven pulley B with outside radius 0.4 m. The distance between the axes of the pulley shafts is 1 m. If the coefficients of friction between the belt and the pulleys are 0.2 at A and 0.15 at B determine at which pulley the belt first slips. What is the maximum torque that can be transmitted through the driven pulley if the allowable tension in the belt is 800 N?

Will it always be true in a two-pulley system, with μ the same for both pulleys, that the criterion for slipping is at the smaller pulley?

5.15 A belt is attached at its two ends to two pins at the same level and distance d apart. A uniform cylinder – diameter d and mass m, to the surface of which a small body, mass m, is attached – is placed to lie in the belt with its axis horizontal. If the length of the belt is greater than $\frac{1}{2}\pi d$ and μ between belt and cylinder is 0.2, determine the angular position θ of the small body from the vertical line through the centre of the cylinder at which the cylinder will just slip in the belt. (*Hint*: Consider the equilibrium of the pulley with the attached mass; relate the tensions.)

5.16 An electric motor drives a pulley of radius 0.2 m through a clutch. The pulley is fitted with a belt with angle of lap 150°, and μ between the belt and the pulley is 0.25. The clutch plate has inner and outer radii 0.3 m and 0.6 m, and μ is 0.2. Determine the required end load on the clutch if the clutch is to slip just before the belt, when the tension on the tight side of the belt is 500 N.

5.17 A screwed shaft moving in a fixed guide has a mean diameter 0.5 m and lead 0.25 m, $\mu = 0.05$. For an axial load of magnitude 1000 N determine the torque required (a) to just sustain the load in equilibrium and (b) to move the shaft against the load. (*Hint*: For (a) which way is motion impending?)

5.18 A screwed shaft moving in fixed guide carries an end load of 500 N. The thread has a mean diameter 0.15 m and a lead 0.05 m; μ is 0.2. Find the magnitude and sense of the torque on the shaft required to (a) move the shaft against the load, (b) move the shaft in the direction of the load.

What should the mean diameter be in order that the torque should be zero in case (b)?

5.19 A body mass 1000 kg is to be lifted by means of a screw jack consisting of a screwed spindle moving in a fixed guide. The body cannot rotate and is carried on a thrust plate, which acts as a thrust bearing for the top face of the spindle. The top face of the spindle is made 0.3 diameter and the pressure may be assumed to be uniform. The thread has a mean diameter 0.1 m and a lead 0.02 m; μ at both sliding surfaces is 0.2

(a) Find the torque required to be applied to the spindle to raise the load.
(b) Will the load stay in position if this external torque is removed?
(c) What should the lead be if the load is to be on the point of descending without assistance when the torque is removed?

(*Hint*: Consider the friction torques (in magnitude and sense) at both thread and thrust bearing.)

5.20 In the thrust bearing of figure 5.14, $r_1 = 0.1$ m, $r_2 = 0.3$, $\alpha = 60°$ and $\mu = 0.05$. The end load is 10 000 N. Obtain a conservative estimate of the friction torque.

5.21 A flat thrust bearing is to have the inner radius one-third that of the outer radius. It carries a load of 20 000 N and the pressure p (assumed uniform) is to be 50 000 N/m^2. Determine the radii and the torque required to overcome friction. Assume $\mu = 0.08$.

If the bearing wears until pr = constant what will then be the maximum pressure and the torque required to overcome friction?

5.22 A rope passing over a pulley is used to lift a body mass 20 kg. What force, applied vertically downwards on the other end of the rope, is required in the following cases: (a) the pulley is locked on its axle and the rope slips on the pulley, with $\mu = 0.3$; (b) the pulley rotates on its axle, the clearance being sufficient for contact to be made along a single horizontal line. The ratio of the outside radius of the pulley to the radius of the axle is 10:1 and μ is again 0.3. (*Hint*: For (b) consider the equilibrium of the pulley noting that the reaction of the axle on the pulley is tangential to the friction circle.)

5.23 The wheel of the container in figure 5.16 has outside radius R_o, and axle radius r_a; the coefficient of friction between the wheel periphery and its contact surface is μ_o and at the axle is μ_a. Find the relationship between μ_o and μ_a, assuming $r_a \ll R_o$, in order that slip will first occur at the wheel periphery.

5.24 The carrier in figure 5.21 has impending motion up the plane. Its total mass

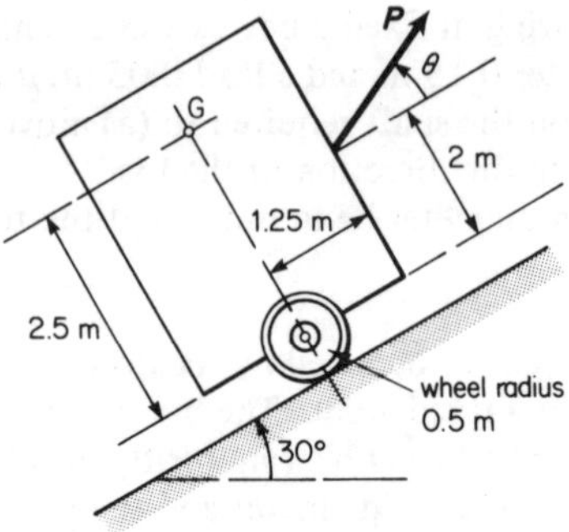

Figure 5.21

is 100 kg, its centre of gravity is at G; the friction circle of the axle is of 0.1 m radius and μ at the road surface is 0.5.

(a) In the impending motion will slip take place at the road surface or at the axle?
(b) What is the magnitude of P and the value of θ?

(*Hint*: Decide the direction of the reaction of the road on the wheel; consider the equilibrium of the carrier.)

6 Virtual Work

The methods used in chapter 4 for the complete analysis of trusses and frames involved successive applications of the conditions for equilibrium either for a particle or for a rigid body. In this chapter we consider a further interpretation of the state of equilibrium that has important implications. In particular, a method is provided for enabling selected unknown forces in connected bodies to be determined without having to apply the conditions for equilibrium to each body separately.

6.1 Work

Before going on to discuss the principle of *virtual work* it is necessary to define the *work* of a force. In chapters 10 and 13 the discussion of work will be much enlarged in the context of *power* and *energy* but for our present purposes the basic definition which follows is sufficient.

6.1.1 Work of a Force

In figure 6.1a a particle is moving along a path under the influence of a force **F**.

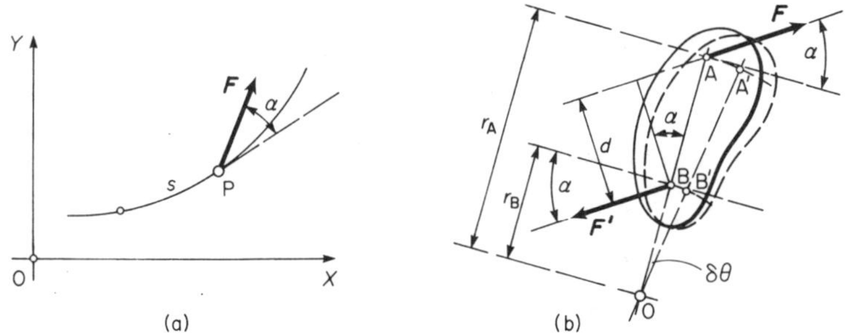

Figure 6.1

The *work* of the force for a small displacement δs of the particle from P to P′ is defined to be $F \cos \alpha \delta s$ or $F_s \delta s$ where $F_s = F \cos \alpha$ is the component of **F** in the direction of the tangent to the path of the particle at P. The total work of the force on the particle as it moves from position s_1 to position s_2 along its path is symbolised by U_{1-2}, and it follows that

$$U_{1-2} = \int_{s_1}^{s_2} F_s \, ds \tag{6.1}$$

Work is positive if F_s has the same sense as δs; its unit is the joule (J), which is equivalent to 1 N m.

It is easily demonstrated (see section 10.5) that the work of a force $\boldsymbol{F}$ can be calculated by reference to its components in arbitrary x- and y-directions, since

$$F \cos \alpha \, \delta s = F_s \delta s = F_x \delta x + F_y \delta y$$

and

$$U_{1-2} = \int_{x_1}^{x_2} F_x \, \mathrm{d}x + \int_{y_1}^{y_2} F_y \, \mathrm{d}y$$

the initial and final positions of the particle being (x_1, y_1) and (x_2, y_2).

6.1.2 Work of a Torque or Couple

The preceeding definition is sufficient to allow the total work of a set of forces acting on a rigid body to be calculated. However, where torques or couples are applied to a rigid body, it is convenient to be able to state the work of a couple in terms of its moment. A couple is indicated in figure 6.1b by two equal unlike parallel forces $\boldsymbol{F}$ and $\boldsymbol{F}'$ a distance d apart such that the moment of the couple $M = Fd$. It is required to determine the work of the couple as the body turns through an angle $\delta\theta$ about a point O so that the line AB moves to A′B′, its length remaining unchanged. If $\delta\theta$ is small, AA′ and BB′ are both perpendicular to AB, and make equal angles with the common direction of $\boldsymbol{F}$ and $\boldsymbol{F}'$. The total work of the couple is then

$$\begin{aligned}
\delta U &= F \cos \alpha \times \mathrm{AA}' - F' \cos \alpha \times \mathrm{BB}' \\
&= F \cos \alpha \, (\mathrm{AA}' - \mathrm{BB}') \\
&= F \cos \alpha \, (r_{\mathrm{A}} \, \delta\theta - r_{\mathrm{B}} \, \delta\theta) \\
&= F \, \delta\theta \times \mathrm{AB} \cos \alpha \\
&= Fd \, \delta\theta \\
&= M \, \delta\theta
\end{aligned}$$

The total work of the couple as the body rotates from θ_1 to θ_2 is

$$U_{1-2} = \int_{\theta_1}^{\theta_2} M \, \mathrm{d}\theta \qquad (6.2)$$

This is positive if M and $\delta\theta$ have the same sense. The unit is again the joule, with θ measured in radians.

It is a simple matter to demonstrate that if the body moves in such a way that the orientation of AB is unchanged, that is, the body does not rotate, the couple does no work; the work of a couple or torque is therefore associated only with rotation of the body.

6.2 Principle of Virtual Work

In chapters 3 and 4 the forces acting on the particles of a particle system were broadly classified as external and internal, depending on their source relative to the system boundary. For the purposes of the present discussion, different criteria are adopted to classify the forces; they again fall into two categories, now referred to as *applied forces* and *forces of constraint*. Constraints are geometrical restrictions imposed on the motion of a system; constraint forces are those external forces associated with such constraints when the system is subjected to applied forces, and also those internal forces required to maintain the geometrical configuration of the system particles. The distinction is brought out further in the course of the discussion.

Suppose now a single particle subject to a given set of forces undergoes an arbitrary, infinitesimally small displacement δs whose effect on the magnitudes and directions of the forces acting on the particle is negligible. The cause of the displacement need not be questioned since it is arbitrary and is not necessarily related to the forces acting. Such a supposed displacement is termed a *virtual displacement.* Arising from the displacement there will be work $F_s\,\delta s$ associated with each individual force and the total work $\Sigma\delta U = \Sigma F_s\,\delta s$ for all the forces can in principle be calculated. Since the work arises from a virtual displacement it is called *virtual work*.

If the particle is in equilibrium under the action of the force set, then the total work is zero in a virtual displacement. This follows since the sum of the virtual works of the forces is equal to the virtual work of their resultant, and the resultant is zero if the particle is in equilibrium.

For a particle system the virtual work for the forces, both external and internal, on all the particles can again, in principle, be calculated. The total virtual work is not, in general, zero even if the system is initially in equilibrium, since the internal forces in particular may change during the arbitrary displacements, which are, in general, different for each particle.

Now we can limit the virtual displacements to those that are said to be *consistent* with certain constraints. Thus if the system is a *rigid body* the distances between the particles are unchanging; the total virtual work of the internal forces taken in pairs is then zero. Further, the rigid body itself can be subject to constraints on its movement, which limit the number of possible virtual displacements; if the work of a constraint force during such a displacement is zero then the constraint is said to be a *workless* constraint. Typical examples of forces at workless constraints are

(1) the normal reaction at a sliding contact with a fixed smooth surface
(2) the normal reaction at a rolling contact
(3) the reaction at a smooth fixed pin
(4) the pair of equal and opposite reactions at a smooth pinned joint between connected bodies.

It now follows that if a rigid body, or a system of rigid bodies connected by smooth pins, is in equilibrium under a set of applied forces and forces of workless constraints, then the virtual work of the applied forces is zero in a virtual displacement consistent with the constraints, since the work of the constraint forces is zero. If friction is present it is treated as an applied force.

The converse of this statement can be shown to be a condition for equilibrium. It is then referred to as the *principle of virtual work* and can be stated as follows: a rigid body or system of connected rigid bodies is in equilibrium if and only if the work of the applied forces is zero in any arbitrary virtual displacement that is consistent with workless constraints.

If $F_{app,s}$ is the component of an applied force in the direction of a virtual displacement, magnitude δs, of its point of application, then the condition for equilibrium of the system can be stated in the form

$$\Sigma \delta U = \Sigma F_{app,s}\ \delta s = 0 \tag{6.3}$$

When applied couples or torques are present $\Sigma \delta U$ is understood to include the virtual work of such couples or torques.

6.3 Applications: Connected Bodies

The principle of virtual work, when expressed in equation form, can be used to calculate unknown forces acting on systems that are known to be, or are required to be, in equilibrium. For this purpose the virtual displacements at the salient points are expressed in terms of the changes in those coordinates that fix the configuration of the body or system of connected bodies. The number of such coordinates represents the number of degrees of freedom of the system. For example, in figure 6.2a the angle θ fixes the configuration of the slider - crank

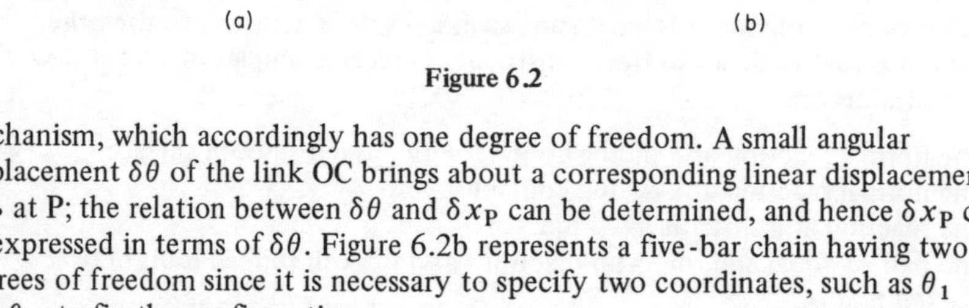

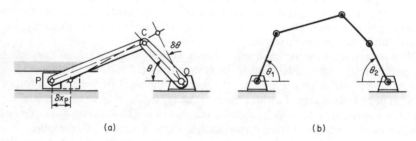

Figure 6.2

mechanism, which accordingly has one degree of freedom. A small angular displacement $\delta\theta$ of the link OC brings about a corresponding linear displacement δx_P at P; the relation between $\delta\theta$ and δx_P can be determined, and hence δx_P can be expressed in terms of $\delta\theta$. Figure 6.2b represents a five-bar chain having two degrees of freedom since it is necessary to specify two coordinates, such as θ_1 and θ_2, to fix the configuration.

In the case of a truss supported on a roller and pin joint, or a framework pinned to a support, it might be thought at first sight that no displacement is possible.

However, if one or other of the constraints is relaxed and replaced by a force equivalent to the original constraint force then a virtual displacement is possible in which the equivalent force, now regarded as an applied force, has virtual work associated with it.

In order to set up the virtual-work equation it is again advisable to draw a diagram of the isolated body or system. However, the forces of workless constraints should not be entered since they do not appear in the virtual-work equation. The remaining applied forces are sometimes referred to as *active forces*, and the diagram is then best referred to as an *active-force diagram*.

Worked Example 6.1

For the framework shown in figure 6.3a determine the force in the member CD in terms of the load W and the angle θ. All pins are smooth.

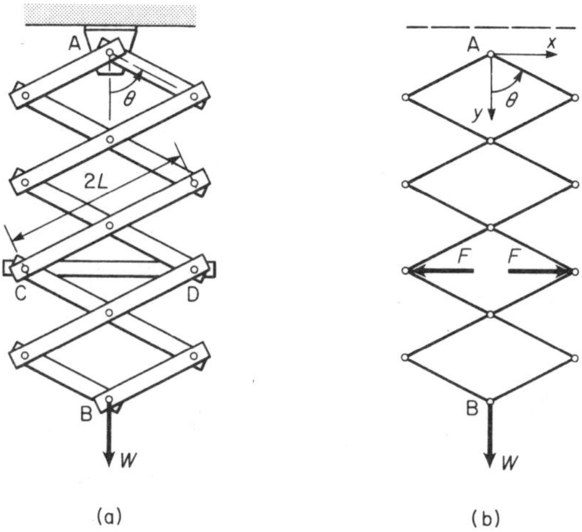

Figure 6.3

Solution

The framework as shown is rigid. Now the member CD supplies equal and opposite forces of magnitude F at joints C and D. If the member is removed and the same forces are applied at C and D the configuration of the frame can be arbitrarily changed. Since the configuration of the frame can be determined by the angle θ only, or by the length AB only, the system has one degree of freedom. An active-force diagram is drawn as shown in figure 6.3b, the constraints at all the pins being workless. Rectangular coordinate directions are indicated with origin at the fixed pin A.

In terms of θ the coordinates of C, D and B are respectively

$$x_C = -L \sin\theta \qquad x_D = L \sin\theta \qquad y_B = 8L\cos\theta$$

The change δx in a typical coordinate x due to a small change $\delta\theta$ is given by $\delta x = (\mathrm{d}x/\mathrm{d}\theta)\delta\theta$. The virtual displacements at C, D and B corresponding to a virtual angular displacement $\delta\theta$ are thus respectively

$$\delta x_{\mathrm{C}} = -(L \cos \theta)\delta\theta$$

$$\delta x_{\mathrm{D}} = (L \cos \theta)\delta\theta$$

$$\delta z_{\mathrm{B}} = -(8L \sin \theta)\delta\theta$$

Applying the principle of virtual work, and using equation 6.3, (note that the force at C is in the negative x-direction)

$$(-F)\delta x_{\mathrm{C}} + F\delta x_{\mathrm{D}} + W\delta z_{\mathrm{B}} = 0$$

therefore

$$(-F) \times (-L \cos \theta)\delta\theta + F \times (L \cos \theta)\delta\theta + W \times (-8L \sin \theta)\delta\theta = 0$$

and

$$F = 4W \tan \theta$$

In the above example the relaxation of a constraint in order to make a virtual displacement possible had the effect of converting an otherwise rigid frame into a mechanism. The principle of virtual work is thus directly applicable to mechanisms to determine the relation between input and output forces and torques; a velocity diagram (as described in chapter 11) for the mechanism is then the most convenient method of relating displacements at the salient points.

6.4 Connected Bodies: Friction and Elastic Members

In the application of the virtual-work principle to connected bodies, the forces considered to be active forces were so chosen as to make certain constraints workless. If external friction were present at a constraint then it was to be treated as an active force. A difficulty arises here since the magnitude and direction of such a friction force is usually not known; even if it were known to be limiting friction then it would depend on the normal reaction, which would have to be determined first, usually by separation of the members. The main advantages of the virtual-work method are then lost. The same remarks would apply to problems involving friction at internal constraints. Worked example 6.2 illustrates a case in which an initial assumption has been made on the magnitude of the friction force at an internal constraint. The essential fact is that the net work of the pair of equal and opposite friction forces at a connection at which relative movement is taking place, is always negative because the frictional force always opposes the relative motion.

Suppose the system contains an elastic member. By this we mean one whose deformation δ is proportional to the magnitude F of the force exerted by it, such that $F = k\delta$, where k is termed the *elastic constant*, with units N/m. If this elastic member, for example a spring having undeformed length L_0, has length L then its action is equivalent to two equal and opposite forces each having magnitude F_e, where $F_e = k(L - L_0)$. If the spring is extended ($L > L_0$) then F_e is positive and the two forces are directed inwards. F_e therefore has a positive numerical value if the member is in tension and conversely a negative value if in compression. If in the virtual displacement of the system the length of the member changes from L to $L + \delta L$, where δL may be a positive or negative quantity, then the work of the pair of elastic forces is in general $- F_e \delta L$. This quantity must be included in the virtual-work equation $\Sigma \delta U = 0$ which now may be written

$$\Sigma \delta U_{\text{external active forces}} + (- F_e \delta L) = 0 \tag{6.4}$$

In the application of equation 6.4, $F_e = k(L - L_0)$ and δL is the incremental change in L arising from the change in a chosen reference coordinate such as θ.

Worked Example 6.2

In the framework shown in figure 6.4a members AB, AC each have length L, and

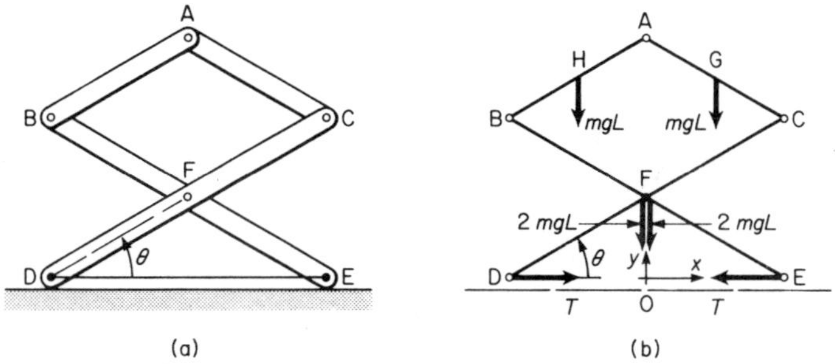

Figure 6.4

members BE, CD each have length $2L$ with the hinge F at their midpoints. The four members have mass m per unit length. The ends D and E rest on a smooth horizontal surface and are joined by an inextensible cord whose length can be varied. The angle θ is initially made less than 30°.

When the cord is slowly tightened an internal friction couple having moment $mgL^2/2$ is brought into play at each hinge. Determine the tension in the cord when it is tightened sufficiently to make $\theta = 30°$.

The cord is now further tightened and then slowly slackened until θ is again 30°. What is now the tension in the cord?

You may assume the friction couples are insufficient to hold the frame in equilibrium.

Solution

Remove the cord and replace it by equal and opposite forces of magnitude T. The active-force diagram is given in figure 6.4b. Choose point O on the axis of symmetry as the origin of x- and y-coordinates. The coordinates of the relevant points G, H, F, D and E are set down, together with the displacements corresponding to a change $\delta\theta$ in the angle θ.

$$x_E = L\cos\theta \qquad \delta x_E = -L\sin\theta\,\delta\theta$$
$$x_D = -L\cos\theta \qquad \delta x_D = L\sin\theta\,\delta\theta$$
$$y_F = L\sin\theta \qquad \delta y_F = L\cos\theta\,\delta\theta$$
$$y_G = 2.5L\sin\theta \qquad \delta y_G = 2.5L\cos\theta\,\delta\theta$$
$$y_H = 2.5L\sin\theta \qquad \delta y_H = 2.5L\cos\theta\,\delta\theta$$

For a change $\delta\theta$ in θ the relative angular movement of the members meeting at A, B, C and F is $2\delta\theta$ in each case. For both positive and negative numerical values of $\delta\theta$ the friction work at a hinge is negative with magnitude $\frac{1}{2}mgL^2 \times 2|\delta\theta|$. When the cord is tightened $\delta\theta$ is a positive quantity. The virtual-work equation for equilibrium at angle θ is $\Sigma F_{\text{app},s}\,\delta s = 0$, thus

$$(-T)\times\delta x_E + T\times\delta x_D + (-mgL)\times\delta y_G + (-mgL)\times\delta y_H + (-4mgL)\times\delta y_F - 4\times\tfrac{1}{2}mgL^2\times 2|\delta\theta| = 0$$

$$2TL\sin\theta\,\delta\theta - 5mgL^2\cos\theta\,\delta\theta - 4mgL^2\cos\theta\,\delta\theta - 4mgL^2|\delta\theta| = 0$$

and

$$T = \frac{mgL}{2\sin\theta}(9\cos\theta + 4)$$

When $\theta = 30°$

$$T = \tfrac{1}{2}mgL(9\sqrt{3} + 8)$$

When the cord is slackened $\delta\theta$ takes a negative sign but the friction work is unchanged. The virtual-work equation now becomes

$$2TL\sin\theta(-\delta\theta) - 9mgL^2\cos\theta(-\delta\theta) - 4mgL^2|\delta\theta| = 0$$

and

$$T = \frac{mgL}{2\sin\theta}(9\cos\theta - 4)$$

When $\theta = 30°$

$$T = \tfrac{1}{2}mgL(9\sqrt{3} - 8)$$

6.5 Systems with Two Degrees of Freedom

In the two preceeding worked examples the system in each case had one degree of freedom and the components of the virtual displacements at all relevant points could be expressed in terms of the small change $\delta\theta$ in one selected coordinate such as θ, which fixed the configuration. For systems having two degrees of freedom the configuration depends on two coordinates such as θ_1 and θ_2, either or both of which can be varied *independently* (see for example figure 6.2b) to bring about a virtual displacement of the system. The corresponding components of the virtual displacements of all relevant points are now expressed in terms of the changes in either or both of the coordinates θ_1 and θ_2. The position of some relevant point is first expressed as a function of the two variables θ_1 and θ_2, for example the x-component $x(\theta_1, \theta_2)$, and it then follows that the change in x is given by

$$\delta x = \frac{\partial x}{\partial \theta_1} \delta \theta_1 + \frac{\partial x}{\partial \theta_2} \delta \theta_2$$

with similar expressions for other components of displacement.

Now an equilibrium configuration requires for its description particular values of both θ_1 and θ_2. However, two virtual-work equations can now be written, one in terms of virtual displacements related to $\delta\theta_1$, with θ_2 kept fixed, and the other in terms of virtual displacements related to $\delta\theta_2$, with θ_1 kept fixed; in the first equation virtual-displacement components are quantities such as $(\partial x/\partial\theta_1)\delta\theta_1$ (with $\delta\theta_2 = 0$), and in the second equation the components are quantities such as $(\partial x/\partial\theta_2)\delta\theta_2$ (with $\delta\theta_1 = 0$).

Worked Example 6.3

In figure 6.5a the uniform members AB, BC have masses m_1, m_2 and lengths L_1, L_2 respectively, and can swing freely in the vertical plane. Determine the angles θ_1 and θ_2 at which the members are in equilibrium when the horizontal force P is applied at the end C.

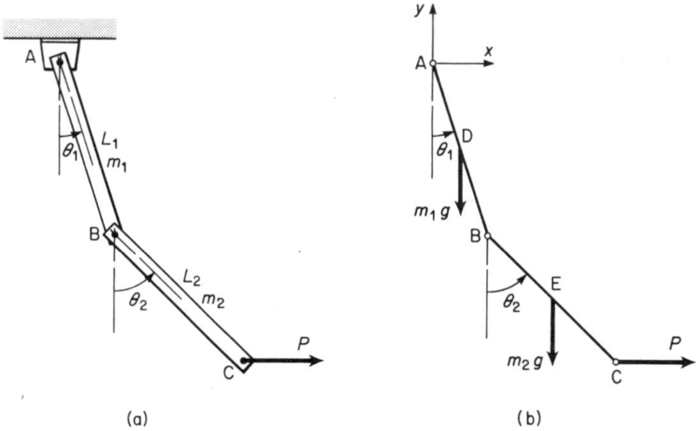

Figure 6.5

Solution

The active forces are shown in figure 6.5b, on which are shown x- and y-coordinate directions.

$$y_D = -\tfrac{1}{2}L_1 \cos\theta_1$$
$$\delta y_D = \tfrac{1}{2}L_1 \sin\theta_1\,\delta\theta_1$$
$$y_E = -L_1\cos\theta_1 - \tfrac{1}{2}L_2\cos\theta_2$$
$$\delta y_E = L_1\sin\theta_1\,\delta\theta_1 + \tfrac{1}{2}L_2\sin\theta_2\,\delta\theta_2$$
$$x_C = L_1\sin\theta_1 + L_2\sin\theta_2$$
$$\delta x_C = L_1\cos\theta_1\,\delta\theta_1 + L_2\cos\theta_2\,\delta\theta_2$$

Applying the virtual-work equation with θ_2 fixed, and $\delta\theta_2 = 0$

$$-m_1g \times \tfrac{1}{2}L_1\sin\theta_1\,\delta\theta_1 - m_2g \times L_1\sin\theta_1\,\delta\theta_1 + P \times L_1\cos\theta_1\,\delta\theta_1 = 0$$

and

$$\tan\theta_1 = \frac{2P}{(m_1 + 2m_2)g}$$

Applying the virtual-work equation with θ_1 fixed, and $\delta\theta_1 = 0$

$$-m_2g \times \tfrac{1}{2}L_2\sin\theta_2\,\delta\theta_2 + P \times L_2\cos\theta_2\,\delta\theta_2 = 0$$

and

$$\tan\theta_2 = \frac{2P}{m_2g}$$

6.6 Summary

(1) The work of a force

$$U_{1-2} = \int_{s_1}^{s_2} F\cos\alpha\,\mathrm{d}s \tag{6.1}$$

where α is the angle between F and δs.
(2) The work of a torque or couple of moment M is

$$U_{1-2} = \int_{\theta_1}^{\theta_2} M\,\mathrm{d}\theta \tag{6.2}$$

(3) Constraints are geometrical restrictions on the motion of a system.
(4) The forces on a system are of two kinds, (a) applied forces, and (b) forces of constraint.
(5) A virtual displacement is an arbitrary infinitesimally small displacement.
(6) Virtual work is the work associated with the forces acting on the particles of the system when the system undergoes a virtual displacement.

(7) A rigid body or system of connected rigid bodies is in equilibrium if and only if the work of the applied forces is zero in any arbitrary virtual displacement that is consistent with workless constraints (principle of virtual work).

$$\Sigma\delta U = \Sigma F_{\text{app},s}\,\delta s = 0 \tag{6.3}$$

(8) For mechanisms, the velocity diagram serves to relate displacements at salient points, and can be used to relate forces and torques using the principle of virtual work.
(9) External friction forces at constraints are treated as applied or active forces. Friction forces at internal constraints occur in pairs of equal and opposite forces, and in a virtual displacement of the system involving relative movement at the constraint, net negative virtual work is associated with those forces.
(10) An elastic member can be replaced by two equal and opposite forces having magnitude F_e. The work of the pair in a virtual displacement is $-F_e\,\delta L$ and the virtual-work equation becomes

$$\Sigma\delta U_{\text{external active forces}} - F_e\,\delta L = 0 \tag{6.4}$$

(11) For two-degree-of-freedom systems displacement components are typically

$$\delta x = \frac{\partial x}{\partial\theta_1}\delta\theta_1 + \frac{\partial x}{\partial\theta_2}\delta\theta_2$$

Two virtual-work equations are now written and solved for equilibrium values of θ_1 and θ_2.

Problems

6.1 For the platform of problem 4.22 determine the magnitude of P required to maintain the platform in position, with $\theta = 30°$, using the principle of virtual work.

6.2 A tripod consists of three legs smoothly jointed at the apex, each leg having length L and inclined at the same angle to the vertical. The mid-points of the legs are joined by three inextensible cords each having length $L/2$. The tripod stands on a smooth horizontal floor and a body mass m is hung from the apex. If the weights of the legs are neglected show that the tension in each cord is $[\sqrt{2}/(3\sqrt{3})]\,mg$.

6.3 In figure 6.6 AC and CB are pairs of members length L pinned at A, C and B. Outward movement of B compresses the spring BD, which has a spring constant k and is undeformed when $\theta = \theta_0 = 60°$. Show that when a body mass m is hung from C equilibrium is attained when $2\sin\theta - \tan\theta = mg/2kL$. Neglect the weights of the members.

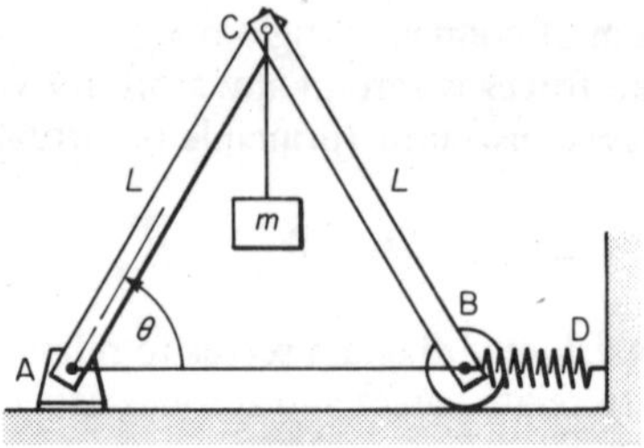

Figure 6.6

6.4 In figure 6.7 the uniform members OA, AB each have mass m and length L. The member OA can rotate about a pin fixed to the support and AB can rotate about a pin fixed to OA. Spiral springs are attached to the pins with their outer

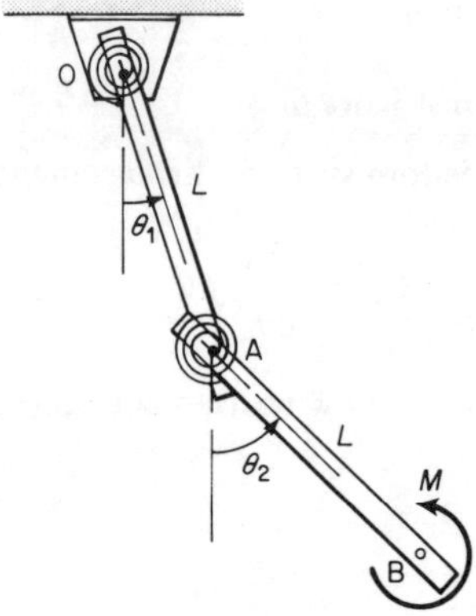

Figure 6.7

ends attached to the adjacent members. Each spring has a constant c, this being the torque exerted per radian of relative angular movement of the adjacent member. When OA, AB are hanging vertically there are no torques exerted by the springs.

If a couple having moment M is applied to member AB, show that if the angular deflections θ_1 and θ_2 from the vertical are small then for equilibrium

$$\theta_1 = \frac{M}{c}\left(\frac{1}{1 + 5p/2 + 3p^2/4}\right)$$

$$\theta_2 = \frac{M}{c}\left(\frac{3p/2 + 2}{1 + 5p/2 + 3p^2/4}\right)$$

where $p = mgL/c$.

6.5 A rigid uniform beam AB, length $2a$, is freely pivoted at its mid-point O. A light rod BC, length b ($b > a$), is pin-jointed to AB at B and the end C is constrained by a frictionless guide to move in the vertical line below the point O. A body mass $2m$ hangs from A and a body mass m hangs from C. If the angle between OB and the upward vertical direction at O is θ($\theta < 90°$) and there is a friction torque having constant magnitude M_f opposing relative angular motion at the joint B, show that the system is in equilibrium for a value of θ given by the equation

$$\frac{\sin\theta\,(b^2/a^2 - \sin^2\theta)^{1/2} + \sin\theta\cos\theta}{(b^2/a^2 - \sin^2\theta)^{1/2} - \cos\theta} = \pm\frac{M_f}{mga}$$

(*Hint*: Start the solution with two angular coordinates. These are not independent and can be related by an equation of constraint.)

7 Centres of Gravity and of Mass: Centroids

In this chapter we discuss certain properties of rigid bodies related to their weight, their mass and their geometrical form. These properties arise in the mechanics of rigid bodies when we are led to define certain quantities that are representative of the body as a whole. For example, we have already used the result that the gravitational forces on the individual particles of a rigid body reduce to a single force acting at some well-defined point in the body; if the location of this point is known in a particular case the analysis is obviously greatly simplified.

At the same time we introduce the idea of continuous distributions of matter and gravitational forces, enabling the summations over particles of a rigid body to be replaced by integration processes.

In the following sections we define and set out methods of obtaining centres of gravity and centres of mass of rigid bodies, and also the centroids of geometrical entities comprising lines, areas and volumes.

7.1 Centre of Gravity

We recall from section 4.4 that the centre of parallel gravitational forces, referred to as the centre of gravity, was a point whose position relative to the particles of a particle system did not depend on the orientation of the system.

To determine the position of the centre of gravity in a thin plane rigid body, lamina or plate, we need only find the line of action of the resultant gravitational force for two orientations of the body, and the centre of gravity is then the point at which the lines of action intersect. This is the basis of the experimental method of determination in which the lamina is suspended in turn at each of two small

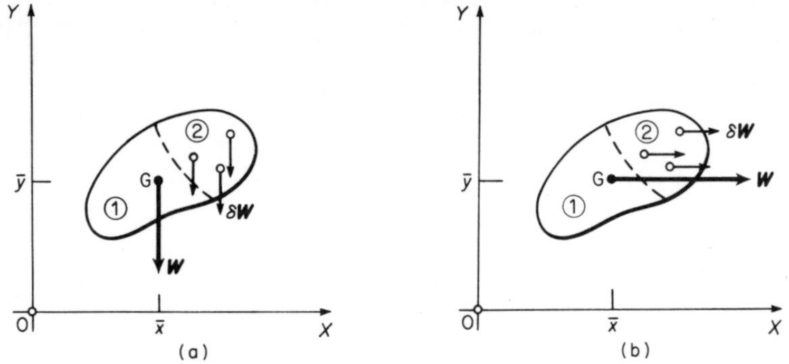

Figure 7.1

holes drilled in the lamina. When the lamina is in equilibrium the centre of gravity lies in the vertical line through the point of suspension, as defined by a plumb line suspended from the same point. If the line is marked on the lamina in both cases, the intersection of the lines locates the centre of gravity.

The analytical method of determination is based on arbitrary orientations of the force set. Thus in figure 7.1 the plane of the lamina is vertical and axes OX, OY are chosen to be parallel to two perpendicular orientations of a set of parallel gravitational forces acting on the particles of the lamina.

In figure 7.1a the resultant $\boldsymbol{W}$ of the elementary gravitational forces, such as $\delta \boldsymbol{W}$, passes through the point G, the centre of gravity. Choosing O as a moment-centre the x-coordinate of G, $\bar{x}$, is given by

$$(\Sigma\delta W) \times \bar{x} = W\bar{x} = \Sigma(\delta W)x$$

and

$$\bar{x} = \frac{\Sigma(\delta W)x}{W} \tag{7.1a}$$

where W is the weight of the lamina.

By similar consideration of figure 7.1b we obtain

$$\bar{y} = \frac{\Sigma(\delta W)y}{W} \tag{7.1b}$$

If the lamina is subdivided into two or more parts as indicated in the figure, then if the subscripts 1, 2, 3, . . . refer to these parts

$$\bar{x} = \frac{(\Sigma(\delta W)x)_1 + (\Sigma(\delta W)x)_2 + \ldots}{W_1 + W_2 + \ldots}$$

$$= \frac{W_1\bar{x}_1 + W_2\bar{x}_2 + \ldots}{W_1 + W_2 + \ldots} \tag{7.2a}$$

and

$$\bar{y} = \frac{W_1\bar{y}_1 + W_2\bar{y}_2 + \ldots}{W_1 + W_2 + \ldots} \tag{7.2b}$$

where $\bar{x}_1, \bar{x}_2, \ldots, \bar{y}_1, \bar{y}_2, \ldots$ refer to the centres of gravity of the parts. If a lamina can be subdivided into parts whose centres of gravity are known then equations 7.2 enable the centre of gravity of the whole lamina to be located.

For a thin wire shaped into a plane curve the centre of gravity is located in the same way as for a thin plate.

For three-dimensional bodies three coordinates are required to locate the centre of gravity. If mutually perpendicular axes OX, OY, OZ are chosen, as in figure 7.2, the orientations of the gravitational forces can be made parallel to the OX- and OY-axes in turn. The coordinates $\bar{x}$ and $\bar{y}$ are determined from moments about the OZ-axis for the two orientations (figure 7.2a). The coordinate $\bar{z}$ is then determined

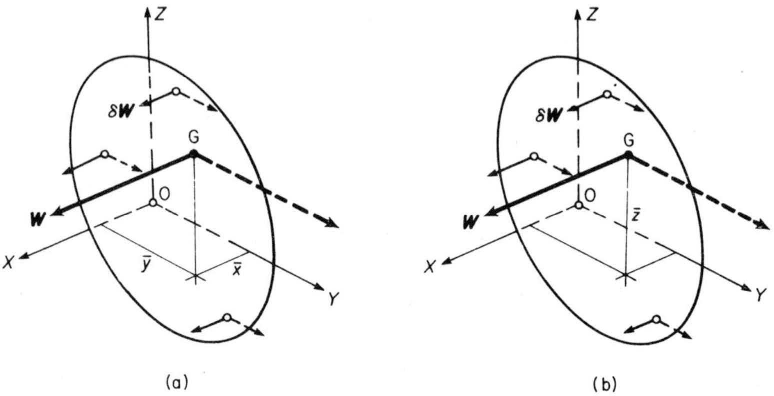

Figure 7.2

from the moments of either the x- or y-directed forces about the OY-or OX-axes respectively (figure 7.2b). The three coordinates are then

$$\bar{x} = \frac{\Sigma(\delta W)x}{W} \qquad \bar{y} = \frac{\Sigma(\delta W)y}{W} \qquad \bar{z} = \frac{\Sigma(\delta W)z}{W} \tag{7.3}$$

However, in many engineering applications a plane of symmetry can usually be found. The centre of gravity lies in this plane and the OX- and OY-axes can be chosen to coincide with it; only $\bar{x}$ and $\bar{y}$ then need to be determined.

It may now be noted that although the orientations of the gravitational field have been chosen in accordance with the physical significance of the centre of gravity as a centre of parallel forces, the expressions for $\bar{x}$, $\bar{y}$ and $\bar{z}$ given by equations 7.3, although obtained as a result of moments of forces about respective axes, merely contain products of *magnitudes* of the weights of the particles (their directions being irrelevant) and their distances from a particular plane. Having chosen axes OX, OY and OZ we therefore repeat equations 7.3

$$\bar{x} = \frac{\Sigma(\delta W)x}{W} \qquad \bar{y} = \frac{\Sigma(\delta W)y}{W} \qquad \bar{z} = \frac{\Sigma(\delta W)z}{W}$$

where x, y and z are now the distances of the particles from the YOZ, ZOX- and XOY-planes respectively.

In the application of equations 7.3 we replace the particle model by one for which a continuous distribution of matter is assumed. The particle is now replaced by an element of line, area or volume. It is now appropriate to use the idea of intensity of gravitational force or weight intensity, signified by the symbol w. This represents weight per unit length, weight per unit area or weight per unit volume, as appropriate, for the material of wires, thin plates or three-dimensional bodies respectively. Thus the particle weight δW is now replaced by $w\delta L$, $w\delta A$ or $w\delta V$ for an element, as appropriate. The summations can now be written as integrations and typical expressions for the x-coordinate of the centre of gravity are

$$\bar{x} = \frac{\int_L wx \, dL}{\int_L w \, dL} \qquad \bar{x} = \frac{\int_A wx \, dA}{\int_A w \, dA} \qquad \bar{x} = \frac{\int_V wx \, dV}{\int_V w \, dV} \tag{7.4}$$

for wires, plates and three-dimensional bodies respectively. Corresponding expressions can be written for $\bar{y}$ and $\bar{z}$ in each case.

The choice of suitable axes and line, area or volume elements should be such as to simplify and expedite the integration. The examples considered in the next section indicate the method of approach.

7.2 Centres of Gravity: Standard Cases

(1) *The uniform triangular plate* (figure 7.3). Place an axis OY along any side, say

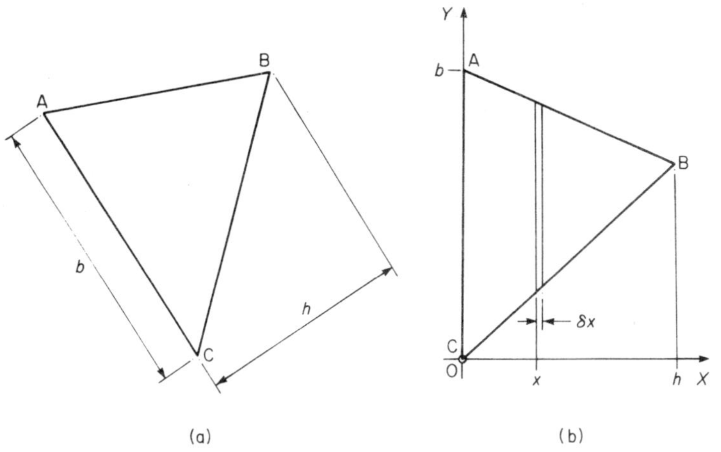

Figure 7.3

AC, and choose an element of plate, width δx, parallel to and a distance x from the axis OY. All parts of the element are therefore at a distance x from the axis. The length of the element is $b \times (h-x)/h$ and its area is $b(h-x)\delta x/h$, where h is the altitude in relation to side AC. If w is the weight per unit area, taken as being uniform, the weight of the element is

$$\delta W = wb(h-x)\frac{\delta x}{h}$$

and the weight of the plate is

$$W = w \times \tfrac{1}{2}bh$$

From the second of equations 7.4

$$(w \times \tfrac{1}{2}bh) \times \bar{x} = \int_0^h \frac{wb(h-x)x\,\mathrm{d}x}{h}$$

$$= \frac{wb}{h}\left[\frac{hx^2}{2} - \frac{x^3}{3}\right]_0^h$$

$$= \frac{wbh^2}{6}$$

and

$$\bar{x} = \frac{h}{3}$$

If a y-axis is chosen to lie along any other side a similar result follows, and we conclude that the centre of gravity is at a distance equal to one-third of the corresponding altitude from any side. This is equivalent to stating that the centre of gravity is located at the intersection of the medians. This result indicates that the centre of gravity could have been found without integration by noting that the centres of gravity of strips parallel to one side lie along the median to that side, implying in turn that the centre of gravity of the triangle lies on that median. By taking strips parallel to the other sides we are led to the result that has been stated.

(2) *The uniform circular wire* (figure 7.4). Place an axis OX along the axis of sym-

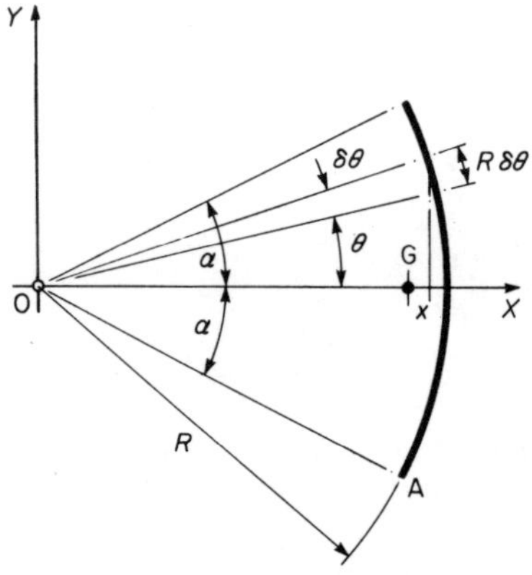

Figure 7.4

metry of the wire arc **AB** with the origin at the centre. Choose an element of wire at an angular distance θ from the axis subtending an angle $\delta\theta$ at the centre O. The length of the element is $R\delta\theta$. If w is the weight per unit length, taken as being uniform, then

$$\delta W = wR\delta\theta \qquad W = 2wR\alpha \qquad x = R\cos\theta$$

From the first of equations 7.4

$$(2wR\alpha) \times \bar{x} = \int_{-\alpha}^{+\alpha} wR^2 \cos\theta \, d\theta$$

$$= wR^2 \, [\sin\theta]_{-\alpha}^{+\alpha}$$

$$= 2wR^2 \sin\alpha$$

and

$$\bar{x} = \frac{R\sin\alpha}{\alpha}$$

It follows that for a semicircular-shaped wire $\bar{x} = 2R/\pi$

(3) *The uniform sector-shaped plate* (figure 7.5). With the axis OX along the axis of symmetry choose a sector-shaped element of the plate with vertex angle $\delta\theta$ at the origin. If $\delta\theta$ is small the element is effectively a triangular plate with weight δW acting at its centre of gravity G_e distance $2R/3$ from the vertex at O. The resultant

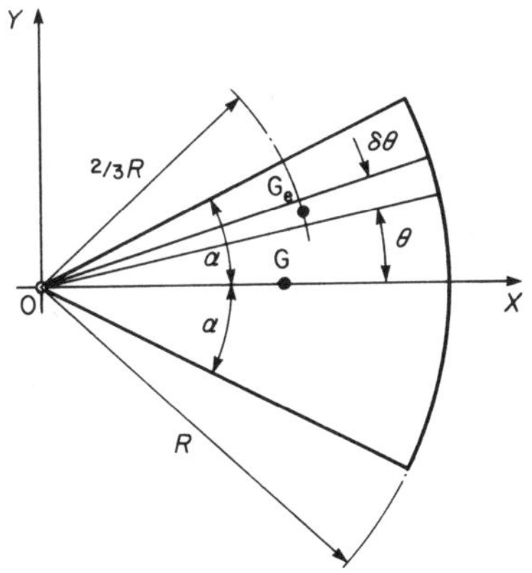

Figure 7.5

gravitational force on all such elements therefore acts through the centre of gravity of an equivalent circular wire having radius $2R/3$, and it follows that for the plate

$$\bar{x} = \frac{2R\sin\alpha}{3\alpha}$$

It follows that for a semicircular-shaped plate

$$\bar{x} = \frac{4R}{3\pi}$$

Worked Example 7.1

Locate the position of the centre of gravity of the uniform thin plate shown in figure 7.6.

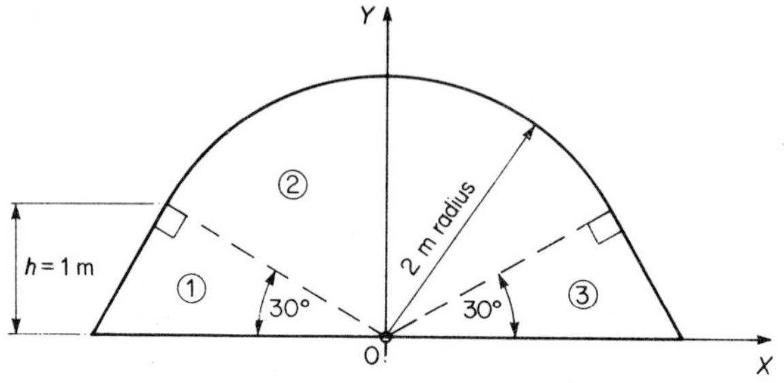

Figure 7.6

Solution

Choose axes OX and OY as indicated. Since OY is an axis of symmetry the centre of gravity must lie on this line, that is, $\bar{x} = 0$. In order to determine $\bar{y}$ divide the plate into standard shapes, (1) and (3) being triangles and (2) a sector.

$$A_1 = A_3 = \tfrac{1}{2} \times 2 \times \frac{2}{\sqrt{3}} = \frac{2}{\sqrt{3}}\ \text{m}^2$$

$$A_2 = \tfrac{1}{3}\pi 2^2 = \frac{4\pi}{3}\ \text{m}^2$$

$$\bar{y}_1 = \bar{y}_3 = \tfrac{1}{3}\ \text{m}$$

$$\bar{y}_2 = \tfrac{2}{3} \times \frac{2 \sin(\pi/3)}{\pi/3} = \frac{2\sqrt{3}}{\pi}\ \text{m}$$

If w is the weight per unit area, then from the result expressed by equation 7.2b

$$\bar{y} = \frac{w(A_1\bar{y}_1 + A_2\bar{y}_2 + A_3\bar{y}_3)}{w(A_1 + A_2 + A_3)}$$

$$= \frac{(2/\sqrt{3}) \times (1/3) + (4\pi/3) \times (2\sqrt{3}/\pi) + (2/\sqrt{3}) \times (1/3)}{(2/\sqrt{3}) + (4\pi/3) + (2/\sqrt{3})}$$

$$= 0.83\ \text{m}$$

Worked Example 7.2

Locate the centre of gravity of the block shown in figure 7.7a.

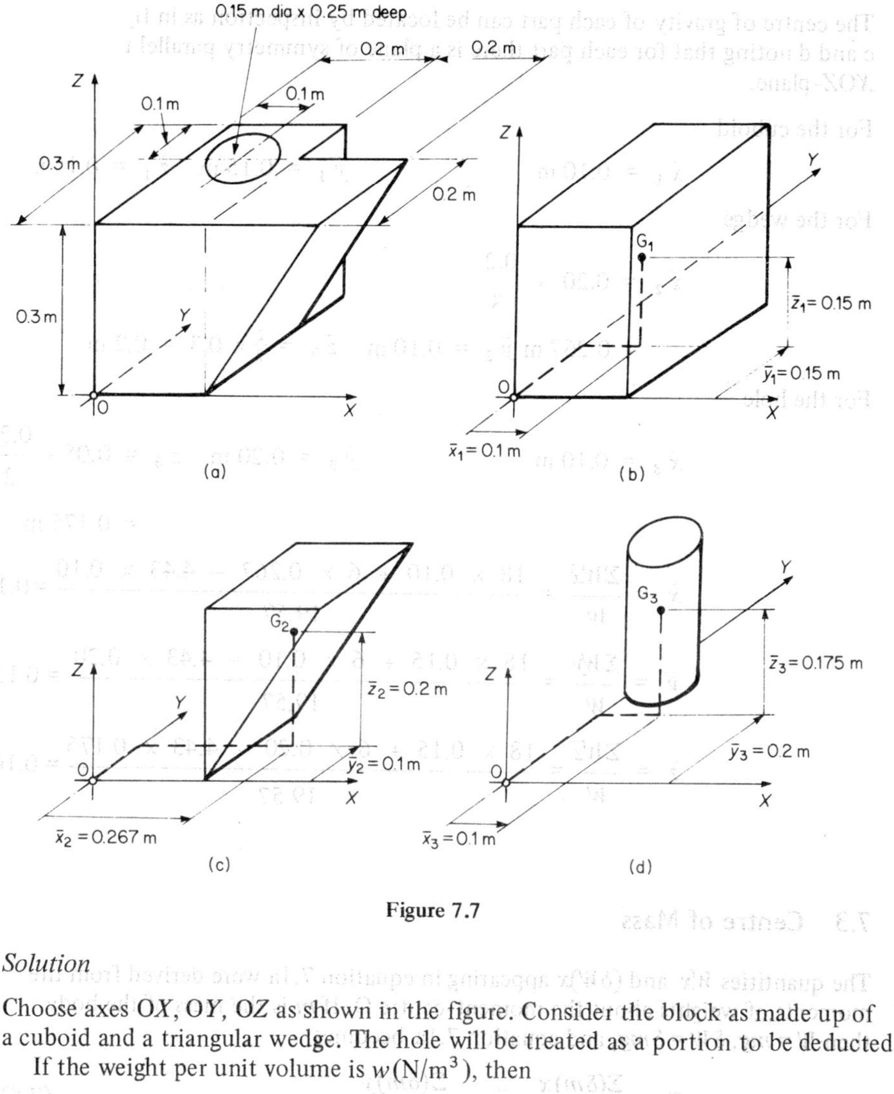

Figure 7.7

Solution

Choose axes OX, OY, OZ as shown in the figure. Consider the block as made up of a cuboid and a triangular wedge. The hole will be treated as a portion to be deducted.

If the weight per unit volume is $w(\mathrm{N/m^3})$, then

$$\text{weight of cuboid} = W_1 = w \times 3 \times 3 \times 2 \times 10^{-3} = 18 \times 10^{-3} w\ \mathrm{N}$$

$$\text{weight of wedge} = W_2 = w \times \tfrac{1}{2} \times 2 \times 2 \times 3 \times 10^{-3} = 6 \times 10^{-3} w\ \mathrm{N}$$

weight of material to be deducted at hole

$$= W_3 = w \times \frac{\pi}{4} \times 1.5^2 \times 2.5 \times 10^{-3} = 4.43 \times 10^{-3} w\ \mathrm{N}$$

$$\text{total weight} = W_1 + W_2 - W_3 = 19.57 \times 10^{-3} w\ \mathrm{N}$$

The centre of gravity of each part can be located by inspection as in figures 7.7b, c and d noting that for each part there is a plane of symmetry parallel to the *XOZ*-plane.

For the cuboid

$$\bar{x}_1 = 0.10 \text{ m} \qquad \bar{y}_1 = 0.15 \text{ m} \quad \bar{z}_1 = 0.15 \text{ m}$$

For the wedge

$$\bar{x}_2 = 0.20 + \frac{0.2}{3}$$

$$= 0.267 \text{ m} \; \bar{y}_2 = 0.10 \text{ m} \quad \bar{z}_2 = \tfrac{2}{3} \times 0.3 = 0.2 \text{ m}$$

For the hole

$$\bar{x}_3 = 0.10 \text{ m} \qquad \bar{y}_3 = 0.20 \text{ m} \quad \bar{z}_3 = 0.05 + \frac{0.25}{2}$$

$$= 0.175 \text{ m}$$

$$\bar{x} = \frac{\Sigma W\bar{x}}{W} = \frac{18 \times 0.10 + 6 \times 0.267 - 4.43 \times 0.10}{19.57} = 0.151 \text{ m}$$

$$\bar{y} = \frac{\Sigma W\bar{y}}{W} = \frac{18 \times 0.15 + 6 \times 0.10 - 4.43 \times 0.20}{19.57} = 0.123 \text{ m}$$

$$\bar{z} = \frac{\Sigma W\bar{z}}{W} = \frac{18 \times 0.15 + 6 \times 0.20 - 4.43 \times 0.175}{19.57} = 0.160 \text{ m}$$

7.3 Centre of Mass

The quantities $W\bar{x}$ and $(\delta W)x$ appearing in equation 7.1a were derived from the moments of weights about the moment-centre O. If m is the mass of the body then $W = mg$, $\delta W = \delta mg$, and equation 7.1a becomes

$$\bar{x} = \frac{\Sigma(\delta m)x}{m} \qquad \bar{y} = \frac{\Sigma(\delta m)y}{m} \tag{7.5}$$

after cancelling the gs (assuming g is constant throughout the body).

The numerators now represent products of particle masses and distances from the axes OY and OX respectively. The point $(\bar{x}, \bar{y})$ is no longer a centre of gravitational force but a point that is representative of the particle mass distribution. This point is called the *centre of mass* or the *mass-centre*.

To take account of continuous distributions of mass we now introduce the quantity mass per unit volume, symbolised by ρ. For an element of a three-dimensional body the particle mass δm is now replaced by $\rho\delta V$ and the integral forms for $\bar{x}$, $\bar{y}$ and $\bar{z}$ now become

$$\bar{x} = \frac{\int_V \rho x \, dV}{\int_V \rho \, dV} \qquad \bar{y} = \frac{\int_V \rho y \, dV}{\int_V \rho \, dV} \qquad \bar{z} = \frac{\int_V \rho z \, dV}{\int_V \rho \, dV} \tag{7.6}$$

The corresponding expressions for $\bar{x}$ in the case of wires having cross-sectional area a and thin plates having thickness t are

$$\bar{x} = \frac{\int_L \rho a x \, dL}{\int_L \rho a \, dL} \qquad \bar{x} = \frac{\int_A \rho t x \, dA}{\int_A \rho t \, dA} \tag{7.7}$$

with corresponding expressions for $\bar{y}$ and $\bar{z}$

If the integrations are carried out for the standard cases already discussed then, provided ρ is constant, the coordinates of the centre of mass are in each case identical to those of the centre of gravity. However it should be carefully noted that the mass-centre is a property of the body that is quite independent of the existence of gravitational forces.

7.4 Centroids of Lines, Surfaces and Volumes

If in the expressions for $\bar{x}$ in equations 7.6 and 7.7 the quantities ρ, a and t are constant they can be cancelled, and we obtain the following expressions

$$\bar{x} = \frac{\int_L x \, dL}{L} \qquad \bar{x} = \frac{\int_A x \, dA}{A} \qquad \bar{x} = \frac{\int_V x \, dV}{V} \tag{7.8}$$

and correspondingly expressions for $\bar{y}$ and $\bar{z}$. Equations 7.8 now relate to the length of a wire, the surface area of a thin plate and the volume of a three-dimensional body respectively. These are geometrical properties and the values of $\bar{x}$, $\bar{y}$ and $\bar{z}$ are dependent only on the shapes of the geometical line, surface or volume concerned. The point $(\bar{x}, \bar{y}, \bar{z})$ as defined by equations such as 7.8 is called the *centroid* of the particular geometrical entity, and the quantities $L\bar{x}$, $A\bar{x}$, $V\bar{x}$ are described as first moments of line, area or volume respectively with respect to the YOZ-plane. Of these the one most used is $A\bar{x}$, the first moment of a plane area, and, although not employed in this book, it is nevertheless conveniently associated with the discussions in this chapter. Facility in the determination of centroids of area is an essential prerequisite to the study of internal forces.

It is evident that for a thin flat plate having uniform thickness and uniform density the coordinates of the centre of gravity and the centre of mass coincide with those of the centroid of the area of the plate. However, it must be stressed that the centroid is a property of the geometrical surface only, and that its location is dependent only on the shape of the surface.

7.5 Theorems of Pappus and Guldinus

Using the ideas of the preceding section two useful results can be obtained. These are associated with the name of Pappus of Alexandria (third century A.D.) and Paul Guldinus (1577 - 1643) and are useful in connection with the geometrical properties of bodies possessing axes of symmetry.

(1) In figure 7.8a a surface, area A, is generated by revolving a line length L about

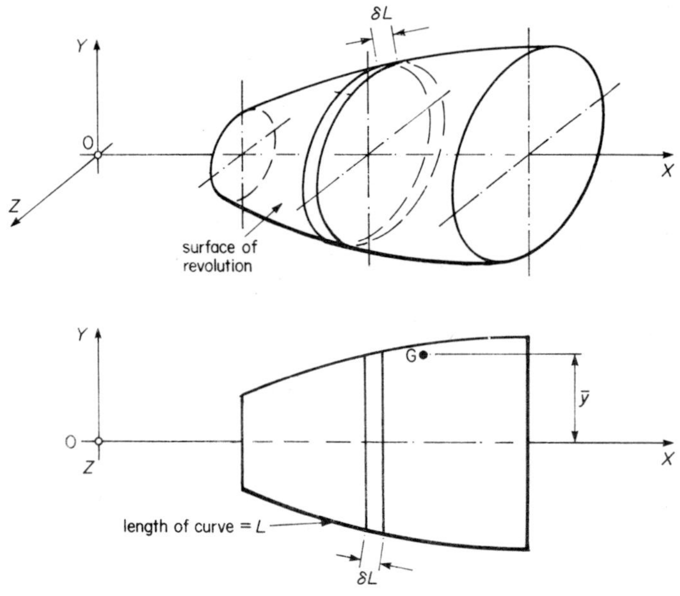

Figure 7.8a

the axis OX, which it does not intersect. The area swept out by an element of the line δL at a distance y from the axis is clearly $\delta L \times 2\pi y$. The total area of the surface of revolution is then

$$A = \int_L 2\pi y \, \mathrm{d}L = 2\pi \int_L y \, \mathrm{d}L$$

$$= 2\pi L\bar{y}$$

$$= L \times 2\pi\bar{y} \qquad (7.9)$$

where $\bar{y}$ is the distance of the centroid G of the line from the axis OX. It follows that the area of the surface of revolution is equal to the product of the length of the line and the distance travelled by the centroid of the line in one revolution.

(2) In figure 7.8b a torus-shaped volume V is generated by revolving a closed curve, enclosing area A, about the axis OX. The curve again should not cross the OX axis.

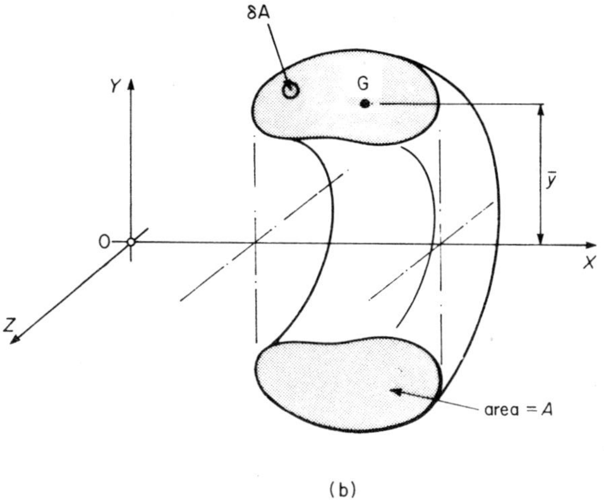

Figure 7.8b

The volume swept out by an element of area δA at a distance y from the axis is $\delta A \times 2\pi y$. The total volume swept out is then

$$V = \int_A 2\pi y \, dA = 2\pi \int_A y \, dA$$
$$= 2\pi A\bar{y}$$
$$= A \times 2\pi\bar{y} \tag{7.10}$$

where $\bar{y}$ is the distance of the centroid of the area from the axis. It follows that the volume generated by a closed curve is equal to the product of the area enclosed by the curve and the distance travelled by the centroid of the area in one revolution.

These two theorems, which will be referred to as the first and second theorems of Pappus, are not only useful for calculating areas and volumes of the kind described, but can be useful for locating centroids, if the areas or volumes of revolution are known together with the lengths or enclosed areas of the appropriate curve.

Worked Example 7.3

Locate (a) the centroid of a circular arc radius R and (b) the centroid of a circular sector radius R, if the angle subtended at the centre is 90° in each case.

Solution

(a) The circular arc is placed in the first quadrant of axes OX and OY with its centre at O. If the arc, length $\frac{1}{2}\pi R$, is revolved about the axis OX, a hemispherical surface is generated whose area is $2\pi R^2$. If $\bar{y}$ is the y-coordinate of the centroid of the arc then using the first theorem of Pappus

$$2\pi R^2 = \tfrac{1}{2}\pi R \times 2\pi\bar{y}$$

$$\bar{y} = \frac{2R}{\pi}$$

similarly, by revolving the curve about the axis OY

$$\bar{x} = \frac{2R}{\pi}$$

The centroid is therefore on the axis of symmetry of the arc at a distance $2\sqrt{2R}/\pi$ from the centre. This result can be compared with that obtained in section 7.2 for the centre of gravity of a circular wire.

(b) If the sector corresponding to the circular arc is revolved about the axis OX the volume generated is that of a hemisphere, namely $2\pi R^3/3$. The area of the sector is $\pi R^2/4$ therefore by the second theorem of Pappus

$$\frac{2}{3}\pi R^3 = \frac{1}{4}\pi R^2 \times 2\pi\bar{y}$$

and

$$\bar{y} = \frac{4}{3}\frac{R}{\pi}$$

similarly

$$\bar{x} = \frac{4}{3}\frac{R}{\pi}$$

The centroid is now at a distance $4\sqrt{2R}/3\pi$ from O on the axis of symmetry, a result which can again be compared with that given previously in section 7.2.

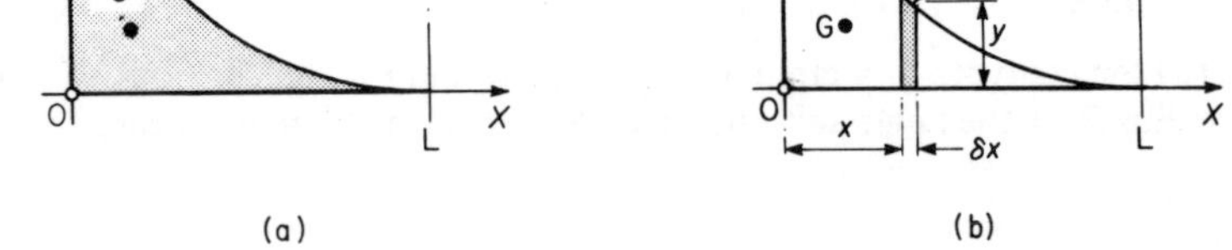

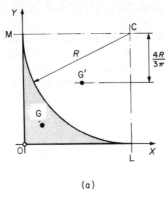

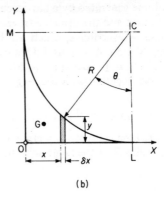

Figure 7.9

Worked Example 7.4

Locate the centroid of the shaded area shown in figure 7.9a using the second theorem of Pappus, and verify the result by integration.

Solution

Revolve the shaded area about the axis OX. The volume of revolution for the shape LOM is not known. For the sector LCM

$$\text{volume of revolution} = \frac{1}{4}\pi R^2 \times 2\pi\left(R - \frac{4R}{3\pi}\right)$$

$$= \frac{1}{2}\pi^2 R^3\left(1 - \frac{4}{3\pi}\right)$$

For the square OLCM

$$\text{volume of revolution} = \pi R^3$$

therefore for the shape LOM

$$\text{volume of revolution} = \pi R^3 - \frac{1}{2}\pi^2 R^3\left(1 - \frac{4}{3\pi}\right)$$

$$= \pi R^3\left(\frac{5}{3} - \frac{\pi}{2}\right)$$

For the shape LOM

$$\text{area} = R^2 - \frac{1}{4}\pi R^2$$

$$= R^2\left(1 - \frac{\pi}{4}\right)$$

By the second theorem of Pappus

$$\pi R^3\left(\frac{5}{3} - \frac{\pi}{2}\right) = R^2\left(1 - \frac{\pi}{4}\right) \times 2\pi\bar{y}$$

therefore

$$\bar{y} = \frac{10 - 3\pi}{4 - \pi} \times \frac{R}{3}$$

and similarly

$$\bar{x} = \bar{y}$$

For the integration method choose strips parallel to the axis OY as in figure 7.9b

$$\text{area of shape LOM} = \int_0^R y \, dx$$

$y = R(1 - \cos\theta)$, $x = R(1 - \sin\theta)$, $dx = -R\cos\theta \, d\theta$ and when θ is used as the variable the lower and upper limits become $\pi/2$ and 0 respectively. Therefore

$$\text{area} = -\int_{\pi/2}^{0} R^2(1 - \cos\theta)\cos\theta \, d\theta$$

$$= R^2\left(1 - \frac{\pi}{4}\right)$$

The details of the integration, which is straightforward, are omitted. From the definition of the centroid

$$\text{area} \times \bar{x} = \int_0^R yx \, dx$$

$$= -\int_{\pi/2}^{0} R^3(1 - \cos\theta)(1 - \sin\theta)\cos\theta \, d\theta$$

$$= R^3\left(\frac{5}{6} - \frac{\pi}{4}\right)$$

the details of the integration again being omitted. Therefore

$$\bar{x} = \frac{5/6 - \pi/4}{1 - \pi/4} \times R$$

$$= \frac{10 - 3\pi}{4 - \pi} \times \frac{R}{3}$$

confirming the previous result.

7.6 Summary

(1) The centre of gravity of a uniform thin plate lying in the XOY-plane is at the point $\bar{x}, \bar{y}$ given by

$$\bar{x} = \frac{\int wx \, dA}{\int w \, dA} \qquad \bar{y} = \frac{\int wy \, dA}{\int w \, dA}$$

where w is the weight per unit area.

For a thin wire lying in the XOY-plane

$$\bar{x} = \frac{\int wx \, dL}{\int w \, dL} \qquad \bar{y} = \frac{\int wy \, dL}{\int w \, dL}$$

where w is the weight per unit length.

For three-dimensional bodies

$$\bar{x} = \frac{\int wx \, dV}{\int w \, dV} \qquad \bar{y} = \frac{\int wy \, dV}{\int w \, dV} \qquad \bar{z} = \frac{\int wz \, dV}{\int w \, dV}$$

where w is the weight per unit volume and x, y, z are distances from the YOZ, ZOX and XOY-planes respectively.

(2) The mass-centre of a three-dimensional body is at the point $\bar{x}, \bar{y}, \bar{z}$ given by

$$\bar{x} = \frac{\int \rho x \, dV}{\int \rho \, dV} \qquad \bar{y} = \frac{\int \rho y \, dV}{\int \rho \, dV} \qquad \bar{z} = \frac{\int \rho z \, dV}{\int \rho \, dV}$$

where ρ is the mass per unit volume.

For thin plates, thickness t

$$\int \rho \, dV = \int \rho t \, dA$$

For wires, cross-sectional area a

$$\int \rho \, dV = \int \rho a \, dL$$

(3) The centroid of a plane area in the XOY-plane with respect to the x- and y-axes respectively is given by

$$\bar{y} = \frac{\int y \, dA}{A} \qquad \bar{x} = \frac{\int x \, dA}{A}$$

$A\bar{y}$, $A\bar{x}$ are the first moments of area with respect to the x- and y-axes respectively.

(4) Composite bodies and areas can usually be subdivided into standard-shaped bodies and areas whose properties are known. Axes should be chosen, where possible, such that there are planes of symmetry parallel to one or other of the chosen axes.

(5) The total area of revolution of a line length L is $A = L \times 2\pi\bar{y}$, where $\bar{y}$ is the distance of the centroid of the line from the axis of revolution.

(6) The total volume of revolution of an area A is $V = A \times 2\pi\bar{y}$, where $\bar{y}$ is the distance of the centroid of the area from the axis of revolution.

Problems

Unless otherwise stated the mass density ρ is constant. The thickness of plates is to be taken as constant.

7.1 Locate the centre of gravity of the thin plate shown in figure 7.10.

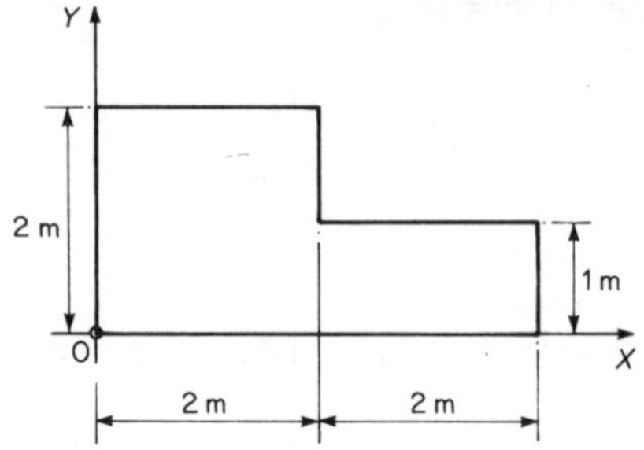

Figure 7.10

7.2 A thin rectangular plate 5 m by 3 m has a hole 2 m diameter cut in it such that the centre is 1.5 m from a short edge. Determine the distance of the mass-centre from the same edge.

7.3 A box has a base 0.2 m by 0.4 m and its sides are 0.5 m high. It is made of thin sheet steel, the bottom being reinforced by having a double thickness. Find the distance of the centre of gravity above the base when the top is (a) on and (b) off the box.

7.4 A wire frame is in the form of a triangle ABC with AB = BC = 0.1 m and CA = 0.15 m. If it is hung from the point A determine the angle that AB makes with the vertical. (*Hint:* The centre of gravity will be directly below A; do not make the mistake of treating the frame as a solid triangular plate.)

7.5 Determine the position of the centroid of the area in figure 7.11.

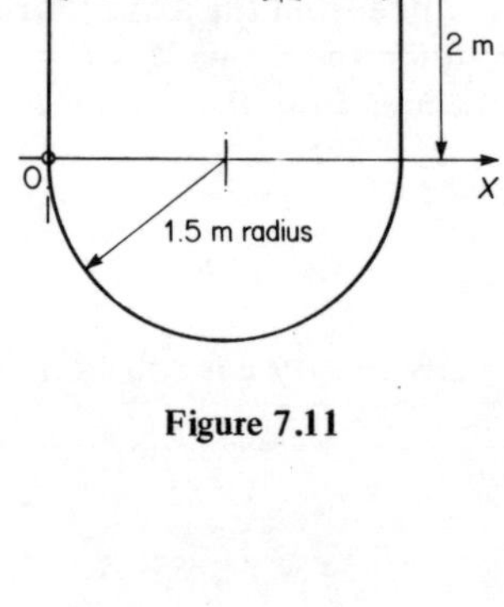

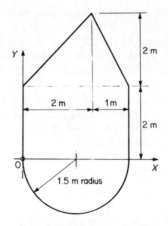

Figure 7.11

7.6 Locate the centre of gravity of the body shown in figure 7.12.

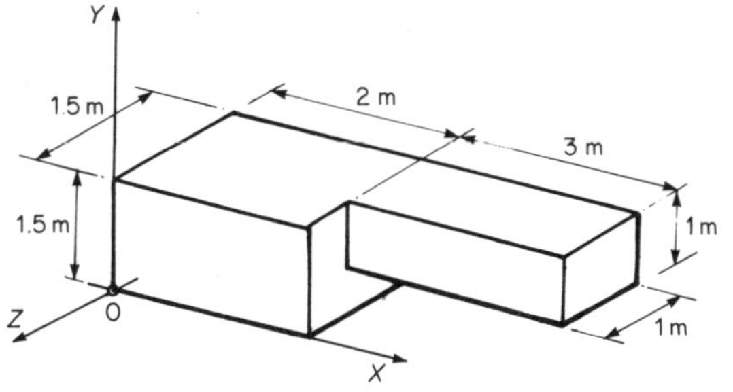

Figure 7.12

7.7 A body consists of a semicircular plate, radius R, on the end of a rectangular plate – width $2R$ and length L – the edge of length $2R$ coinciding with the diameter of the semicircle. The length L is to be such that when the body is stood in the vertical plane with its curved edge standing on a horizontal plane, it is to be in equilibrium for any point of contact. (*Hint:* For equilibrium note the line of action of the normal reaction of the plane and the weight of the body.)

7.8 Show that the mass-centre of a wire length L bent into an arc of radius R is at a distance $2R^2 \sin(L/2R)/L$ from the centre of the arc.

A frame consists of a wire bent into a rectangular shape 0.3 m by 0.2 m plus a length of the same wire bent into a semicircle 0.3 m diameter fixed to a 0.3 m side. Find the distance of the mass-centre of the frame from the 0.3 m base.

7.9 A 5 m length of wood tapers uniformly from a rectangular section 0.4 m by 0.3 m at one end to 0.2 m by 0.15 m at the other. Find (a) its weight in terms of ρ its mass per unit volume and (b) the distance of the centre of gravity from the thicker end. (*Hint:* Consider a lamina distance x from one end and obtain an expression for its area in terms of x.)

7.10 A conical body, radius 1 m at the base and height 3 m is cast in a mould, but the metal runs so badly during moulding that the density varies from 5000 kg/m^3 at the apex to 1000 kg/m^3 at the base. Assuming that the density varies linearly with height from apex to base determine (a) the mass of the body and (b) the distance of its mass-centre from the base. (*Hint:* Obtain an expression for the mass density at any point and consider a circular lamina distance x from the apex.)

7.11 By using the theorems of Pappus determine the surface area and volume of the body of revolution formed by revolving the line in figure 7.13 around the axis OY.

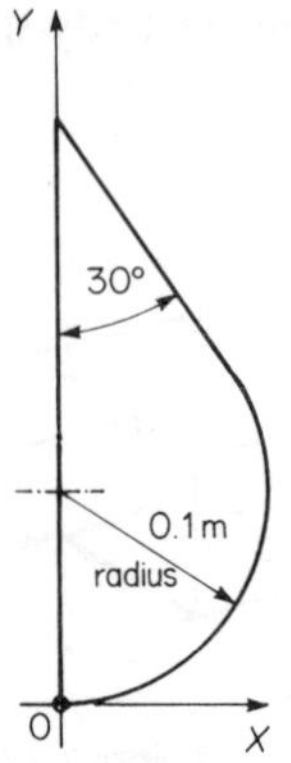

Figure 7.13

7.12 Use the theorems of Pappus to determine the position of the centre of gravity of the thin plate of figure 7.10. (*Hint*: The volumes of revolution can be calculated directly.)

7.13 Verify the formula for the volume of a sphere ($4\pi R^3/3$) by calculating the swept volume of a semicircular area.

7.14 The equation of a parabolic curve is $y = Kx^2$. Determine the distance from the y-axis of the centroid of the area lying between the curve, the x-axis and the ordinate at $x = R$. Hence find its volume of revolution about the y-axis.

When a cylindrical can partly filled with water is rotated at a uniform speed ω rad/s about its vertical axis, the height of the water surface at radius r is $y = \omega^2 r^2/2g$, measured relative to the water surface at the axis. If such a can having radius 0.2 m, has depth equal to twice its radius and is initially half filled with water, determine the speed at which water will just begin to spill from the can. (*Hint*: Equate water volumes.)

8 Kinematics of a Particle

Dynamics is the study of systems that are not in equilibrium. The set of forces acting on the system under consideration reduces to either a single resultant force or a resultant couple, and the system is no longer in a state of rest or constant speed in a straight line. In our discussion of the statics of systems, making use of Newton's first law, we assumed the particle or the particle system to be at rest, and for the most part disregarded the possibility of constant speed in a straight line, since we did not have at our disposal the necessary criteria for establishing the nature of motion in general. However, if the system is not in equilibrium, then motion of some kind is certainly taking place and the study of that motion becomes all-important.

This study will have two aspects, namely the description of the motion itself, the 'how' of the motion, which is named *kinematics*, together with an examination of the relations between the motion and the associated set of forces, the 'why' of the motion, named *kinetics.*

In this chapter we discuss the kinematics of a particle. We first define the terms that are used to describe motion in mathematical terms. The mathematical descriptions that follow from the definitions can take different forms, the form selected being that which is appropriate to the problem in hand. We also consider in detail some simple types of motion that occur frequently in engineering situations.

8.1 Rectilinear Motion of a Particle

A particle whose motion is confined to a straight line is said to be undergoing *rectilinear motion.*

8.1.1 Definitions

(1) We shall now take the straight line to be an X-axis and choose any point O on the axis as a reference point. Suppose the particle to be at point P at a certain time t measured from some arbitrary instant (figure 8.1a). Then OP = x, where x varies with t. We say that x is a function of t, and express this dependence by writing $x = x(t)$. The *position* of the particle at time t is defined to be the vector **OP** = $\mathbf{x}$ having magnitude x.

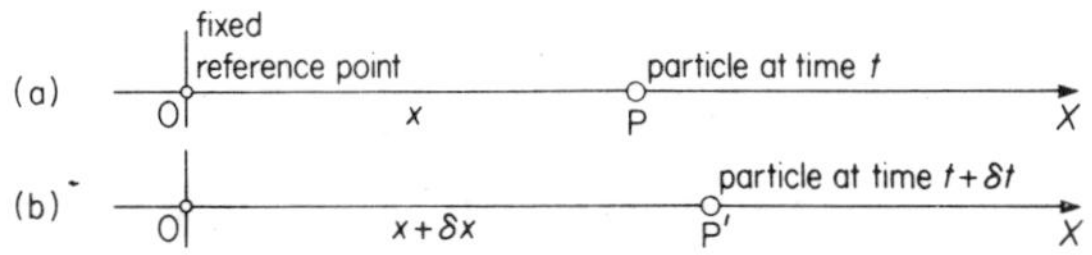

Figure 8.1

(2) At time $t + \delta t$ the particle will be at some point P′, where OP′ = $x + \delta x$ (figure 8.1b). The change of position, or the *displacement* of the particle in the time inteval δt, is defined to be the vector **PP′** having magnitude δx.
(3) The average rate of change of position, or the average rate of displacement, in the time interval δt, is given by the ratio **PP′**/δt, the value of which will depend upon the magnitude of δt. The smaller the magnitude of δt the closer P′ is to P and the above ratio has a limiting value as $\delta t \to 0$.

The rate of change of position of the particle at the point P, or the rate of displacement of the particle at the point P, is the *velocity* of the particle at the point P and is defined to be

$$\lim_{\delta t \to 0} \frac{\mathbf{PP'}}{\delta t} = \frac{\mathrm{d}x}{\mathrm{d}t}$$

a vector having magnitude $\mathrm{d}x/\mathrm{d}t$. Velocity is symbolised as $\boldsymbol{v}$, with magnitude v.
(4) If in the time interval δt during which the particle moves from P to P′ the magnitude of the velocity changes from v to $v + \delta v$, then we can obtain the average rate of change of velocity, the magnitude of which will be $\delta v/\delta t$. Again, as P′ approaches P with reduction in δt, the ratio $\delta v/\delta t$ attains a limiting value as $\delta t \to 0$. The rate of change of velocity of the particle at the point P, or the *acceleration* of the particle at the point P, is defined to be

$$\lim_{\delta t \to 0} \frac{\delta \boldsymbol{v}}{\delta t} = \frac{\mathrm{d}\boldsymbol{v}}{\mathrm{d}t}$$

a vector having magnitude $\mathrm{d}v/\mathrm{d}t$.

Acceleration is symbolised as $\boldsymbol{a}$, with magnitude a. It follows that

$$a = \frac{\mathrm{d}v}{\mathrm{d}t} = \frac{\mathrm{d}^2x}{\mathrm{d}t^2}$$

If the unit for velocity is m/s then the unit for acceleration is m/s^2.

The quantities we have defined are all vectors, but since the motion was specified as being rectilinear, that is, in a given direction, only the scalar magnitude of a given quantity need be stated. This can be positive or negative; the sign of the quantity then indicates the sense with respect to the positive sense chosen for x.

Summarising for the rectilinear motion of a particle

position is defined by x
change of position, displacement, is defined by δx
rate of change of position, velocity, is defined by $\mathrm{d}x/\mathrm{d}t$
rate of change of velocity, acceleration, is defined by $\mathrm{d}^2x/\mathrm{d}t^2$

A convenient notation is that by which a time derivative is indicated by a dot over the quantity being operated upon; thus $v = \dot{x}$ and $a = \dot{v} = \ddot{x}$.

If the particle path is curved then the motion is not strictly rectilinear. However, the position of the particle can be specified by the distance s measured along the path from some fixed reference point on the path. Following the same procedure as before we obtain the quantity $\mathrm{d}s/\mathrm{d}t$, the positive magnitude of which is called the *speed*, v, of the particle at P. A separate statement is required to describe the sense of the motion in relation to increasing values of s. The quantity $\mathrm{d}^2s/\mathrm{d}t^2$ can also be found, the positive magnitude of which is again given the name acceleration, a, if the speed is increasing; if the speed is decreasing then a is referred to as the *retardation*.

It is stressed that s, $\mathrm{d}s/\mathrm{d}t$ and $\mathrm{d}^2s/\mathrm{d}t^2$ in the preceding paragraph are position and rates of change measured along the path, and are positive quantities. Motion in a curved path, so-called curvilinear motion, will be discussed in section 8.2, and it will be found that these quantities are not sufficient to describe the motion fully.

8.1.2 Applications

The definitions framed above can now be used to describe the motion of a particle undergoing (1) rectilinear motion in a fixed x-direction, and (2) motion in a curved path to the extent that position, speed and acceleration or retardation along the path can be ascertained. We shall use the symbols x, v and a in our equations; if the path is curved then s can be substituted for x and the equations are identical in form for the purpose of determining the magnitudes of $\mathrm{d}s/\mathrm{d}t$ and $\mathrm{d}^2s/\mathrm{d}t^2$. The following paragraphs illustrate the types of analysis that can arise.

(1) If the position of the particle is stated as a particular function of time, $x = x(t)$, then successive differentiations with respect to time will give the velocity and acceleration at specified times and positions.

(2) Usually it is the velocity or the acceleration that is stated and the position of the particle at any time is to be determined. If the velocity or the acceleration is stated as a function of time then a graphical representation of the information can lead to a solution that is more easily handled. The *velocity - time* graph (figure 8.2) is such a representation; this graph displays the variation of velocity with time. Two important properties of the graph can be noted.

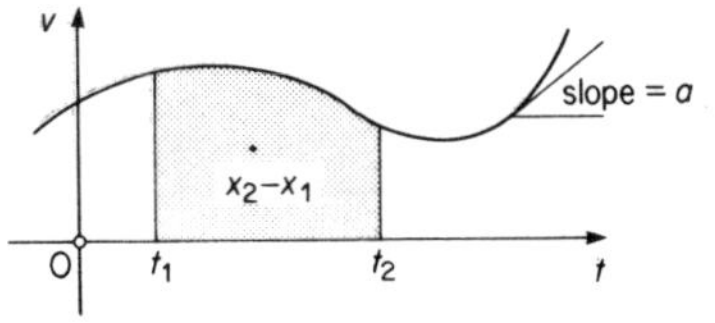

Figure 8.2

(a) Since $a = \mathrm{d}v/\mathrm{d}t$, the slope of the curve at any point represents the acceleration of the particle at the corresponding time.

(b) Since $v = \mathrm{d}x/\mathrm{d}t$, then $v\mathrm{d}t = \mathrm{d}x$, and

$$\int_{t_1}^{t_2} v\mathrm{d}t = \int_{x_1}^{x_2} \mathrm{d}x = x_2 - x_1 \tag{8.1}$$

The integral on the left is therefore the change in position of the particle, $x_2 - x_1$, in the time interval $t_2 - t_1$, and is represented by the area under the curve between the ordinates at t_1 and t_2. The required areas can usually be obtained by calculation, the unit of area being derived from the units adopted for the velocity and time axes. Properties of some standard geometrical shapes are given in the appendix.
(3) The most frequently occurring problems are those in which the acceleration is given, as a function of time, position or velocity, and we need to determine the velocity and position of the particle at some specified instant. It now becomes useful to express the acceleration in other equivalent forms, thus

$$a = \frac{\mathrm{d}v}{\mathrm{d}t} = \frac{\mathrm{d}v}{\mathrm{d}x} \times \frac{\mathrm{d}x}{\mathrm{d}t} = v\frac{\mathrm{d}v}{\mathrm{d}x} = \frac{\mathrm{d}}{\mathrm{d}x}\left(\frac{v^2}{2}\right) \tag{8.2}$$

We can now integrate the given function in the following ways, the choice depending on the manner in which the acceleration is described.

(a) If a is given as a function of time, $a(t)$, we use

$$a(t) = \frac{\mathrm{d}v}{\mathrm{d}t}$$

then

$$\int \mathrm{d}v = \int a(t)\mathrm{d}t$$

and $v(t)$ is determined

$$v(t) = \frac{\mathrm{d}x}{\mathrm{d}t}$$

therefore

$$\int \mathrm{d}x = \int v(t)\mathrm{d}t$$

and x(t) is determined.

(b) If a is given as a function of position, $a(x)$, we use

$$a(x) = v\frac{\mathrm{d}v}{\mathrm{d}x}$$

then

$$\int v\mathrm{d}v = \int a(x)\mathrm{d}x$$

and $v(x)$ is determined

$$v(x) = \frac{dx}{dt}$$

therefore

$$\int \frac{dx}{v(x)} = \int dt$$

and $x(t)$ is determined.

(c) If a is given as a function of velocity, $a(v)$, then we have the alternatives (i) using $a(v) = dv/dt$, then

$$\int \frac{dv}{a(v)} = \int dt$$

and $v(t)$ is determined

$$v(t) = \frac{dx}{dt}$$

therefore

$$\int dx = \int v(t)dt$$

and $x(t)$ is determined.

(ii) using $a(v) = v\ dv/dx$ then

$$\int \frac{v\ dv}{a(v)} = \int dx$$

and $v(x)$ is determined

$$v(x) = \frac{dx}{dt}$$

therefore

$$\int \frac{dx}{v(x)} = \int dt$$

and $x(t)$ is determined.

The above integrations involve arbitrary constants, the evaluation of which require knowledge of the details of the motion at some initial instant or position.

(4) A particularly simple case is that in which the acceleration is constant. If the time t and the position x are taken to be zero when the speed of the particle is u, then from the velocity – time graph (now a sloping line) or by using any of the integration methods discussed, the familiar relations may be derived for the

position x and velocity v at a subsequent time t, namely

$$v = u + at \tag{8.3a}$$

$$x = ut + \tfrac{1}{2}at^2 \tag{8.3b}$$

$$v^2 = u^2 + 2ax \tag{8.3c}$$

(5) An important case of a motion for which $a = a(x)$ is that for which the magnitude of the acceleration a is proportional to the position x measured from some reference point, and the sense is always opposite to that of the position. This implies that if the particle is moving away from the reference point its speed is decreasing, and if moving towards the reference point its speed is increasing. The acceleration a is now expressed as $a = -kx$, but it will be found more convenient to write the proportionality constant as ω^2, and then $a = -\omega^2 x$, which can be written

$$\ddot{x} + \omega^2 x = 0 \tag{8.4}$$

This equation can be solved for $\dot{x}$ and x by the integrations given under section (3) above if we write $\ddot{x}$ in the form $\dot{x}(\mathrm{d}\dot{x}/\mathrm{d}x)$ or $(\mathrm{d}/\mathrm{d}x)(\dot{x}^2/2)$. Integrating once, $\dot{x}^2 = -\omega^2x^2 + C$, and $\dot{x} = \omega\sqrt{(A^2 - x^2)}$ in which the constant C has been replaced by $\omega^2 A^2$, for convenience later.

Putting $\dot{x} = \mathrm{d}x/\mathrm{d}t$ and integrating again

$$\int \frac{\mathrm{d}x}{\sqrt{(A^2 - x^2)}} = \omega\int \mathrm{d}t$$

and

$$\sin^{-1}\left(\frac{x}{A}\right) = \omega t + B$$

or

$$x = A\sin(\omega t + B) \tag{8.5}$$

The values of A and B are determined from prescribed initial conditions, for example, the values of x and $\dot{x}$ when $t = 0$.

The expression for x in equation 8.5 shows that it takes the same value, implying that the particle passes through the same point, at time intervals $2\pi/\omega$ since $\sin[\omega(t + 2\pi/\omega) + B] = \sin(\omega t + 2\pi + B) = \sin(\omega t + B)$. From the expressions for $\dot{x}$ and $\ddot{x}$ we see that there are associated with this point a definite magnitude of the velocity and a definite magnitude and sense of the acceleration. This implies that the motion repeats itself, the time for one cycle or the *periodic time* τ being given by

$$\tau = \frac{2\pi}{\omega} \tag{8.6}$$

The position x attains maximum and minimum values $x_{max/min} = \pm A$. The magnitude A is termed the *amplitude* of the motion. The motion of the particle is therefore confined to the range $x = -A$ and $x = +A$, and is termed *oscillatory*.

Oscillatory motion of this kind which is described by the equation $\ddot{x} + \omega^2 x = 0$ is named *simple harmonic motion* and is the starting point for the study of other oscillatory phenomena of many kinds.

A basic characteristic of simple harmonic motion is the number of cycles of the motion undergone in unit time. Since the frequency $f = 1/\tau = \omega/2\pi$ it follows that the quantity ω entering into the equation of motion 8.4 is given by $\omega = 2\pi f$; ω can be referred to as the *circular frequency*. If unit time is the second then frequency has the unit, cycle/second, which is named the hertz (Hz). The quantity ω, from its association with t in equation 8.6 can be ascribed the unit, rad/s.

Summarising, we have

$$x = A\sin(\omega t + B) \tag{8.7a}$$

$$\dot{x} = A\omega\cos(\omega t + B) \tag{8.7b}$$

$$\ddot{x} = -A\omega^2\sin(\omega t + B) = -\omega^2 x \tag{8.7c}$$

from which we note that the magnitude of the acceleration is a maximum when $x = \pm A$ and is zero when the particle passes through the reference point. Further, when passing through the reference point at $x = 0$ the magnitude of the velocity is a maximum, with a value $A\omega$. Since $A = (1/\omega) \times$ (the magnitude of the particle velocity at $x = 0$) it follows that the amplitude of the motion A depends on the speed of the particle as it passes through the reference point.

This case has been solved analytically to indicate the application of the standard methods of integration described in paragraph (3) to simple harmonic motion. A fuller and perhaps simpler appreciation of the characteristics of this type of motion will emerge after we have discussed the motion of a particle in a circular path.

Worked Example 8.1

A particle is moving in the x-direction with constant acceleration 5 m/s^2. At time $t = 0$ it passes through a certain point with velocity 20 m/s. What is its displacement in the subsequent 5 s?

At the instant $t = 5$ s the acceleration changes to -10 m/s^2, this acceleration remaining constant until the particle again passes through its initial position. At what value of t does this occur?

What is the extreme position of the particle in the x-direction and at what value of t does the particle reach this position?

What is the velocity of the particle as it passes through its initial position?

Solution

The velocity – time graph of the motion is drawn as in figure 8.3. This consists

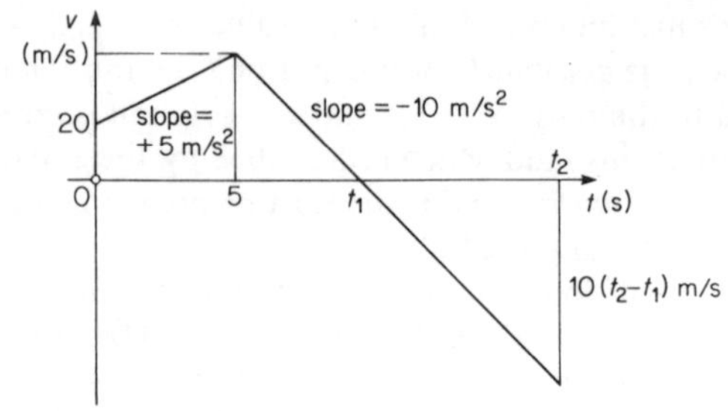

Figure 8.3

of two sloping lines with gradients 5 m/s² and − 10 m/s² respectively. When $t = 5$ s, velocity = 20 + (5 × 5) = 45 m/s.

Displacement in 5 s = area under graph between $t = 0$ and $t = 5$ s

$$= \frac{1}{2}(20 + 45) \times 5$$

$$= 162.5 \text{ m}$$

Between $t = 5$ s and t_1 s displacement is positive since area is increasing. For $t > t_1$ s we have negative area indicating negative displacement. Eventually at time t_2 s net area is zero, therefore net displacement is zero and the particle has regained its initial position.

From the diagram for $5 < t < t_1$

$$t_1 - 5 = \frac{45}{10} = 4.5$$

therefore

$$t_1 = 9.5$$

The particle attains its extreme position after 9.5 s.

Displacement in 9.5 s = area under graph between $t = 0$ and $t = 9.5$ s

$$= 162.5 + \frac{1}{2} \times 4.5 \times 45$$

$$= 162.5 + 101.25$$

$$= 263.75 \text{ m}$$

and this is the extreme position reached by the particle.

Area under graph between t_1 s and t_2 s = − 263.75 m

$$= -\frac{1}{2}(t_2 - 9.5) \times 10(t_2 - 9.5)$$

$$(t_2 - 9.5)^2 = \frac{2 \times 263.75}{10}$$

$$t_2 - 9.5 = 7.26$$

and

$$t_2 = 16.76$$

The particle returns to its initial position after 16.76 s.

$$\begin{aligned}\text{Velocity at time } t_2 &= -10(t_2 - 9.5)\\ &= -10 \times 7.26\\ &= -72.6 \text{ m/s}\end{aligned}$$

Worked Example 8.2

A particle moving along the x-axis has acceleration a, given by $a(v) = -kv$, where k is a constant. When $t = 0$, $v = v_0$ and $x = x_0$. Obtain

(a) equations for velocity and position as functions of time
(b) an equation for velocity in terms of position.

Solution

(a) Since we are given $a(v)$ and we require v and x as functions of t we choose

$$a(v) = \frac{\mathrm{d}v}{\mathrm{d}t} = -kv$$

then

$$\frac{\mathrm{d}v}{v} = -k\,\mathrm{d}t$$

Integrating

$$\ln v = -kt + C_1$$

When $t = 0$, $v = v_0$, therefore

$$\ln v_0 = C_1$$

Eliminating C_1

$$\ln\left(\frac{v}{v_0}\right) = -kt$$

and

$$v = v_0 \mathrm{e}^{-kt}$$

With $v = \mathrm{d}x/\mathrm{d}t$, $\mathrm{d}x = v_0\mathrm{e}^{-kt}\,\mathrm{d}t$. Integrating

$$x = -\frac{v_0}{k}\mathrm{e}^{-kt} + C_2$$

When $t = 0, x = x_0$, therefore

$$x_0 = -\frac{v_0}{k} + C_2$$

Eliminating C_2

$$x = x_0 + \frac{v_0}{k}(1 - e^{-kt})$$

(b) Since we require v in terms of x we choose

$$a(v) = v\frac{dv}{dx} = -kv$$

then

$$dv = -k\,dx$$

Integrating

$$v = -kx + C_3$$

When $v = v_0, x = x_0$, therefore

$$v_0 = -kx_0 + C_3$$

Eliminating C_3

$$v = v_0 - k(x - x_0)$$

Using the previous solution for $x - x_0$ we verify that

$$v = v_0 - k \times \frac{v_0}{k}(1 - e^{-kt})$$

$$= v_0 e^{-kt}$$

as before.

Worked Example 8.3

A particle moving in the positive x-direction has acceleration $a(v) = (100 - 4v^2)$ m/s^2, where v is the speed of the particle in m/s.

(a) For what values of speed is the speed of the particle (i) increasing, (ii) decreasing?
(b) If the speed of the particle is initially zero what is its maximum speed? Is this a mathematical maximum or is it a final steady value? If $t = 0$ when the particle is at rest, at what time t is the maximum speed attained? What is the speed at $t = 0.05$ s?
(c) In case (b) what is the time interval and the displacement of the particle as the speed changes from 1 m/s to 3 m/s?
(d) If the speed of the particle is initially 8 m/s what is its ultimate speed?

Solution

(a) $a(v) = 100 - 4v^2$; $a(v)$ is positive for $0 < v < 5$ m/s and negative for $v > 5$ m/s. Thus (i) for speeds less than 5 m/s speed is increasing and (ii) for speeds greater than 5 m/s speed is decreasing. *Note* :negative values of v are not considered since particle is moving in the positive x-direction.

(b) The speed v will have a maximum value when $dv/dt = 0$ and d^2v/dt^2 is negative. Since $dv/dt = a(v) = 100 - 4v^2$, $dv/dt = 0$ when $v = 5$ m/s, and $d^2v/dt^2 = -8v(dv/dt)$, which is also zero at $v = 5$ m/s. Thus $v = 5$ m/s is not a mathematical maximum. However when v attains the value 5 m/s, $a(v) = 0$; the speed therefore remains constant and is a final steady value.

Since we are interested in the time, we retain the form

$$a(v) = \frac{dv}{dt} = 100 - 4v^2$$

and

$$\frac{dv}{4(25 - v^2)} = dt$$

Integrating this equation gives a relation between v and t. We can use partial fractions or use the standard form

$$\int \frac{dx}{a^2 - x^2} = \frac{1}{2a} \ln \left(\frac{a + x}{a - x}\right) \qquad x < a$$

with appropriate limits.

$$\frac{1}{4} \int_0^v \frac{dv}{5^2 - v^2} = \int_0^t dt$$

$$t = \frac{1}{4} \times \frac{1}{10} \left[\ln \frac{5 + v}{5 - v}\right]_0^v$$

$$= \frac{1}{40} \times \ln \frac{5 + v}{5 - v}$$

As $v \rightarrow 5$, $t \rightarrow \infty$, therefore the steady value of speed is attained after infinite time.

To obtain the speed after time t we rearrange the expression to make v explicit

$$\frac{5 + v}{5 - v} = e^{40t}$$

from which

$$v = \frac{5(e^{40t} - 1)}{1 + e^{40t}}$$

when $t = 0.05$ s

$$v = \frac{5(e^2 - 1)}{1 + e^2} = 3.81 \text{ m/s}$$

(c) We use the same integration as in (b) but amend the limits, putting $v = 1$ m/s at $t = t_1$ and $v = 3$ m/s at $t = t_2$. Then

$$t_2 - t_1 = \frac{1}{40} \left[\ln \frac{5 + v}{5 - v} \right]_1^3$$

$$= \frac{1}{40} \left(\ln \frac{8}{2} - \ln \frac{6}{4} \right)$$

$$= \frac{1}{40} \ln \frac{8}{3} = 0.0245 \text{ s}$$

For the displacement we require the form

$$a(v) = v \frac{dv}{dx} = 100 - 4v^2$$

and

$$\frac{v \, dv}{4(25 - v^2)} = dx$$

then

$$x_2 - x_1 = \frac{1}{4} \int_1^3 \frac{v \, dv}{25 - v^2} = -\frac{1}{8} \left[\ln (25 - v^2) \right]_1^3$$

$$= -\frac{1}{8} (\ln 16 - \ln 24)$$

$$= \frac{1}{8} \ln \frac{3}{2} = 0.0506 \text{ m}$$

(d) If the initial speed is 8 m/s then, by similar considerations as those in (b) above, particle speed reaches a final steady value 5 m/s.

8.2 Curvilinear Motion of a Particle

The particle now moves along a plane curved path C (figure 8.4) but the definitions framed in the previous section still formally apply, provided we recognise

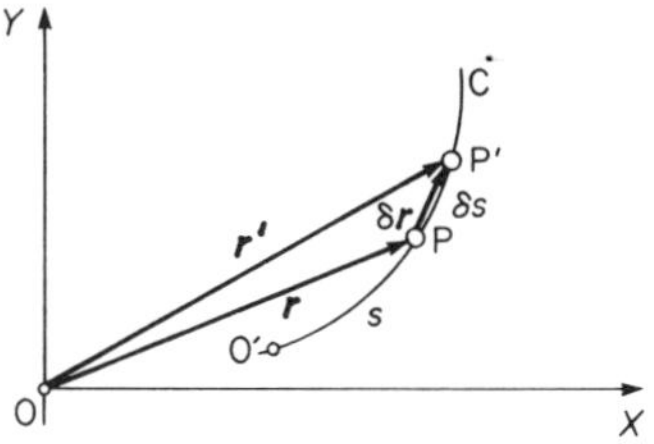

Figure 8.4

that, because of the curvilinear motion the directions of the position, displacement, velocity and acceleration vectors no longer coincide with each other nor do they remain fixed. For particular applications it is necessary to determine the components of these vectors in specified directions.

8.2.1 Position and Velocity Components

We again choose a fixed reference direction, which we take as an X-axis, and select a reference point O on the axis. If the particle at time t is at point P of its path we again define its *position* as being **OP**, having magnitude and direction. We shall call this the *position vector* $\boldsymbol{r}$.

In a time interval δt the particle will have moved to a new position $\mathbf{OP'} = \mathbf{r'}$. and the change of position, the *displacement*, is given vectorially by **PP′** which we signify by $\delta\boldsymbol{r}$ such that $\boldsymbol{r}' = \boldsymbol{r} + \delta\boldsymbol{r}$.

The ratio $\delta\boldsymbol{r}/\delta t$ represents the average rate of change of position, a vector having the direction of $\delta\boldsymbol{r}$.

The average speed of the particle during the same time interval is $\delta s/\delta t$ where s is the distance of P measured along the path from some reference point O′ on the path, and δs is the length of the curved segment PP′.

The *velocity* of the particle at P at time t can now be formally defined to be

$$\lim_{\delta t \to 0} \frac{\delta \boldsymbol{r}}{\delta t} = \frac{\mathrm{d}\boldsymbol{r}}{\mathrm{d}t}$$

Since in the limiting process the magnitude of $\delta\boldsymbol{r}$ tends to δs, and the direction of $\delta\boldsymbol{r}$ tends to that of the tangent at P, we interpret $\mathrm{d}\boldsymbol{r}/\mathrm{d}t$ as being a velocity vector $\boldsymbol{v}$ with magnitude $v = \mathrm{d}s/\mathrm{d}t$ and direction that of the tangent at **P**.

The motion of the particle is therefore characterised by a stretching and a swinging of the position vector $\boldsymbol{r}$, the particle thereby tracing a path to which the velocity vector $\boldsymbol{v}$ is always tangential.

(1) Rather than proceed using vector symbolism exclusively we can use the *polar* coordinates of P to indicate the position of the particle (figure 8.5). These are particularly appropriate since the magnitude of the position vector is conveniently given by r and its direction by θ.

The velocity $\boldsymbol{v}$ can now be expressed in terms of time rates of change of r and

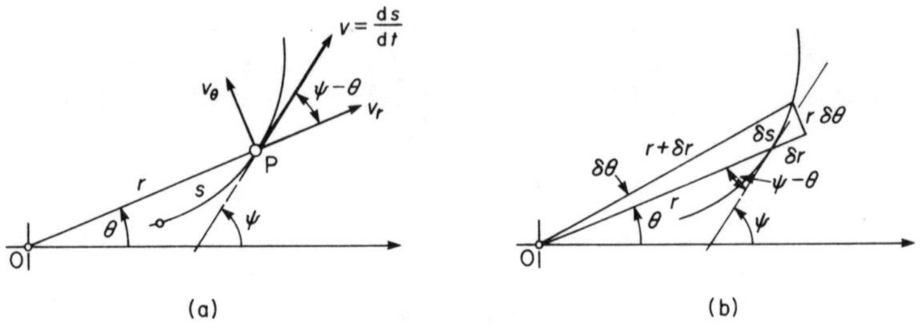

Figure 8.5

θ, by resolving v into its components in the radial and transverse directions shown in figure 8.5a, from which $v_r = v\cos(\psi - \theta)$ and $v_\theta = v\sin(\psi - \theta)$. In figure 8.5b, $\delta\theta$ is small and in the limit $\cos(\psi - \theta) = \mathrm{d}r/\mathrm{d}s$ and $\sin(\psi - \theta) = r\,\mathrm{d}\theta/\mathrm{d}s$.

$$v_r = \frac{\mathrm{d}s}{\mathrm{d}t}\cos(\psi - \theta) = \frac{\mathrm{d}s}{\mathrm{d}t} \times \frac{\mathrm{d}r}{\mathrm{d}s} = \frac{\mathrm{d}r}{\mathrm{d}t} = \dot{r} \tag{8.8a}$$

$$v_\theta = \frac{\mathrm{d}s}{\mathrm{d}t}\sin(\psi - \theta) = \frac{\mathrm{d}s}{\mathrm{d}t} \times \frac{r\mathrm{d}\theta}{\mathrm{d}s} = \frac{r\,\mathrm{d}\theta}{\mathrm{d}t} = r\dot{\theta} \tag{8.8b}$$

(2) We can also indicate the position of the particle by specifying the *rectangular* coordinates of P, and in a similar manner the velocity v may be expressed in terms of the time rates of change of x and y by resolving in the x- and y-directions (figure 8.6a).

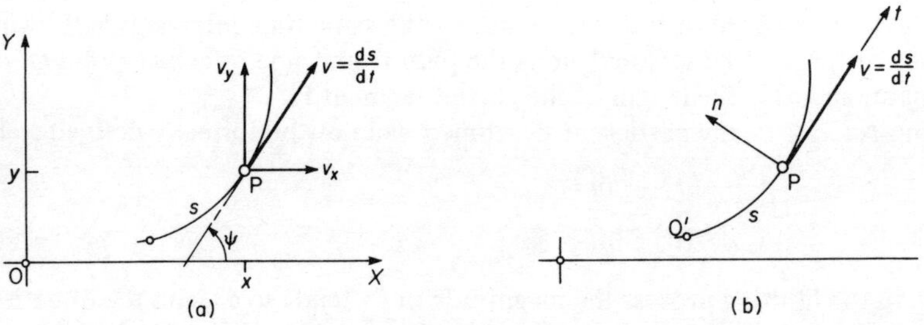

Figure 8.6

From the figure

$$v_x = \frac{\mathrm{d}s}{\mathrm{d}t}\cos\psi = \frac{\mathrm{d}s}{\mathrm{d}t} \times \frac{\mathrm{d}x}{\mathrm{d}s} = \frac{\mathrm{d}x}{\mathrm{d}t} = \dot{x} \tag{8.9a}$$

$$v_y = \frac{\mathrm{d}s}{\mathrm{d}t}\sin\psi = \frac{\mathrm{d}s}{\mathrm{d}t} \times \frac{\mathrm{d}y}{\mathrm{d}s} = \frac{\mathrm{d}y}{\mathrm{d}t} = \dot{y} \tag{8.9b}$$

(3) Further, we can indicate the position of the particle by specifying the distance of P from some reference point O′ on the path, measured along the path. We call this the *intrinsic* coordinate of P (figure 8.6b). At the point P we have two perpendicular directions, namely, the t-direction, that of the tangent, and the n-direction, that of the normal.

We can now write immediately

$$v_t = \frac{ds}{dt} = \dot{s} \tag{8.10a}$$

$$v_n = 0 \tag{8.10b}$$

8.2.2 Acceleration Components: The Hodograph

The acceleration of the particle is the rate of change of its velocity and again can be formally defined as being $d\boldsymbol{v}/dt$. However, we note that as the particle moves the velocity vector is changing in both magnitude and direction, and, in the same way as the velocity itself was the result of a stretching and swinging of the position vector, so the acceleration is now the result of the stretching and swinging of a velocity vector in the plane of the motion. In figure 8.4 we were able to see the stretching and swinging of the position vector since the path itself was the result of the movement of the end point of this vector. In the case of the velocity we should draw a separate diagram of the velocity vector, that is, a representation of the velocity at P, and observe how it too stretches and swings.

This is conveniently done by choosing another plane, the so-called *hodograph plane*, in which the behaviour of the velocity vectors can be observed. While the actual particle traverses the actual path in what we shall now call the *physical plane*, a second particle is made to move in the hodograph plane in such a way that its position vector is always equal to the velocity vector of the actual particle. The path traced out by the second particle, the hodograph particle, is called the *hodograph* of the motion of the actual particle. Thus in figure 8.7, if the actual

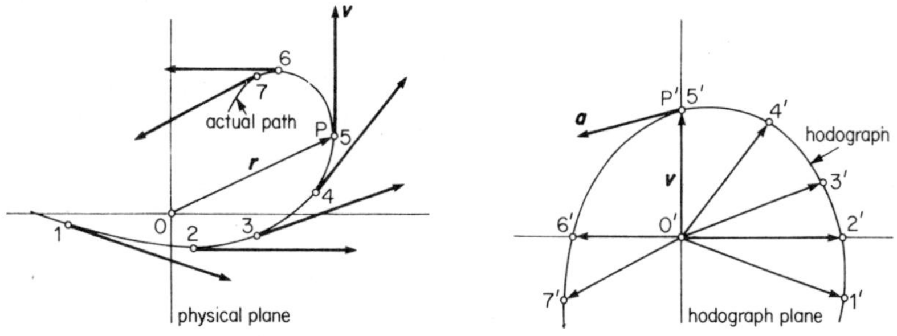

Figure 8.7

particle, when at the point 5 in the physical plane, has velocity v then the hodograph particle is at the point 5′ in the hodograph plane, the vector **O′P′** having magnitude v and the direction of $\boldsymbol{v}$.

We can now state generally that if the actual particle is at a point P and the hodograph particle is at P′, then

(the velocity of the actual particle in the physical plane) = (the position of the hodograph particle in the hodograph plane)

It follows that

(the rate of change of the velocity of the actual particle) = (the rate of change of the position of the hodograph particle)

or

(the acceleration of the actual particle in the physical plane) = (the velocity of the hodograph particle in the hodograph plane)

(1) Consider now our particle, which is at the point (r, θ) and moving with velocity magnitude $\mathrm{d}s/\mathrm{d}t$ in the direction of the tangent to its path, as in figure 8.8a. The hodograph of the motion will be of the form shown in figure 8.8b. Then as the

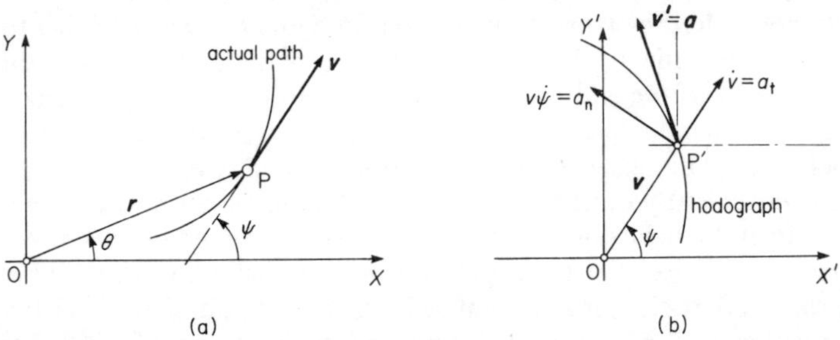

Figure 8.8

position vector $\boldsymbol{r}$ of the actual particle stretches and swings, the position vector $\boldsymbol{v}$ of the hodograph particle, magnitude v, also stretches and swings, and the velocity $\boldsymbol{v}'$ of the hodograph particle is equal in magnitude and direction to the acceleration $\boldsymbol{a}$ of the actual particle at all times.

By comparison with the radial and transverse components of the velocity of a particle in the physical plane, as already discussed (see equations 8.8), the velocity $\boldsymbol{v}'$ of the hodograph particle has a radial component $\dot{v}$ and a transverse component $v\dot{\psi}$. We note that the radial component $\dot{v}$ is in the direction of the tangent at P to the actual path, and the component $v\dot{\psi}$ is in the direction of the normal at P. We have therefore, for the actual particle

$$a_\mathrm{t} = \dot{v} = \frac{\mathrm{d}v}{\mathrm{d}t} \tag{8.11a}$$

$$a_\mathrm{n} = v\dot{\psi} = v\frac{\mathrm{d}\psi}{\mathrm{d}s} \times \frac{\mathrm{d}s}{\mathrm{d}t} = v^2\frac{\mathrm{d}\psi}{\mathrm{d}s}$$

since $\mathrm{d}s/\mathrm{d}t = v$. Since (from coordinate geometry) $\mathrm{d}s/\mathrm{d}\psi = \rho$, the radius of curvature of the path at P

$$a_\mathrm{n} = \frac{v^2}{\rho} \tag{8.11b}$$

a_t and a_n are respectively the tangential and normal components of the acceleration of the particle at P.

(2) We can also resolve the velocity of the hodograph particle in the x- and y-directions and obtain rectangular components of acceleration. Thus, referring to figure 8.8b

$$a_x = \dot{v}\cos\psi - v\dot{\psi}\sin\psi$$

Noting that $v\cos\psi = \mathrm{d}x/\mathrm{d}t$

$$a_x = \frac{\mathrm{d}}{\mathrm{d}t}\left(\frac{1}{\cos\psi}\,\frac{\mathrm{d}x}{\mathrm{d}t}\right)\cos\psi - \left(\frac{1}{\cos\psi}\,\frac{\mathrm{d}x}{\mathrm{d}t}\right)\frac{\mathrm{d}\psi}{\mathrm{d}t}\sin\psi$$

$$= \left(\frac{1}{\cos\psi} \times \frac{\mathrm{d}^2x}{\mathrm{d}t^2} + \frac{\sin\psi}{\cos^2\psi} \times \frac{\mathrm{d}\psi}{\mathrm{d}t} \times \frac{\mathrm{d}x}{\mathrm{d}t}\right)\cos\psi$$

$$- \frac{\sin\psi}{\cos\psi} \times \frac{\mathrm{d}x}{\mathrm{d}t} \times \frac{\mathrm{d}\psi}{\mathrm{d}t}$$

$$= \frac{\mathrm{d}^2x}{\mathrm{d}t^2} = \ddot{x} \tag{8.12a}$$

It can be verified that

$$a_y = \dot{v}\sin\psi + v\dot{\psi}\cos\psi$$

$$= \frac{\mathrm{d}^2y}{\mathrm{d}t^2} = \ddot{y} \tag{8.12b}$$

a_x and a_y are respectively the x and y-components of the acceleration of the particle at P.

(3) To obtain the radial and transverse components of acceleration we can again resolve the hodograph particle velocity in the direction of and perpendicular to the direction of the actual particle position vector; but it will be found more convenient to use two hodographs, one in plane 1 to follow the radial velocity component $\dot{r}$, and the other in plane 2 to follow the transverse component $r\dot{\theta}$ (figure 8.9).

In hodograph plane 1 the hodograph particle shown has a radial velocity component $\mathrm{d}(\dot{r})/\mathrm{d}t = \ddot{r} = a_{r1}$ and a transverse velocity component $(\dot{r})\mathrm{d}\theta/\mathrm{d}t = \dot{r}\dot{\theta} = a_{\theta 1}$. In hodograph plane 2 the hodograph particle shown has a radial velocity

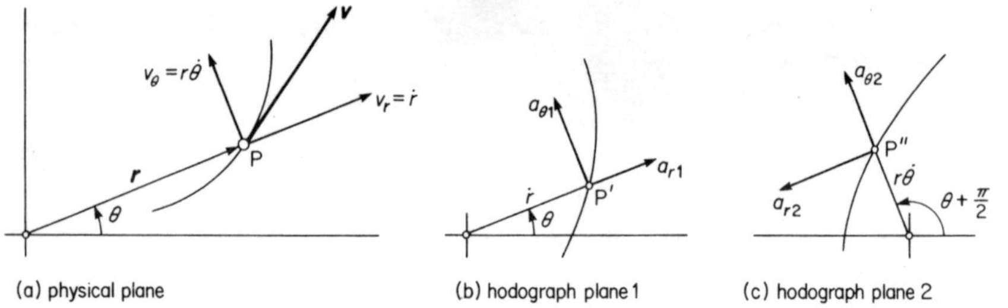

Figure 8.9

component $\mathrm{d}(r\dot{\theta})/\mathrm{d}t = \dot{r}\dot{\theta} + r\ddot{\theta} = a_{\theta 2}$ and a transverse velocity component $(r\dot{\theta})\mathrm{d}(\theta + \pi/2)/\mathrm{d}t = r\dot{\theta}^2 = a_{r2}$. The subscripts r and θ refer to directions in the physical plane.

If we now add corresponding components we obtain

$$a_r = a_{r1} - a_{r2} = \ddot{r} - r\dot{\theta}^2 \tag{8.13a}$$

$$a_\theta = a_{\theta 1} + a_{\theta 2} = \dot{r}\dot{\theta} + \dot{r}\dot{\theta} + r\ddot{\theta} = r\ddot{\theta} + 2\dot{r}\dot{\theta} \tag{8.13b}$$

a_r and a_θ are respectively the radial and transverse components of the acceleration of the particle at P.

The expressions we have derived can be summarised as in figure 8.10.

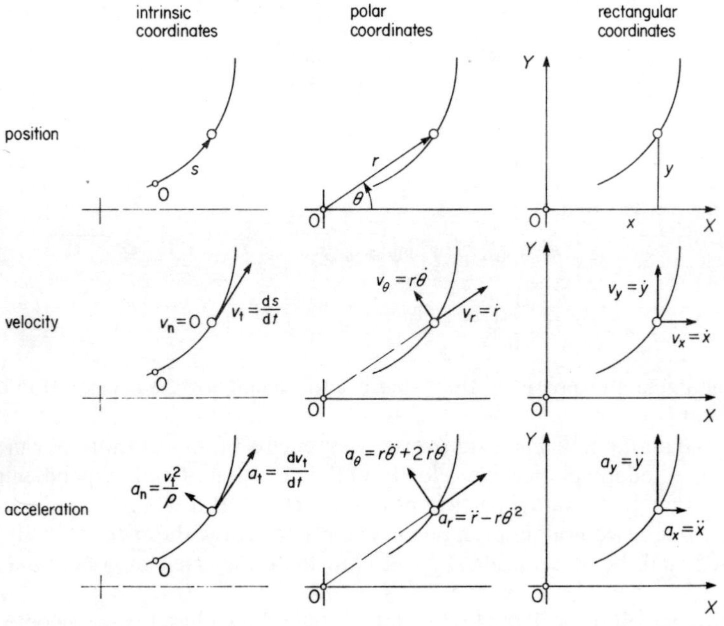

Figure 8.10

The choice of coordinate system will depend on the problem under consideration, and the solution in turn will be considerably simplified by a correct choice. We shall find later that the choice will depend to a large extent on the type of force acting on the particle. Typical choices would be (a) for vehicle motion – intrinsic coordinates, (b) for orbital motion – polar coordinates, (c) for motion under unidirectional forces such as gravity – rectangular coordinates.

Worked Example 8.4

A particle travels along the parabolic path $y = x^2$ with speed $v = 2 + 2t$ (m, s units), the time t being zero as the particle passes through the origin. Determine the time at which the particle passes through the point (1,1) and the x- and y-components of its velocity and acceleration at that instant.

Solution

If the tangential direction at any point makes an angle ψ with the x-axis then

$$v_x = \frac{dx}{dt} = v \cos \psi$$

$$\tan \psi = \frac{dy}{dx} = 2x$$

and

$$\cos \psi = \frac{1}{\sqrt{(1 + 4x^2)}}$$

therefore

$$\frac{dx}{dt} = \frac{2 + 2t}{\sqrt{(1 + 4x^2)}}$$

$$\int_0^1 \sqrt{(1 + 4x^2)}dx = \int_0^{t_1} (2 + 2t)dt$$

$$\tfrac{1}{2}[x\sqrt{(1 + 4x^2)} + \tfrac{1}{2} \ln \tfrac{1}{2}\{2x + \sqrt{(1 + 4x^2)}\}]_0^1 = [2t + t^2]_0^{t_1}$$

$$\tfrac{1}{2}\sqrt{5} + \tfrac{1}{4}\ln(2 + \sqrt{5}) = 2t_1 + t_1^2$$

$$t_1^2 + 2t_1 - 1.479 = 0$$

$$t_1 = 0.575 \text{ (positive value)}$$

therefore

$$v_1 = 2 + 2t_1 = 3.150 \text{ m/s}$$

at the point (1,1), $\cos\psi = 1/\sqrt{5}$, $\sin\psi = 2/\sqrt{5}$, therefore

$$v_{x1} = v_1 \cos\psi = 3.15/\sqrt{5} = 1.408 \text{ m/s}$$

$$v_{y1} = v_1 \sin\psi = 6.30/\sqrt{5} = 2.817 \text{ m/s}$$

From above, $v_x = \mathrm{d}x/\mathrm{d}t = v\cos\psi$, therefore $\sqrt{(1+4x^2)}\,\mathrm{d}x/\mathrm{d}t = 2 + 2t$. Differentiating

$$\sqrt{(1+4x^2)}\frac{\mathrm{d}^2x}{\mathrm{d}t^2} + \frac{4x}{\sqrt{(1+4x^2)}}\left(\frac{\mathrm{d}x}{\mathrm{d}t}\right)^2 = 2$$

At the point (1, 1)

$$\sqrt{5}\frac{\mathrm{d}^2x}{\mathrm{d}t^2} + \frac{4}{\sqrt{5}}\left(\frac{\mathrm{d}x}{\mathrm{d}t}\right)^2 = 2$$

therefore

$$a_x = \frac{\mathrm{d}^2x}{\mathrm{d}t^2} = \frac{1}{\sqrt{5}}\left(2 - \frac{4 \times 1.408^2}{\sqrt{5}}\right) = -0.692 \text{ m/s}^2$$

From above, $v_y = \mathrm{d}y/\mathrm{d}t = v\sin\psi$

$$\sin\psi = \frac{2x}{\sqrt{(1+4x^2)}} = \frac{2\sqrt{y}}{\sqrt{(1+4y)}} = \frac{2}{\sqrt{(1/y+4)}}$$

therefore $\sqrt{(1/y+4)}\,\mathrm{d}y/\mathrm{d}t = 4 + 4t$. Differentiating

$$\sqrt{\left(\frac{1}{y}+4\right)}\frac{\mathrm{d}^2y}{\mathrm{d}t^2} + \frac{-1/y^2}{2\sqrt{(1/y+4)}}\left(\frac{\mathrm{d}y}{\mathrm{d}t}\right)^2 = 4$$

At the point (1, 1)

$$\sqrt{5}\frac{\mathrm{d}^2y}{\mathrm{d}t^2} - \frac{1}{2\sqrt{5}}\left(\frac{\mathrm{d}y}{\mathrm{d}t}\right)^2 = 4$$

therefore

$$a_y = \frac{\mathrm{d}^2y}{\mathrm{d}t^2} = \frac{1}{\sqrt{5}}\left(4 + \frac{2.817^2}{2\sqrt{5}}\right) = 2.582 \text{ m/s}^2$$

8.3 Angular Motion of a Line

In our discussion of the motion of a particle in a curved path we found that it was characterised by the stretching and swinging of position and velocity vectors. The swinging was conveniently referred to the changes in the angles θ and ψ, and we encountered quantities such as $\dot{\theta}$ and $\dot{\psi}$ in our description of the motion of a particle. To accommodate such angle changes we introduce further definitions relating to the angular motion of a line.

Suppose OP is such a line (figure 8.11a), which is rotating about a fixed point O. Choose a fixed direction OX. Then we define the angular position of OP to be

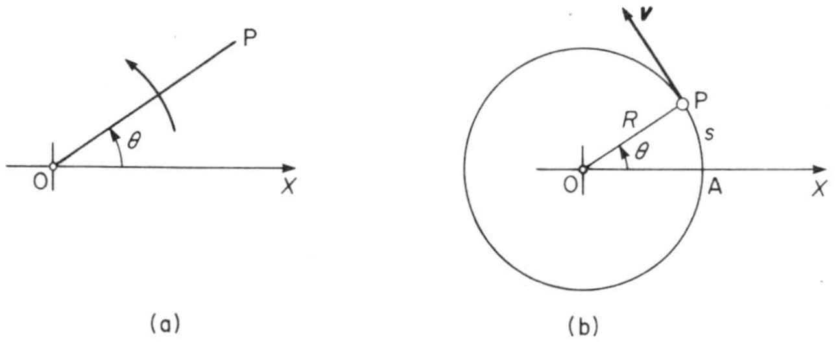

Figure 8.11

the angle θ, measured in the anticlockwise sense from OX. The angular velocity of the line, Ω, is then defined to be $\mathrm{d}\theta/\mathrm{d}t = \theta$, and the angular acceleration of the line, α, is defined to be $\mathrm{d}^2\theta/\mathrm{d}t^2 = \ddot{\theta}$, with the anticlockwise direction being taken as the positive sense in both cases. If the unit for measurement of angle is the radian, then the units for Ω and α are respectively rad/s and rad/s^2.

It is evident that the relationships between position x, speed v and acceleration a that followed from the definitions in the case of rectilinear motion can be formally repeated for angular motion, with the substitution of θ, Ω and α respectively for x, v and a. For example, if α is constant, we can write

$$\Omega = \Omega_0 + \alpha t \tag{8.14a}$$

$$\theta = \Omega_0 t + \tfrac{1}{2}\alpha t^2 \tag{8.14b}$$

$$\Omega^2 = \Omega_0{}^2 + 2\alpha\theta \tag{8.14c}$$

where t is measured from the instant when $\theta = 0$ and $\Omega = \Omega_0$.

Similarly we have a corresponding angular simple harmonic motion expressed by the equation

$$\ddot{\theta} + \omega^2\theta = 0 \tag{8.15}$$

Note that whereas previously we were discussing the motion of a material particle we are now referring to the angular motion of a geometrical line. We shall later have occasion to refer to the motion of a geometrical point and the angular motion of a material body. It is meaningless, however, to speak of the angular velocity of a particle.

If a particle is moving in a circular path then we obtain particularly simple relations between linear and angular motions. In figure 8.11b suppose the particle is moving in a circular path having radius R. If s is the arc length AP measured from the reference axis and θ is the angular position of OP, then for the particle at P we have

$$s = R\theta \qquad (8.16a)$$

and

$$v = \dot{s} = R\dot{\theta} = R\Omega \qquad (8.16b)$$

the direction of the velocity being tangential to the path, or at right angles to the radius OP.

The acceleration of the particle at P has two components, a_t and a_n where (from figure 8.10)

$$a_t = \frac{dv}{dt} = R\ddot{\theta} = R\alpha \qquad \text{(tangential to the path)} \qquad (8.17)$$

$$a_n = \frac{v^2}{\rho} = \frac{v^2}{R} = R\Omega^2 \qquad \text{(towards the centre)} \qquad (8.18)$$

a_t is termed the tangential acceleration component and a_n is termed the centripetal acceleration component.

Although for purposes of definition we chose the reference direction in figure 8.11a to pass through O, we could equally well have chosen a parallel reference direction $O'X'$ (figure 8.12), and the definitions of θ, Ω and α would have been

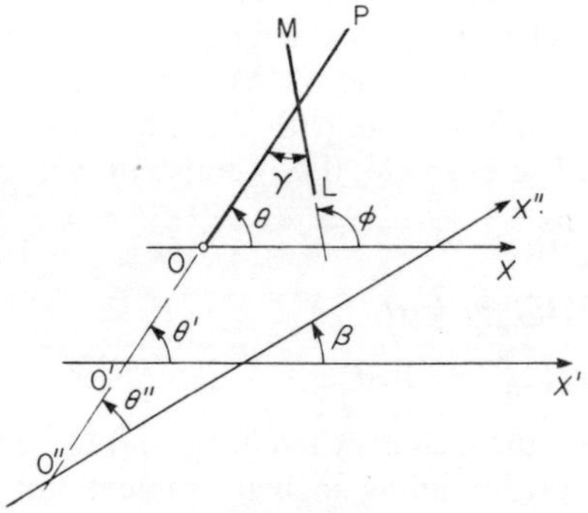

Figure 8.12

unchanged. Had we chosen any other reference direction $O''X''$ inclined at a fixed angle β to the direction OX, then since $\theta'' = \theta' - \beta = \theta - \beta$ the angular position of line OP would have been different, but after differentiating

$$\dot{\theta}'' = \dot{\theta} = \Omega$$

$$\ddot{\theta}'' = \ddot{\theta} = \alpha$$

and Ω and α would have been unchanged in value.

We go further and select another line LM as shown which rotates with OP in such a way that the angle γ between the lines is fixed. Then if the angular position of LM is ϕ with respect to the direction OY

$$\phi = \theta + \gamma$$

and again, after differentiating

$$\dot{\phi} = \dot{\theta} = \Omega$$

$$\ddot{\phi} = \ddot{\theta} = \alpha$$

The same result is true for all lines that are in a fixed relation to one another. It follows that all line segments of a rotating plane have the same angular velocity and the same angular acceleration. If a plane 2 is rotating relative to another plane 1, we can now associate with its rotation an angular velocity Ω and an angular acceleration α relative to plane 1, without specifying any particular axis of rotation or any particular line in plane 2. However, we must still define θ, Ω and α by reference to the angular motion of some selected line in plane 2 and a reference direction in plane 1.

8.4 Simple Harmonic Motion

We now return to the particular type of motion that we described in section 8.1.2 as being simple harmonic, and review its characteristics in the light of the motion of a particle in a circular path. In figure 8.13 a particle B is moving at constant

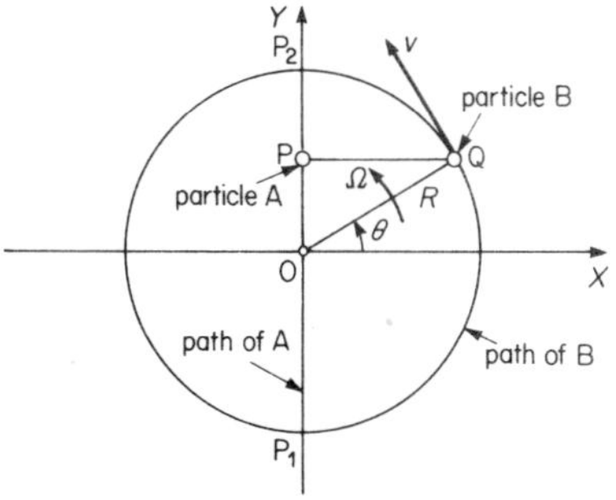

Figure 8.13

speed v in a circular path radius R. At the same time a particle A is moving along the axis OY in such a manner that the line PQ joining the points occupied by the particles is always perpendicular to OY. Time t will be measured from the instant the particles cross the axis OX, this being also the reference direction for the angular position θ of the line OQ.

Since v is constant and $\Omega = v/R$, then Ω is constant, and $\theta = \Omega t$. We can then write for particle A when at the point P at time t

$$y = R \sin \Omega t \tag{8.19a}$$

$$\dot{y} = R\,\Omega \cos \Omega t \tag{8.19b}$$

$$\ddot{y} = -R\,\Omega^2 \sin \Omega t = -\Omega^2 y \tag{8.19c}$$

For the motion of particle A we have therefore, at all points of its path $\ddot{y} + \Omega^2 y = 0$, and the particle is moving with simple harmonic motion (SHM) between the points P_1 and P_2. The amplitude of the motion is R, and the periodic time τ for one cycle of the motion is the time taken for particle B to traverse the circular path, or for the line OQ to rotate through the angle 2π, that is $2\pi/\Omega$.

Usually the motion of a particular particle is already known or can be shown to conform to an equation of the form $\ddot{y} + \Omega^2 y = 0$, and the acceleration $\ddot{y}$ and the position y are known at some instant. From equations 8.19 it can be seen that

$$\tau = \frac{2\pi}{\Omega} = 2\pi \sqrt{\left(\frac{y}{\ddot{y}}\right)}$$

$$= 2\pi \sqrt{\left(\frac{\text{position}}{\text{corresponding acceleration}}\right)} \tag{8.20}$$

without regard to signs.

If the time t is measured from the instant that OQ is in the angular position ϕ, our expressions now become

$$y = R \sin (\Omega t + \phi) \tag{8.21a}$$

$$\dot{y} = R\,\Omega \cos (\Omega t + \phi) \tag{8.21b}$$

$$\ddot{y} = -R\,\Omega^2 \sin (\Omega t + \phi) \tag{8.21c}$$

The angle ϕ is called the *phase angle*. Comparing equations 8.21 with equations 8.7 previously derived analytically, we identify A with the radius R of the so-called auxiliary circle; we identify ω with the angular velocity Ω of the line OQ, and we identify B with the phase angle ϕ.

Finally we can plot a position - time curve as in figure 8.14 which shows how y varies sinusoidally with time, each cycle of the motion being repeated at time intervals of $2\pi/\Omega$.

Worked Example 8.5

A particle moves with simple harmonic motion in a straight line, the amplitude being 2 m and the frequency 15 Hz. Determine (a) its positions during one cycle when its velocity has magnitude 100 m/s, (b) its acceleration at these positions, and (c) the time taken to travel 3 m from one end position.

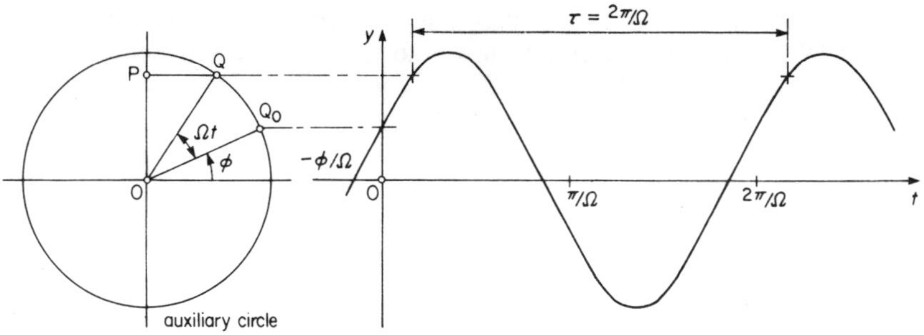

Figure 8.14

Solution

Frequency = $\omega/2\pi$, therefore

$$\omega = 2\pi \times 15 = 94.25 \text{ rad/s}$$

(a) In equations 8.7a

$$x = A \sin(\omega t + B)$$

$$\dot{x} = A\,\omega \cos(\omega t + B)$$

Therefore

$$\cos(\omega t + B) = \frac{\dot{x}}{A\omega} = \frac{100}{2 \times 94.25} \text{ or } \frac{-100}{2 \times 94.25}$$

(since the velocity changes direction during each cycle).

$$\cos(\omega t + B) = \pm 0.5305$$

and

$$(\omega t + B) = 58^\circ, 122^\circ, 238^\circ \text{ and } 302^\circ$$

$$\sin(\omega t + B) = 0.8477, 0.8477, -0.8477 \text{ and } -0.8477$$

and

$$x = 2\sin(\omega t + B) = 1.695 \text{ m}, 1.695 \text{ m}, -1.695 \text{ m and } -1.695 \text{ m}$$

when the velocity will be 100 m/s, − 100 m/s, − 100 m/s and 100 m/s.

(b)

$$\ddot{x} = -A\omega^2 \sin(\omega t + B) = -2 \times 94.2^2 \sin(\omega t + B)$$

$$= -15.06 \text{ km/s}^2, -15.06 \text{ km/s}^2, 15.06 \text{ km/s}^2 \text{ and } 15.06 \text{ km/s}^2$$

when

$$(\omega t + B) = 58^\circ, 122^\circ, 238^\circ \text{ and } 302^\circ$$

(c) The required time is best obtained by using an auxiliary circle similar to figure 8.13 with $OP_1 = 2$ m; then after moving 3 m the particle is at the point P where OP = 1 m. The angle θ is then $\sin^{-1} 1/2 = 30°$, and angle $P_1OQ = 120°$. For one cycle, $\tau = 1/f = 1/15$ s. Therefore, for the part cycle

$$\text{time taken} = \frac{120}{360} \times \frac{1}{15}$$
$$= 1/45 \text{ s}$$

8.5 Relative Motion of Particles

We shall shortly be discussing many-particle systems, in particular the motion of rigid bodies. As the first step towards that discussion we now consider the motion of two particles moving in a plane. We first confine their motion to a straight line and then allow the particles to move independently in curved paths.

8.5.1 Rectilinear Motion

Consider two particles A and B moving independently in the same straight line (figure 8.15).

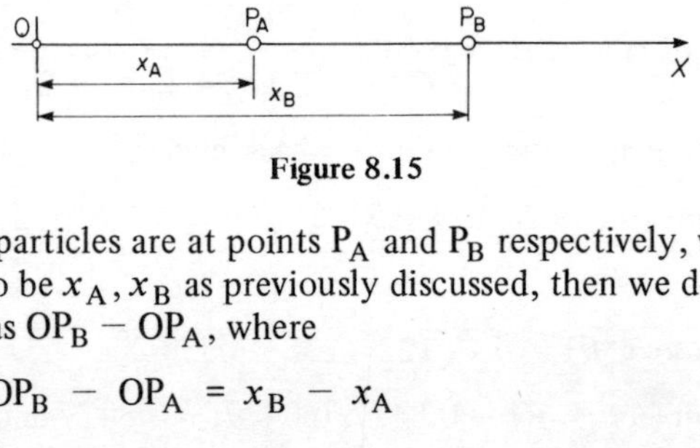

Figure 8.15

If at time t the particles are at points P_A and P_B respectively, where their positions are defined to be x_A, x_B as previously discussed, then we define the position of B *relative to* A as $OP_B - OP_A$, where

$$OP_B - OP_A = x_B - x_A$$

symbolised as ${}_Ax_B$.

Alternatively we can refer to the *relative position* of B with respect to A. ${}_Ax_B$ can also be regarded as the position of B as viewed by an observer travelling with A who relates the position of B to himself as a reference point.

The definitions of *relative velocity* and *relative acceleration* of B with respect to A follow immediately, their magnitudes being respectively

$${}_Av_B = v_B - v_A = \dot{x}_B - \dot{x}_A$$

$${}_Aa_B = a_B - a_A = \ddot{x}_B - \ddot{x}_A$$

the sense again being based on the choice of a positive x-direction.

The subscript notation for relative position, velocity and acceleration should be carefully noted since its use will be found to facilitate the solution of relative motion problems. A significant extension of the notation can be made if we imagine a third particle to be stationed at the reference point O, its position being x_O, which is now zero. An equivalent expression for x_B is now $x_B - x_O$, which can be written as a position relative to O in the form ${}_Ox_B$. Similarly we can write $x_A = {}_Ox_A$. We then have the position of B relative to A

$$ {}_Ax_B = x_B - x_A = {}_Ox_B - {}_Ox_A $$

and, after rearranging

$$ {}_Ox_B = {}_Ox_A + {}_Ax_B \tag{8.22a} $$

The ordering of the subscripts in this equation should be particularly noted, since the recognition of this ordering enables us to write five further relations each of which is equivalent to equation 8.22a. For example we can write two of these as follows

$$ {}_Ox_A = {}_Ox_B + {}_Bx_A $$

$$ {}_Ax_O = {}_Ax_B + {}_Bx_O $$

with proper regard to signs.

If we differentiate equation 8.22a twice we obtain

$$ {}_Ov_B = {}_Ov_A + {}_Av_B \tag{8.22b} $$

$$ {}_Oa_B = {}_Oa_A + {}_Aa_B \tag{8.22c} $$

in which the cyclic ordering of the subscripts is apparent. Again these relationships can be written in equivalent ways by proper re-arrangement of subscripts.

8.5.2 Curvilinear Motion: Translating Axes

Consider a particle B that is moving along a path in a plane 2, which is attached to some other particle A (figure 8.16a). B is then said to be moving relatively to A,

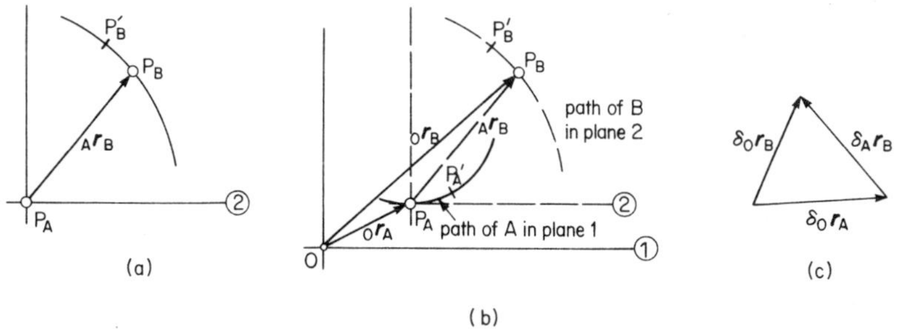

Figure 8.16

and P_A is a reference point for the position of B. Then the position of B relative to A is ${}_A\boldsymbol{r}_B$, the velocity of B relative to A is ${}_A\boldsymbol{v}_B$ tangential to the path and the acceleration of B relative to A is ${}_A\boldsymbol{a}_B$, not usually tangential to the path.

Suppose now that particle A is moving along a path in a plane 1 and that plane 2 attached to particle A moves with it but is not allowed to rotate figure 8.16b. Plane 2 is said to be in *translation* relative to plane 1. This implies that a pair of axes fixed in plane 2 retain their orientations as P_A moves along the path of particle A.

From the figure we see immediately that at time t

$$ {}_O\boldsymbol{r}_B = {}_O\boldsymbol{r}_A + {}_A\boldsymbol{r}_B \tag{8.23a} $$

where ${}_O\boldsymbol{r}_B$ is the position of B in plane 1.

In the subsequent short time interval δt, particle A moves to $P_A{}'$ in plane 1, and particle B moves to $P_B{}'$ in plane 2, the displacements being respectively $\delta {}_O\boldsymbol{r}_A$ and $\delta {}_A\boldsymbol{r}_B$. The total displacement of B in plane 1 is given by the vector addition of these two displacements (figure 8.16c) and $\delta {}_O\boldsymbol{r}_B = \delta {}_O\boldsymbol{r}_A + \delta {}_A\boldsymbol{r}_B$. Dividing by δt, then in the limit as $\delta t \to 0$

$$ {}_O\boldsymbol{v}_B = {}_O\boldsymbol{v}_A + {}_A\boldsymbol{v}_B \tag{8.23b} $$

in which the velocity vectors are in the same directions as the displacement vectors. Thus, if the paths of A in plane 1 and B in plane 2 are known, ${}_O\boldsymbol{v}_A$ and ${}_A\boldsymbol{v}_B$ are tangential to those paths and, by vector addition, we obtain the velocity of B in plane 1, ${}_O\boldsymbol{v}_B$, in magnitude and direction (figure 8.17a).

Without setting out the argument in detail it can be seen that a corresponding relationship can be set down for the accelerations (figure 8.17b).

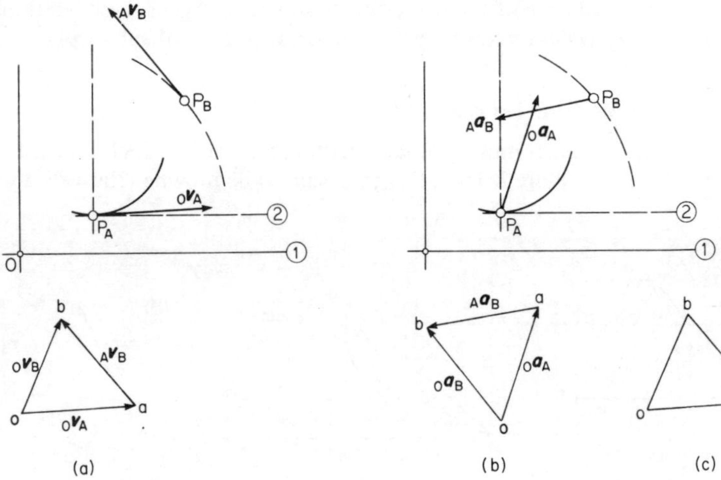

Figure 8.17

The directions of ${}_O\boldsymbol{a}_A$ and ${}_A\boldsymbol{a}_B$ are not tangential to the paths unless the paths are straight lines.

Figure 8.17 illustrates a further facility afforded by the subscript notation. The ends of the vectors in the vector triangles carry lower-case letters, which correspond to the vector being represented. Thus the vector ${}_A v_B$ is represented by the line segment a ——— b, the direction being that from a to b. The same line segment also represents the vector ${}_B v_A$, the direction now being that from b to a. The vector addition of the relative velocities in figure 8.17a can now be displayed as in figure 8.17c, which not only displays the relationship of equation 8.23b but also displays the remaining five equivalent relationships that can be written, for example

$$ {}_B v_O = {}_B v_A + {}_A v_O $$

The same principle is applicable to the graphical summation of relative accelerations.

It is again emphasised that plane 2 must not be allowed to rotate. If rotation also occurred then the expressions for the acceleration of B relative to O, ${}_O a_B$, would contain additional terms arising from the rotation of the path of B. These additional terms are considered in the section that follows.

Worked Example 8.6

At time 14.00 h a ship B is sailing on a N-E course at 12 km/h when it sights another ship A at a position 10 km N 60°E, which is sailing on an E-W course at 15 km/h. Ship B continues on its course until ship A is directly ahead; it then changes course in order to rendezvous with ship A as soon as possible.

(a) Determine the time at which B changes course and its distance from A at that time,
(b) the new course taken by B,
(c) the time at which the two ships meet.

Plot the path of B relative to A and verify that the time taken by B to traverse this relative path agrees with the time determined in (c) above. Disregard the curvature of the surface of the Earth.

Solution

Figure 8.18a gives a *space diagram* to illustrate the solution. If we take plane 1 to be the Earth's surface, then, relative to B, A is moving in plane 2 attached to B, plane 2 being defined by the N-S, E-W directions at B.

Then

$$ {}_E v_A = {}_E v_B + {}_B v_A $$

$$ 15 \leftarrow = 12 \angle 45^\circ + {}_B v_A $$

in which ${}_B v_A$ is not yet known. (Subscript E refers to the Earth.)

This equation is represented graphically in the vector diagram, figure 8.18b, from which ${}_B v_A$ can be determined by measurement (or calculation) of the magnitude and direction of the line ba. Similarly ${}_A v_B$ is represented by ab. The path of A relative to B and parallel to ${}_B v_A$ is then indicated in the space diagram.

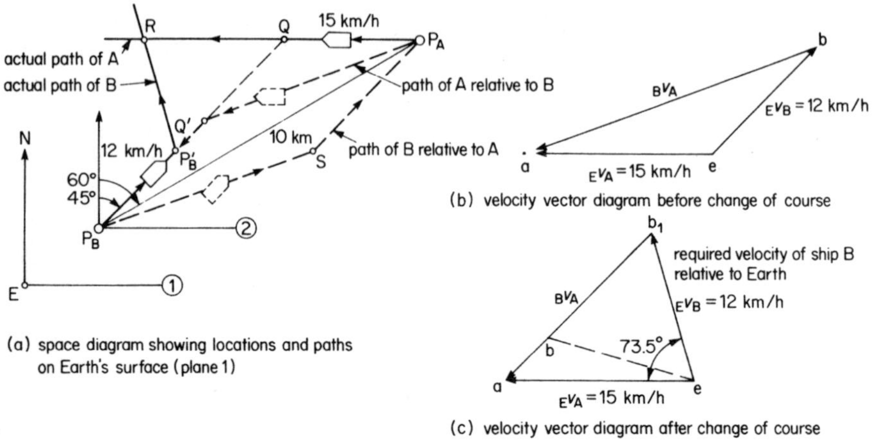

Figure 8.18

As viewed from B, A travels along its relative path and is directly ahead of B when it reaches point Q′ on the relative path, or alternatively, is at point Q on the actual path

$$\text{Elapsed time} = \frac{P_A Q'}{{}_B v_A} \text{ or } \frac{P_A Q}{{}_E v_A} = \frac{3.66}{15} \text{ h} = 14.6 \text{ min}$$

and B changes course at 14 h 14.6 min.

Distance of A from B at this time is $P_B Q' = 4.13$ km, by measurement.

During these 14.6 min, ship B will have travelled to a point P_B', a distance 14.6 (min) × 12 (km/h) = 2.93 km. From this point onwards, in order that the ships may meet it is necessary that the velocity of A relative to B be in the direction QP_B' (or $Q'P_B$) and we write

$$_E v_A = {}_E v_B + {}_B v_A$$

$$15 \leftarrow\; = 12 \angle ? + {}_B v_A \angle -135^\circ$$

This equation contains sufficient information to construct the vector diagram, figure 8.18c, and thereby obtain the required direction for $_E v_B$. We find that two directions are possible, and we select by inspection the one that gives the shorter time, namely eb_1, where $\angle aeb_1 = 73.5^\circ$.

The new course for ship B is therefore N 16.5° W and the ships meet at the point R (see figure 8.18a).

The total elapsed time is

$$\frac{P_A R}{{}_E v_A} = \frac{7.45}{15} \text{ h} = 29.8 \text{ min}$$

and the ships meet at 14 h 29.8 min.

The path of B as viewed from A is as shown in the space diagram. From P_B to S, ${}_A v_B$ is given by ab in the vector diagram, figure 8.18b, and from S to P_A, ${}_A v_B$ is given by ab_1 in the vector diagram figure 8.18c.

The total time for B to travel the relative path is

$$\frac{P_BS}{ab} + \frac{SP_A}{ab_1} = \frac{6.12}{25.0} + \frac{4.13}{16.2}\ \text{h}$$

$$= (14.7 + 15.3)\ \text{min}$$

$$= 30\ \text{min}$$

(The difference between this value and the previous value of 29.8 is due to the graphical construction.)

8.5.3 Curvilinear Motion: Rotating Axes

We now allow plane 2 of the previous section to rotate about an axis through particle A, the axis being perpendicular to the plane. For convenience in description we now signify the point occupied by particle A in plane 2 by O′ (figure 8.19a). Initially we shall assume particle A to be at rest in plane 1 but later we

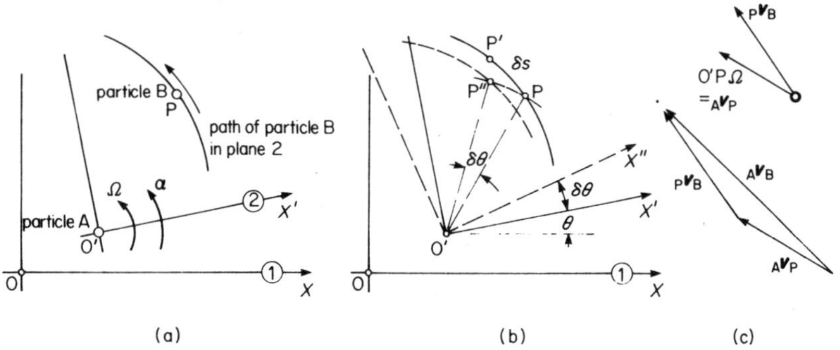

Figure 8.19

shall allow it to move, carrying with it the rotating plane 2. A fixed reference direction OX is taken through O, and the angular position of plane 2 is then defined by an axis $O'X'$ fixed in plane 2 and rotating with it with angular velocity Ω and angular acceleration α relative to the axis OX, which is stationary in plane 1. Particle B travels as before along its path which is now fixed in the rotating plane 2 and rotates with it.

We wish to determine (1) the velocity and (2) the acceleration of particle B as seen by an observer stationed with particle A at O′, who refers all motion to the reference direction defined by OX.

(1) Velocity

At time t the particle B will be at some point P on its path, the point P being that point of plane 2 with which particle B happens to be coincident at time t. This we call the *coincident point* of particle B. In the subsequent time interval δt the particle will have moved in its path to a neighbouring point P′ (figure 8.19b). We have already discussed this motion and concluded that the velocity of the particle B relative to P has magnitude $\mathrm{d}s/\mathrm{d}t = {}_{\mathrm{P}}v_{\mathrm{B}}$ and is tangential to the path. However, due to the rotation of the plane, the point P moves in a circular path with radius O′P to P″, through a distance $(\mathrm{O'P})\delta\theta$ in time δt. It has itself therefore, a velocity of magnitude $(\mathrm{O'P})\,\mathrm{d}\theta/\mathrm{d}t = (\mathrm{O'P})\Omega$ relative to $\mathrm{O}X$, in a direction perpendicular to O′P. The total velocity of particle B relative to $\mathrm{O}X$ therefore has two components, one tangential to the path at P with magnitude ${}_{\mathrm{P}}v_{\mathrm{B}}$ and the other perpendicular to O′P with magnitude $(\mathrm{O'P})\Omega = {}_{\mathrm{A}}v_{\mathrm{P}}$.

We can now write, for the total velocity ${}_{\mathrm{A}}\boldsymbol{v}_{\mathrm{B}}$ of particle B as viewed from particle A at O′

$${}_{\mathrm{A}}\boldsymbol{v}_{\mathrm{B}} = {}_{\mathrm{A}}\boldsymbol{v}_{\mathrm{P}} + {}_{\mathrm{P}}\boldsymbol{v}_{\mathrm{B}} \tag{8.24}$$

These components are shown in figure 8.19c.

(2) Acceleration

(i) When particle B is at the coincident point P it will have two components of acceleration as viewed in plane 2, one tangential to the path with magnitude $\mathrm{d}({}_{\mathrm{P}}v_{\mathrm{B}})/\mathrm{d}t$ and the other normal to the path with magnitude $({}_{\mathrm{P}}v_{\mathrm{B}})^2/\rho$ as previously discussed. For convenience we write ${}_{\mathrm{P}}v_{\mathrm{B}}$ as v, it being understood that v is the speed of particle B in its path. We then write the above components in the simpler forms $\mathrm{d}v/\mathrm{d}t$ and v^2/ρ. These are shown in figure 8.20b.

(ii) Due to the rotation of the plane the coincident point P is travelling in a circular path and therefore has itself two components of acceleration, one perpendicular to O′P with magnitude $\mathrm{O'P}\alpha$ and the other directed towards O′ with magnitude $(\mathrm{O'P}\Omega)^2/\mathrm{O'P} = \mathrm{O'P}\Omega^2$. These are shown in figure 8.20a.

It might be concluded that the total acceleration of particle B as viewed from particle A is the vector sum of these four components. Such a conclusion would be premature since we have not taken into account all the velocity changes that are occurring.

(iii) The two acceleration components referred to in (i) above are the result of the changes in magnitude and direction respectively of the velocity $\boldsymbol{v}$, the change in direction being due to *curvature* of the path. We have not yet taken into account the *rotation* of the path as a whole. Because of this rotation the orientation of the path at P is changing, and therefore the velocity vector $\boldsymbol{v}$ swings through an additional angle $\delta\theta$ in the time interval δt. The angle $\delta\theta$ is the angular displacement of plane 2 in time δt and is additional to the angle $\delta\psi$ previously used to obtain the component v^2/ρ arising from the curvature of the path. Particle B therefore undergoes an additional velocity change having magnitude $v\delta\theta$ in time δt, giving rise to an additional acceleration component having magnitude (as δt tends to zero) $v\,\mathrm{d}\theta/\mathrm{d}t = v\Omega$ in a direction perpendicular to that of $\boldsymbol{v}$, that is, along

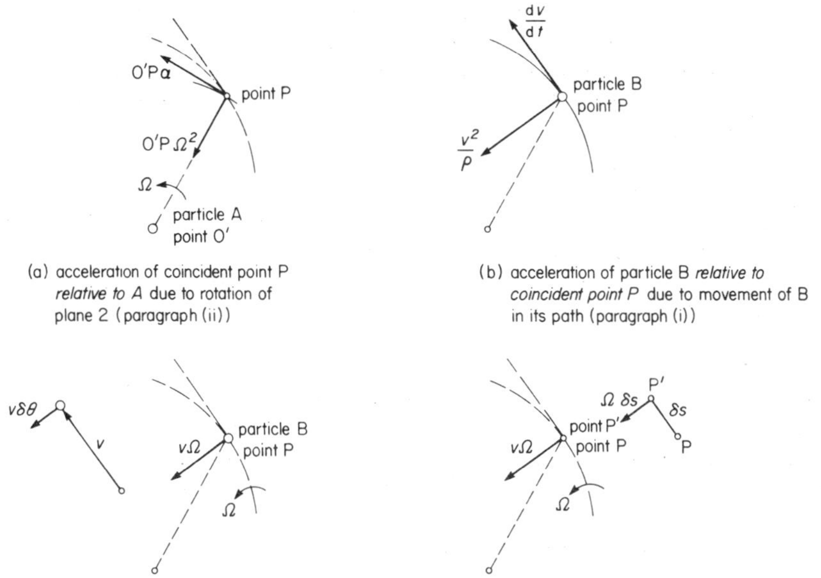

(a) acceleration of coincident point P *relative to A* due to rotation of plane 2 (paragraph (ii))

(b) acceleration of particle B *relative to coincident point P* due to movement of B in its path (paragraph (i))

(c) acceleration of particle B *relative to coincident point P* due to rotation of the path (paragraph (iii))

(d) acceleration of coincident point P′ of particle B *relative to coincident point P* due to rotation of the path (paragraph (iv))

Figure 8.20

the normal to the path at P. This component is shown in figure 8.20c.

(iv) In paragraph (ii) above we have taken into account the acceleration of the point P arising from the rotation of plane 2. We recall that P is that point of plane 2 with which particle B happens to be coincident at time t. It is also that point of the plane with respect to which the particle B has the acceleration components referred to in paragraph (i) and is the point at which those components are defined. Any other point would serve equally well provided it was not moving relative to P. However, since plane 2 is now rotating, all other points of plane 2 have different velocities and accelerations, and in its motion the particle B moves to coincident points whose velocities are different from that of P.

Now after a short time interval δt the coincident point of B will be P′ as noted above (figure 8.19b) and P′ has a velocity relative to P arising out of the rotation of the segment PP′ (assumed straight) through an angle $\delta\theta$, with plane 2. This relative velocity has magnitude $\Omega\delta s$ (with δs small) in a direction perpendicular to PP′, that is, along the normal to the path at P. In time δt the particle has therefore acquired a further velocity $\Omega\delta s$ relative to P, and accordingly has an additional acceleration component normal to its path (in the limit as δt tends to zero) of magnitude $(\Omega\delta s)/\delta t = (\mathrm{d}s/\mathrm{d}t)\Omega = v\Omega$. This component is shown in figure 8.20d. (There will also be a centripetal component of magnitude $(\Omega\delta s)^2/\delta s = \Omega^2\delta s$ directed towards P, but this becomes zero as we proceed to the limit.)

The rotation of the path – see paragraph (iii) – and the change in the coincident point – paragraph (iv) – together give rise to an acceleration component of magnitude $2v\Omega$ directed along the normal to the path. The recognition of this component was due to G. Coriolis (1792 - 1843) and is accordingly referred to as the Coriolis component.

The total acceleration of particle B is the resultant of the components shown in figure 8.20: the resultant of the components shown in figure 8.20a we symbolise as ${}_{A}a_{P}$; the resultant of the components shown in figure 8.20b we symbolise as ${}_{P}a'_{B}$; the Coriolis component resulting from the components shown in figures 8.20c and d we symbolise as ${}_{P}a''_{B}$, or simply a_C, and we write, for the total acceleration of particle B as viewed from particle A

$$ {}_{A}a_{B} = {}_{A}a_{P} + {}_{P}a'_{B} + a_{C} \tag{8.25} $$

where P is the coincident point of particle B and a_C is the Coriolis component (magnitude $2v\Omega$) of the acceleration of B with respect to P.

Note that the direction of the Coriolis component of the total acceleration is given by rotating the direction of the tangential velocity through 90° in the same sense as that of the angular velocity Ω of plane 2.

To complete the analysis of the motion of particle B we now allow particle A (with the attached rotating plane 2) to move in plane 1 with velocity ${}_{O}v_{A}$ and acceleration ${}_{O}a_{A}$. The complete expressions for the velocity and acceleration of particle B in plane 1 now become

$$ {}_{O}v_{B} = {}_{O}v_{A} + {}_{A}v_{P} + {}_{P}v_{B} \tag{8.26a} $$

$$ {}_{O}a_{B} = {}_{O}a_{A} + {}_{A}a_{P} + {}_{P}a'_{B} + a_{C} \tag{8.26b} $$

For example, consider the case of a particle moving around the perimeter of a stationary circular disc radius R, with constant speed v in an anticlockwise direction. We have already found that at any point of its path the particle has acceleration v^2/R directed towards the centre of the disc. Suppose the disc is now made to rotate in its plane about an axis through the centre in an anticlockwise direction with angular velocity Ω, which we choose to be equal to v/R. The peripheral speed of the coincident point of the particle at any time is then $R\Omega = v$ and its acceleration is centripetal with magnitude v^2/R. One might be tempted to say that the total acceleration of the particle has magnitude $2v^2/R$, but this is evidently incorrect since the particle is clearly moving in a circular path with speed $2v$ and the acceleration has therefore magnitude $(2v)^2/R = 4v^2/R$. The mistake arises because of the omission of the Coriolis acceleration component, magnitude $2v\Omega = 2v^2/R$, which in this case is also directed towards the centre of the disc.

Worked Example 8.7

In figure 8.21a a small body B can slide along the rod OQ, which is rotating about the end O with angular velocity $\dot{\theta}$ and angular acceleration $\ddot{\theta}$, in the anticlockwise sense. When at the point P on the rod the velocity and acceleration of the body relative to the rod are respectively $\dot{r}$ and $\ddot{r}$, where $r = OP$. The body thereby traces

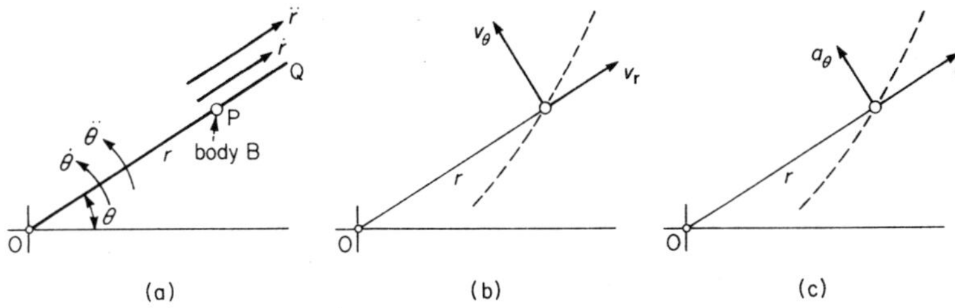

Figure 8.21

a curved path in the reference plane XOY. Obtain expressions for the radial and transverse components of the velocity and acceleration of the body in the reference plane.

Solution

From equation 8.24

$$ {}_{O}v_{B} = {}_{O}v_{P} + {}_{P}v_{B} $$

in which ${}_{O}v_{P} = r\dot{\theta}$ in the transverse direction ${}_{P}v_{B} = \dot{r}$ in the radial direction, therefore ${}_{O}v_{B}$ has the components $v_r = \dot{r}$ and $v_\theta = r\dot{\theta}$. These components are shown in figure 8.21b, which should be compared with the appropriate diagram in figure 8.10.

From equation 8.25

$$ {}_{O}a_{B} = {}_{O}a_{P} + {}_{P}a'_{B} + a_{C} $$

in which ${}_{O}a_{P}$ has two components, $r\ddot{\theta}$ in the transverse direction and $-r\dot{\theta}^2$ in the radial direction

$$ {}_{P}a'_{B} = \ddot{r} \text{ in the radial direction} $$

$$ a_{C} = 2\dot{r}\dot{\theta} \text{ in the transverse direction} $$

therefore ${}_{O}a_{B}$ has the components $a_r = \ddot{r} - r\dot{\theta}^2$ and $a_\theta = r\ddot{\theta} + 2\dot{r}\dot{\theta}$. These components are shown in figure 8.21c, which should be compared with the appropriate diagram in figure 8.10. Note that the path of the body in the rotating plane is a straight line, and that the Coriolis component is perpendicular to this line.

Worked Example 8.8

The vertical plate shown in figure 8.22a is hinged at one corner and has a narrow slot LM cut in it; the centreline of LM is an arc of a circle with centre Q. A small peg B can slide in the slot. At the instant when the bottom edge of the plate passes through the horizontal its angular velocity is 10 rad/s anticlockwise and its angular acceleration is 50 rad/s^2 clockwise. At the same instant the peg is at the point P of the slot and moving towards L with speed 5 m/s and acceleration 20 m/s^2 relative to the slot.

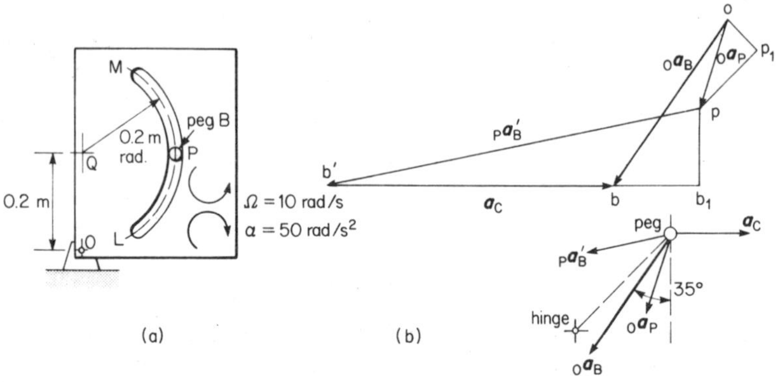

Figure 8.22

Determine the total acceleration of the peg in the vertical plane.

Solution

$$_{O}a_{B} = {_{O}a_{P}} + {_{P}a'_{B}} + a_{C}$$

$_{O}a_{P}$ has two components, $OP\alpha$ perpendicular to OP and $OP\Omega^2$ along PO

$$OP\alpha = (0.2 \times \sqrt{2}) \times 50 = 10\sqrt{2}\ \text{m/s}^2 \angle -45^\circ$$

$$OP\Omega^2 = (0.2 \times \sqrt{2}) \times 10^2 = 20\sqrt{2}\ \text{m/s}^2 \angle -135^\circ$$

These components are represented by op_1 and p_1p in the graphical solution shown in figure 8.22b, their resultant being represented by **op**.

$_{P}a'_{B}$ has two components, a_t tangential and a_n normal to the slot at P.

$$a_t = 20\ \text{m/s}^2 \downarrow$$

$$a_n = \frac{{_{P}v_{B}}^2}{QP} = \frac{5^2}{0.2} = 125\ \text{m/s}^2 \leftarrow$$

These components are represented by pb_1 and b_1b' and their resultant by **pb′**. a_C is perpendicular to $_{P}v_{B}$ and the direction is given by rotating $_{P}v_{B}$ in the same sense as Ω

$$a_C = 2\,{_{P}v_{B}}\Omega = 2 \times 5 \times 10 = 100\ \text{m/s}^2 \rightarrow$$

and a_C is represented by **b′b**.

We now determine the resultant of these vector components, this being represented by the vector **ob**. The total acceleration of the peg is thus found to be $61\ \text{m/s}^2 \angle -125^\circ$.

8.6 Summary

(1) If the position of a particle on the axis OX at time t is $x(t)$ then its velocity $v = \mathrm{d}x/\mathrm{d}t$ and its acceleration $a = \mathrm{d}v/\mathrm{d}t = \mathrm{d}^2x/\mathrm{d}t^2$, the positive sense being that of increasing x. These may also be written $v = \dot{x}$; $a = \dot{v} = \ddot{x}$.
(2) If s is a distance measured along a curved path, the velocity and acceleration along this path are $\mathrm{d}s/\mathrm{d}t$ and $\mathrm{d}^2s/\mathrm{d}t^2$ respectively.
(3) a may be expressed as

$$a = \frac{\mathrm{d}v}{\mathrm{d}t} = v\frac{\mathrm{d}v}{\mathrm{d}x} = \frac{\mathrm{d}}{\mathrm{d}x}\left(\frac{v^2}{2}\right) \qquad (8.2)$$

(4) Simple harmonic motion is defined by the equation

$$\frac{\mathrm{d}^2x}{\mathrm{d}t^2} + \omega^2 x = 0 \qquad (8.4)$$

where the periodic time

$$\tau = \frac{2\pi}{\omega} \qquad (8.6)$$

and the frequency

$$f = \frac{\omega}{2\pi}$$

The solution to the differential equation 8.4 is

$$x = A \sin(\omega t + B) \qquad (8.5)$$

A and B being determined from the initial conditions.
(5) In plane curvilinear motion the pairs of velocity and acceleration components for the respective coordinate directions are
polar

$$v_r = \dot{r} \qquad v_\theta = r\dot{\theta} \qquad (8.8)$$

$$a_r = \ddot{r} - r\dot{\theta}^2 \qquad a_\theta = r\ddot{\theta} + 2\dot{r}\dot{\theta} \qquad (8.13)$$

rectangular

$$v_x = \dot{x} \qquad v_y = \dot{y} \qquad (8.9)$$

$$a_x = \ddot{x} \qquad a_y = \ddot{y} \qquad (8.12)$$

intrinsic

$$v_\mathrm{t} = \dot{\mathrm{s}} \qquad v_\mathrm{n} = 0 \qquad (8.10)$$

$$a_\mathrm{t} = \ddot{\mathrm{s}} \qquad a_\mathrm{n} = v_\mathrm{t}^2/\rho \qquad (8.11)$$

(6) For a particle moving in a circular path

$$s = R\theta \qquad (8.16)$$

$$v = R\Omega$$

$$a_t = R\alpha \qquad (8.17)$$

$$a_n = R\Omega^2 \qquad (8.18)$$

(7) The relationships between position x, speed v and acceleration a for the case of rectilinear motion can be formally repeated for angular motion of a line, with the substitution of θ, Ω and α for x, v and a respectively.
(8) For relative motion in non-rotating reference frames

$${}_{\text{O}}v_{\text{B}} = {}_{\text{O}}v_{\text{A}} + {}_{\text{A}}v_{\text{B}} \qquad (8.22a)$$

$${}_{\text{O}}a_{\text{B}} = {}_{\text{O}}a_{\text{A}} + {}_{\text{A}}a_{\text{B}} \qquad (8.22b)$$

where ${}_{\text{Y}}v_{\text{X}}$ and ${}_{\text{Y}}a_{\text{X}}$ represent generally the velocity and acceleration of X with respect to, or as seen from, Y.
(9) For problems involving relative motion, relative velocities and relative paths are best determined by graphical interpretation of equation 8.22a.
(10) Relative motion in a rotating reference plane in which a path is rotating at a speed Ω involves a Coriolis component of acceleration $\boldsymbol{a}_{\text{C}}$, having magnitude $2\Omega v$ and direction that of $\boldsymbol{v}$ turned through 90° in the sense of Ω. In equation 8.22a the term ${}_{\text{A}}v_{\text{B}}$ now becomes

$${}_{\text{A}}v_{\text{B}} = {}_{\text{A}}v_{\text{P}} + {}_{\text{P}}v_{\text{B}}$$

and in equation 8.22b the term ${}_{\text{A}}a_{\text{B}}$ now becomes

$${}_{\text{A}}a_{\text{B}} = {}_{\text{A}}a_{\text{P}} + {}_{\text{P}}a'_{\text{B}} + a_{\text{C}}$$

where P is the coincident point of B.

Problems

8.1 The maximum possible acceleration of a train is 1.5 m/s², its maximum speed is 30 m/s and its maximum retardation 4.5 m/s². What is the shortest time, from rest to rest, in which it can cover (a) 300 m (b) 1000 m? (*Hint*: Use a velocity - time graph.)

8.2 A train moves off from station A with constant acceleration 1 m/s² and attains a top speed of v m/s, which it maintains for 9.6 km until it reaches station B, 10.05 km from A. The brakes are then applied, causing a constant retardation a m/s² and as soon as the train comes to rest it is reversed to station B, first with constant acceleration 1m/s², and then at constant retardation of 5 m/s². If it comes to rest at B 422 s after it left A find v and a and the distance the train overshot B before reversing. (*Hint*: Use a velocity - time graph.)

8.3 A train starts from rest and for the first kilometre moves with constant acceleration, for the next 3 km it has constant velocity and for another 2 km it moves with constant retardation to come to rest after 10 min. Draw the velocity – time graph and find (a) the maximum speed and (b) the three time intervals.

8.4 The motion of a particle is described by the equation $s = (10v^2 + 50v)$ m.

(a) Will it have a maximum velocity?
(b) Will it have a maximum acceleration?
(c) Find the time taken for the velocity to change from 2 to 10 m/s.
(d) Find the distance covered in this period.
(e) What is the acceleration at the beginning and end of the period?

(*Hint*: Differentiate the equation and use relationships given in section 8.1.2.)

8.5 The motion of a particle in a straight line is described by $a = (5 - 0.05v^2)$ m/s^2.

(a) What is the final velocity of the particle?
(b) If the initial velocity when $t = 0$ is 4 m/s find the time required and distance covered to reach 8 m/s.
(c) If the initial velocity is 20 m/s what happens to the velocity of the particle?

(*Hint*: Use the relationships given in section 8.1.2.)

8.6 A particle moves in a straight line with velocity $v = (50 - 2t + 0.01t^2)$ m/s, where t is the time in seconds.

Sketch a velocity – time graph and calculate the times when the particle is instantaneously (a) at rest and (b) has no acceleration.
(c) What are the values of acceleration in case (a), and (d) what is the velocity in case (b)?
(e) Find the distances from its initial position at which the particle comes instantaneously to rest.
(f) What is the total distance the particle travels in a negative direction?
(g) Find the values of t at which the particle passes again through its original position.
(h) What is the total length of its path from $t = 0$ to the instant when is passes through its initial position for the last time? (*Hint*: Only definitions of displacement, velocity and acceleration are required.)

8.7 A particle has component velocities $v_y = 20 \sin 20t$ m/s and $v_x = 10 \sin (10t + 45°)$ m/s. At time $t = 0, y = x = 0$.

Find, for $t = 0.1$ s, the absolute velocity and acceleration of the particle and its displacement from the origin. (*Hint*: Find component accelerations and displacements by using relationships in section 8.2.1.)

8.8 A particle has component velocities $v_y = t$ m/s and $v_x = 5 - 3t$ m/s. At time $t = 0, y = x = 0$.

Find equations for x and y and thus deduce an expression for the distance r of the particle from the origin.

(a) Find the farthest distance the particle reaches from the origin in the time period 0 to 2.6 s.
(b) At what time does this occur?
(c) State the significance of the instant $t = 2.5$ s.
(d) Find the absolute velocity and acceleration of the particle when $t = 2.0$ s.

8.9 The position of a particle is described in polar coordinates as $r = 5\theta$. If its velocity along its path is constant at 5 m/s show that after 2 s its θ-coordinate is given by an equation that is satisfied by $\theta = 1.53$ rad; hence give the position of the particle at that instant.

Find also the absolute velocity and absolute acceleration at the same instant. (*Hint*: Deduce an equation for s (see the appendix) in terms of θ, hence, or otherwise, relate $\dot{s}$ and $\dot{\theta}$ and $\ddot{s}$ and $\ddot{\theta}$. Use equations relating to polar coordinates to find the absolute velocity and acceleration.)

8.10 A particle is moving along a horizontal smooth surface with constant velocity V when it impinges tangentially on to an upward curved surface of radius R. If the velocity in the horizontal (x) direction is maintained constant by the application of a suitable force, show that the acceleration in the y direction is

$$a_y = \frac{V^2}{R}\left(\frac{1}{\cos^3\phi}\right)$$

where ϕ is the angle subtended at the centre by the arc along which the particle has travelled. (*Hint*: Write equations for x and y measured from the point of impingement (or use any other set of components); differentiate for velocity and acceleration or use other component sets; use conditions on x values to solve.)

8.11 A particle describes SHM having amplitude 0.5 m and frequency 10 Hz.

(a) Find its maximum velocity and maximum acceleration.
(b) Find its velocity and acceleration when travelling away from the centre and at a distance of 0.2 m from one end of its travel.
(c) Find the time taken to travel from 0.1 m from one end position to the other extreme position and back to within 0.2 m of the original position.

8.12 A rigid link in hinged at one end. It performs simple harmonic motion in an angular fashion with amplitude 30° and period 0.5 s.

(a) Determine the maximum angular velocity of the link.
(b) Find its maximum angular acceleration.
(c) Find its angular velocity and angular acceleration when at 10° to the mean position and travelling away from the centre.

(*Hint*: Write $\theta = A \sin \omega t$, where θ is the angle of the link; differentiate for angular velocity and angular acceleration.)

8.13 A particle moving with SHM passes through two points A and B 0.3 m apart with the same velocity having taken 2 s to pass directly from A to B. After another 4 s it passes through B in the opposite direction. Find the period and amplitude of oscillation.

8.14 If $x = a \sin(\omega t + \phi)$ where a, ω and ϕ are constants, show that $\ddot{x} = -\omega^2 x$. Hence solve $\ddot{x} = -16x$, given that when $t = 0$, $x = 4$ and $\dot{x} = 20$.

8.15 A particle moves along a path so that its rectangular coordinates are $x = 2 \sin \theta$, $y = 3(1 - \cos \theta)$. The motion is such that $x = \sin(20t)$ m, t being the time in seconds. Find (a) the distance from the origin, (b) the absolute velocity and (c) the absolute acceleration of the particle when $t = 2\pi/50$ s. (*Hint*: Deduce equations for velocity and acceleration by differentiation; use the two relationships for x and solve the resulting equations.)

8.16 Two straight roads cross at right angles at an intersection, car A is on one road and B on the other. If car A travelling at 40 m/s passes the intersection 0.5 s after car B and its nearest distance to B (at any time) is 10 m, what is the speed of B?

How far is B from the intersection when A is 100 m before the intersection? (*Hint*: determine the path of B relative to A when the former is at the intersection (the position of A is known at that instant).)

8.17 Boat A leaves port X and sails at 25 km/h due west. Boat B leaves half an hour later from port Y which is 30 km south west of X and sails at 20 km/h north west. At what time (after A leaves port) will they be nearest together and how far apart will they then be?

For what duration of time are the boats within 10 km of each other? (*Hint*: Consider relative motion from the time when B leaves port.)

8.18 A pilot boat leaves port to intercept a tanker which is sailing at 30 km/h on a straight course whose nearest point is 5 km from the port. At the instant the pilot boat leaves port the tanker is 8 km away.

(a) At what minimum speed must the pilot boat sail in order to intercept the tanker?
(b) If the pilot boat sails at 20 km/h for how long is the tanker in a position to be intercepted by the pilot boat.

(*Hint*: (a) relative velocity of the pilot boat to tanker must lie along the 8 km path. (b) There are two possible relative velocities; calculate time to reach the tanker for each.)

9 Kinetics of Particles and Particle Systems I – Equations of Motion

In the last chapter we sought to establish methods of describing the motion of a particle, so giving us the ability to predict the motion of the particle if the equations of motion were known and the initial conditions specified. We now have to consider how to determine these equations of motion.

The characteristics of the motion of an engineering system are entirely determined by the forces acting on the system. On this basis the engineer is able to predict the mechanical behaviour of systems, and to assemble systems that behave in a predetermined manner. The mechanical behaviour of a system must therefore be related at all times to the forces acting, and, furthermore, an analysis of the motion of a system that omits any relevant force, or neglects any such force without justification – based on the principles of mechanics – is bound to be defective.

9.1 Equations of Motion for a Particle

The fundamental relationships between force and motion were first clearly enunciated by Isaac Newton in his three laws of motion. We have already encountered the first and third laws in our study of statics; the first law associated with zero resultant force a state of either no motion or a state of rectilinear motion at constant speed; the third law enabled us to extend the first law for a particle to an assembly of particles, the particle system.

We now consider the implications of the second law, which we state in the following form.

The Second Law

If the resultant force on a particle is not zero then the particle is accelerated, the magnitude of the acceleration being proportional to the magnitude of the resultant force ΣF and the direction being that of the resultant force.

For a given particle at any instant the law can be written

$$\Sigma \boldsymbol{F} = \text{constant} \times \boldsymbol{a} \qquad (9.1)$$

For different particles the proportionality constant has in general a different value for each particle. For a resultant force of given magnitude, the larger the constant the smaller the magnitude of the acceleration produced. The proportionality constant is evidently a measure of some property of a particle that determines the acceleration of the particle in response to a resultant force. This property is called the *inertial mass* of the particle, or simpy the *mass.*

This property is additive in the sense that a single particle made up of two particles has mass that is the sum of the masses of the individual particles. We can therefore choose a particular particle and declare that its mass is to be regarded as a unit of mass. Any other particles can then be ascribed a mass m, and the law can be expressed in a form applicable to all particles as

$$\Sigma F = km\boldsymbol{a} \tag{9.2}$$

in which ΣF the resultant force, m the mass, and $\boldsymbol{a}$ the acceleration can be measured in independent arbitrarily defined units. The constant k takes a numerical value that will now depend only on the units selected.

9.1.1 Coherent Units

To avoid the continual use of a proportionality constant it is convenient to adopt a set of units that will enable k to have a numerical value of unity. If units of length, time and mass are chosen initially then the unit of force can be defined as that force which gives unit acceleration to a particle having unit mass. This is the method adopted for deriving a force unit based on the set of primary SI units.

The SI unit of mass is the kilogram (kg), this being the mass of the so-called International Prototype Kilogram which is in the custody of the Bureau International des Poids et Mesures, at Sèvres, near Paris, and with which other standards can be compared directly or indirectly. Having adopted the kilogram as the unit mass, the unit force is then defined to be that force which, when applied to a body having mass 1 kg, gives it an acceleration of 1 m/s^2. This derived unit of force is named the newton (N), and can be expressed equivalently as 1 kg m/s^2.

The mathematical statement of the second law now becomes

$$\Sigma F = m\boldsymbol{a} \tag{9.3}$$

it being understood that the units of force, mass, length and time are respectively N, kg, m and s (or some other coherent set of units so defined as to enable k to have the value unity).

9.1.2 The Inertial Frame

We now ask if the simple relation between force and acceleration for a given particle is invariably satisfied in practice. The answer is that it is not. The justification of this statement reveals the necessity of supplementing the second law statement by an important condition, which must be fulfilled if the relation $\Sigma F = m\boldsymbol{a}$ is to retain its validity.

We first recall that the acceleration of a particle is defined and measured in some frame of reference that must be specified or otherwise inferred. Suppose an observer is fixed in this frame and that this observer anticipates the acceleration of the particle will be given by $\Sigma F/m$. Another observer moving in the frame with acceleration $\boldsymbol{a}_0$ and referring accelerations to a frame travelling with him might also anticipate that the acceleration of the same particle would be given by $\Sigma F/m$, the mass of the particle and the magnitude of the force as indicated by say a spring balance, both being unchanged. Both observers cannot be correct since the accelerations as viewed by the observers are bound to differ by an amount $\boldsymbol{a}_0$. Unfortunately, it could be that the first observer is also being accelerated in some other frame. Since $\Sigma \boldsymbol{F} = m\boldsymbol{a}$ is not satisfied simultaneously in all reference frames that are being accelerated relative to one another, we have to ask if there is any frame in which the relation is true. We now assert that there is such a frame in which $\Sigma \boldsymbol{F} = m\boldsymbol{a}$ holds, and we call this an *inertial frame of reference*. From an engineering point of view this frame is represented to sufficient accuracy by the surface of the Earth, and henceforth all accelerations associated with the relation $\Sigma \boldsymbol{F} = m\boldsymbol{a}$ will be assumed to have the Earth's surface as a reference.

The equality of $\Sigma \boldsymbol{F}$ and $m\boldsymbol{a}$ in both magnitude and direction (in an inertial frame) has been expressed by the notation $\Sigma \boldsymbol{F} = m\boldsymbol{a}$ in which m is a scalar quantity. Having regard to the method of representing a vector by a directed line segment we go further and state that, for a particle, the vectors $\Sigma \boldsymbol{F}$ and $m\boldsymbol{a}$ are *equivalent* and write $\Sigma \boldsymbol{F} \equiv m\boldsymbol{a}$, implying that the vectors are identical in magnitude, direction and line of action. This is an important statement that underlies much of the succeeding work.

The equivalence of $\Sigma \boldsymbol{F}$ and $m\boldsymbol{a}$ implies the equality of their components in specified directions. Unless graphical methods are employed the vector equation is expressed in component form for calculation purposes. The x- and y-component forms of the equation of motion are then

$$\Sigma F_x = ma_x = m\ddot{x} \tag{9.4a}$$

$$\Sigma F_y = ma_y = m\ddot{y} \tag{9.4b}$$

Similar equations can be written for other pairs of component directions, in particular the directions tangential and normal to the particle path at any instant, and the radial and transverse directions.

9.1.3 Effective Force and Inertia Force

The force $\Sigma \boldsymbol{F}$ has been referred to as the *resultant force*, being the resultant of the set of applied forces acting on the particle. The product $m\boldsymbol{a}$ is now referred to as the *effective force*, and we say that for a particle the resultant force and the effective force are equivalent in an inertial frame (figure 9.1a).

Suppose now that in the equation of motion, or its vector representation, a vector $(m\boldsymbol{a})_{\text{rev}}$, equal in magnitude and line of action to $m\boldsymbol{a}$ but reversed in direction, is added to both sides of the equation. We then have (figure 9.1b)

$$\Sigma \boldsymbol{F} + (m\boldsymbol{a})_{\text{rev}} = \Sigma \boldsymbol{F} - m\boldsymbol{a} = 0 \tag{9.5}$$

The vector $(m\boldsymbol{a})_{\text{rev}}$ is referred to as an *inertia force*, and we conclude that a particle subject to the action of a set of forces that includes the inertia force is in equilibrium, meaning that the resultant of the force set is zero. Since the resultant

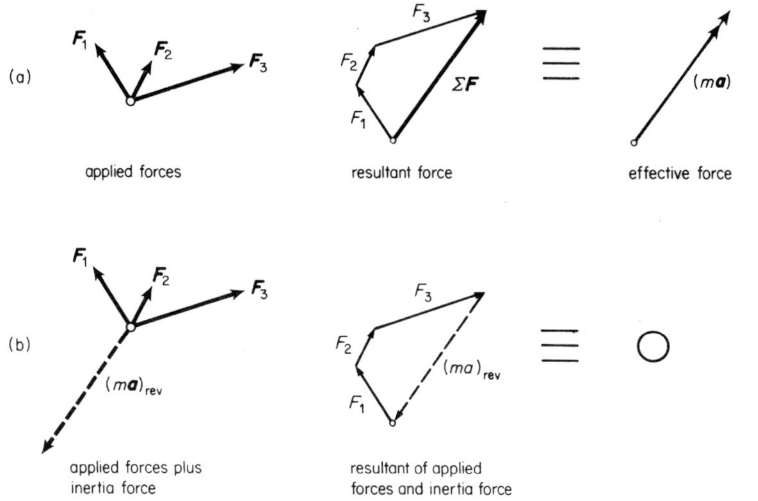

Figure 9.1

of the *applied forces* is not zero and the problem in hand is accordingly one of dynamics, the statement is qualified and the particle is said to be in *dynamic equilibrium.*

It follows that, since the particle is in (dynamic) equilibrium under the action of the applied force set and the inertia force, the principles of statics are applicable. In particular the conditions for equilibrium must be satisfied and they can be used for the determination of unknown forces and accelerations. The inertia force now enters in terms of its components, such as $-ma_x$, $-ma_y$ or $(ma_x)_{rev}$, $(ma_y)_{rev}$.

This method of interpreting the equation of motion may appear to be merely an artifice that, as we shall find later, is very useful in problem solving. However, it will also be found useful in the analysis of motion as observed in non-inertial reference frames by enabling the observer to associate in such frames also a state of zero acceleration with zero resultant force.

9.1.4 Gravitational Force

The magnitudes and directions of the individual applied forces in a force set are prescribed by the problem in hand. Certain forces, which we shall consider more fully later, are determined by the position of the particle in a so-called field of force, that is a region where a certain force is associated with each point. For present purposes such a field of force is exemplified by gravitational forces. A particle near the surface of the Earth experiences a downward vertical force such that, if the particle is released in a vacuum, it falls with constant acceleration. Represented by a vector $\boldsymbol{g}$, it has the same value for all particles. The magnitude of $\boldsymbol{g}$ varies slightly from place to place, but where absolute precision is not demanded it may be taken to be 9.81 m/s^2.

The angle subtended at the centre by the arc length s (figure 9.3b) is

$$\frac{4}{2} \times \frac{360}{2\pi} = 115°$$

The tangential component of acceleration

$$a_t = \frac{dv}{dt} = 0.5 \text{ m/s}^2$$

and the normal component of acceleration

$$a_n = \frac{v^2}{\rho} = \frac{(0.5 \times 4)^2}{2} = 2 \text{ m/s}^2$$

In the free-body diagram, figure 9.3c, $\boldsymbol{R}$ the force of the plate on the particle, is shown acting normally to the path since friction is absent. The sense has been arbitrarily assumed to be inwards.

$\Sigma F_t = ma_t$

$$P - 0.2g \cos 25° = 0.2 \times 0.5$$

$$P = 1.87 \text{ N}$$

$\Sigma F_n = ma_n$

$$R + 0.2g \sin 25° = 0.2 \times 2$$

$$R = 0.32 \text{ N}$$

The assumed direction for $\boldsymbol{R}$ is therefore correct. The particle exerts a force 0.32 N on the plate, directed outwards.

Using the inertia force method the inertia force components are included in the free-body diagram, figure 9.3d. The equations now become

$\Sigma F_t = 0$

$$P - 0.2 \times 9.81 \times \cos 25° - 0.2 \times 0.5 = 0$$

$$P = 1.87 \text{ N}$$

$\Sigma F_n = 0$

$$R + 0.2 \times 9.81 \times \sin 25° - 0.2 \times 2 = 0$$

$$R = 0.32 \text{ N}$$

9.2 Equations of Motion for a Particle System

In engineering applications we have to deal with finite material bodies. Just as in our study of statics, the material bodies of dynamical studies are considered to be made up of a finite number of particles. A finite body is therefore a many-particle system and we have to extend our discussion of the equation of motion to embrace such systems. In doing so we shall find that we can make deductions relating to the system as a whole without having to consider the behaviour of each individual particle.

The motion of a particle system is determined by first applying the equation of motion to each particle separately and then summing all the equations over the particles of the system. Thus, since for each particle, resultant force ≡ effective force, we can write for the system

$$\Sigma(\text{resultant forces}) \equiv \Sigma(\text{effective forces}) \tag{9.6}$$

summed over all particles, by vector addition.

In carrying out the summation we should find that the forces making up the set of forces acting on the system fall into two classes, as follows.

(1) External forces having their source in the surroundings; these can be subdivided into:

(a) contact forces acting at the boundary of the system,
(b) body forces acting on all particles of the system because of its presence in a force field; for example, the weight of the system,

(2) Internal forces, namely the mutual forces between the particles of the system.

Consider, for simplicity, a system consisting of three particles only as in figure 9.4. A typical set of forces is shown acting on the particles of the system.

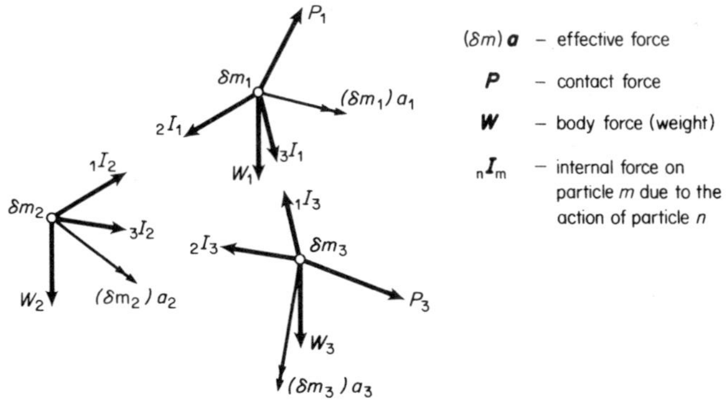

Figure 9.4

Each particle will move independently in accordance with the equation of motion $\Sigma F = (\delta m)a$, where ΣF is the resultant force on the particle and $\boldsymbol{a}$ is the acceleration of the particle. The figure shows the effective force for each particle, the direction of that force being the same as that of the resultant force on that

These two component equations taken together state that, for a particle system, the product of the total mass of the system and the acceleration of the mass-centre is equal in magnitude and direction to the resultant of the external forces. This result can be referred to as the *principle of the motion of the mass-centre* and it implies that the mass-centre of a particle system moves as if it were a particle having mass equal to that of the system and subject to the action of a force that is equal to the resultant of the external force set. Note carefully that we have *not* stated that the line of action of the resultant passes through the mass-centre. Further discussion of this point will appear later in our examination of the motion of rigid bodies. At this stage we emphasise only the role of the mass-centre and the particle-like behaviour of the system in so far as the *magnitude and direction* of the acceleration of its mass-centre are concerned.

For example, a standard problem in elementary mechanics is the flight of a projectile that is fired at a certain angle of elevation. If the drag of the atmosphere is disregarded its path is easily shown to be of parabolic shape under the influence of gravitational forces only acting on its particles. Suppose that at some point in its flight it explodes. The fragments now pursue individual parabolic paths and the analysis of their motions becomes intractable. However, we can still say that the mass-centre of the fragments continues along the same parabolic path that the projectile was originally pursuing (provided the fragments are still in flight), since the resultant gravitational force on the system is still equal to that on the original projectile.

A further observation needs to be made. In the problems relating to this chapter a finite rigid body is treated as a particle in the manner we have described, but no attention is paid to the size of the body or to the exact location of the mass-centre relative to the boundaries of the body. This step requires justification but at this stage we shall use a result that will be demonstrated in chapter 11, namely that if no rotation of the body occurs, any point of the body may be chosen to represent the whole body and all points of the body have the same motion as the mass-centre.

9.3 Solution of Problems

In the solution of problems the drawing of a free-body diagram should be regarded as essential. The free-body diagram should embody all those facts of the problem appertaining to its mechanics and should never be regarded as a rough sketch. The worked example solutions include such diagrams and indicate suitable ways of showing the essential information, namely the mass of the body, the magnitude and direction of the acceleration as given or as assumed, and the forces acting on the body. The effective force vector need not be shown, but if the inertia-force method of solution is adopted then the inertia force should be shown by a dotted arrow acting on the body since it is now included with the applied force set.

9.4 Connected Particles

We have noted that, in general, to determine the motions of the individual particles of a particle system would be very difficult. If the configuration of the particles is fixed in a rigid body the analysis is considerably simplified, as in the case of the non-rotating bodies discussed. The study of the motion of rigid bodies will be taken up fully in later chapters, but meanwhile we can usefully consider certain particle systems whose configurations are such that straightforward solutions are possible.

We can state the situation more precisely. A system containing n particles moving in a plane requires $2n$ independent coordinates to locate the particles at any instant; the system is said to have $2n$ degrees of freedom. It follows that the same number of equations are required to describe the motion at any instant. If n is small, and in addition the particles do not all move independently of each other, the number of degrees of freedom is less than $2n$ and the number of independent equations of motion is reduced. If further the motions of the particles are rectilinear then evidently the analysis of their motions could become straightforward.

Typical examples of simple particle systems are pulley systems of various kinds. To fix ideas we refer to figure 9.6a, illustrating a system consisting of two bodies,

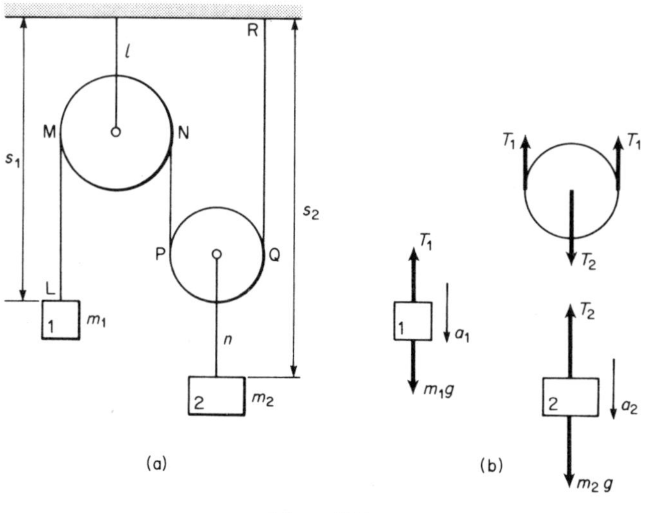

Figure 9.6

two pulleys and inextensible cords. We shall assume the pulleys and cords have negligible mass, the pulleys are frictionless, and the two bodies behave as particles having masses m_1 and m_2 respectively.

We now wish to describe the motion of the system following release from rest, that is, to determine the accelerations a_1 and a_2 of the bodies.

Solution

Only the method of solution is indicated, the complete solution being left as an exercise. Free-body diagrams are shown in figure 9.7b. We note first that there are two cords, therefore we require two constraint equations. In setting up these equations we find that we have to retain a term s_P, the position of the lower pulley, that is not a constant.

These two equations become

$$s_1 + s_2 - 2s_P = \text{constant}$$

$$s_P + s_3 = \text{constant}$$

After differentiating we are left with a term a_P in our two equations. This can be eliminated and we have finally

$$a_1 + a_2 + 2a_3 = 0$$

The solution then proceeds as before, this time with three equations in a_1, a_2, a_3 and T_1, which can be solved in conjunction with the constraint equation. (This problem is referred to later in problem 9.19.)

9.5 Summary

(1) If the resultant force ΣF on a particle is not zero then the particle accelerates in the direction of the force such that

$$\Sigma F = m\boldsymbol{a} \tag{9.3}$$

where ΣF is the resultant force in N, m the mass in kg and $\boldsymbol{a}$ the acceleration in m/s^2.
(2) The equation of motion 9.3 is valid only in an inertial frame. The Earth's surface is assumed to be an inertial frame and all accelerations associated with the equation are relative to the Earth's surface.
(3) The equation of motion 9.3 can be written in component form, for example in the x- and y-directions

$$\Sigma F_x = ma_x \tag{9.4a}$$

$$\Sigma F_y = ma_y \tag{9.4b}$$

where ΣF_x is the component of the resultant force in the x-direction.
(4) The effective force of a particle is $m\boldsymbol{a}$ and is equivalent to ΣF.
(5) The inertia force of a particle is $(m\boldsymbol{a})_{rev}$ or $-m\boldsymbol{a}$, that is, a force having magnitude ma in the opposite sense to that of $\boldsymbol{a}$.
(6) The inertia force and the applied forces acting on the particle have zero resultant and the particle is in dynamic equilibrium.
(7) The gravitational force on, or weight of, a particle mass m is mg where g is the gravitational acceleration normally to be taken as 9.81 m/s^2.

(8) For a particle system having total mass m the motion of the mass centre is given by

$$\Sigma F_x = m\bar{a}_x \tag{9.9a}$$

$$\Sigma F_y = m\bar{a}_y \tag{9.9b}$$

(9) Equations 9.9 are applicable to a rigid body. There is no implication that the resultant of the applied forces passes through the mass-centre.
(10) In problems always draw a free-body diagram.
(11) With connected particles relate the motions of the particles by the constraint(s) imposed. This provides the extra equation(s) required.

Problems

9.1 A body, mass 2 kg, is placed on a weighing machine, which uses a spring as the measuring device. The weighing machine carrying the body is placed in a lift, which accelerates upwards at 5 m/s^2. What weight (in newtons) will the machine indicate?

If a beam-balance-type machine were used what weight would this indicate? (*Hint*: Draw free-body diagrams of the body and the machine; the force exerted by machine on the body is the weight it registers.)

9.2 A body of mass m is released from rest on a rough plane (coefficient of friction μ) inclined at an an angle θ to the horizontal. Obtain an expression for the velocity of the body after it has travelled a distance s down the plane. (*Hint*: Draw a free-body diagram, deduce its acceleration and hence its velocity – see section 8.1.2.)

9.3 Two bodies A and B are connected by a cord, which passes over a frictionless pulley. A, mass m_1, slides on a rough plane (coefficient of friction μ) inclined upwards towards the pulley at θ to the horizontal, the cord being parallel to the plane between A and the pulley. If B, mass m_2, hanging freely downwards from the pulley, accelerates downwards when released derive an expression for its acceleration. (*Hint*: Draw separate free-body diagrams for A and B; insert inertia forces for dynamic equilibrium if this method is used.)

9.4 A particle, mass m, is placed on top of a smooth circular cylinder whose axis is horizontal. If when released it slides from rest, show that it loses contact with the cylinder at a point where the normal to the surface makes an angle $\cos^{-1}(2/3)$ with the vertical. (*Hint*: For a symbolic angle θ insert a symbolic Ω and α; hence decide the component accelerations. Draw a free-body diagram for the particle (including inertia effects if required); solve for α and hence Ω by integration; note the condition on the normal reaction for the particular solution.)

where a_θ is the component of the acceleration of the particle perpendicular to the tube direction in the horizontal plane. Hence show, if at a particular instant the tube has angular velocity ω and angular acceleration α when the particle is a distance r from the centre of rotation, that the equation of motion of the particle along the tube may be written

$$\left(\frac{d^2r}{dt^2}\right)^2 - 2\omega^2 r\frac{d^2r}{dt^2} - 4\mu^2\omega^2\left(\frac{dr}{dt}\right)^2 - 4\mu^2\alpha\omega r\frac{dr}{dt}$$

$$+ (\omega^4 - \mu^2\alpha^2)r^2 = \mu^2 g^2$$

If $\mu = 0$ and the particle has no radial velocity when $t = 0$ and $r = r_0$, show that at time t

$$r = \frac{r_0}{2}(e^{\omega t} + e^{-\omega t})$$

(*Hint*: Draw free-body diagrams in the vertical and horizontal planes, to include all inertia forces based upon assumed accelerations. Note that the normal reaction is affected by its weight and the inertia force due to a_θ; see appendix for the solution of the resulting differential equation.)

9.13 A particle, mass m, hangs on a massless rod of length 2 m that is hinged to one end of a massless arm 1 m long so that it can swing in a vertical plane passing through the arm; this arm rotates in the horizontal plane about its other end. If the arm has a constant angular velocity of 2 rad/s find the angle θ that the rod makes with the vertical.

It is suggested that the equation for θ be solved graphically. (*Hint*: Use a free-body diagram inserting the inertia force.)

9.14 A particle, mass m, is projected along the horizontal section AB of the smooth surface ABCDEFG, shown in figure 9.11, so that it approaches B with velocity V.

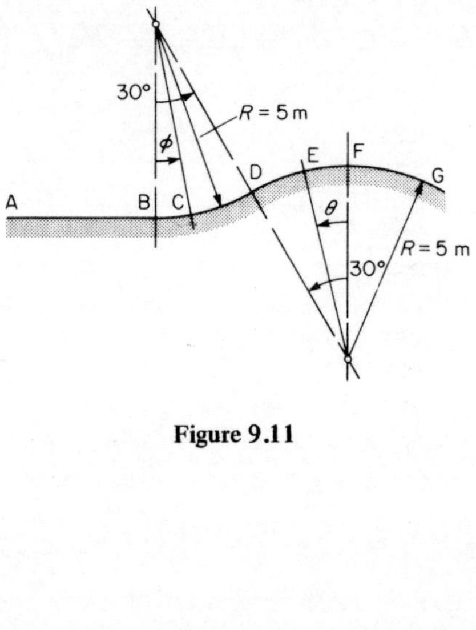

Figure 9.11

Show that when the particle is at C its motion can be described by

$$\ddot{\phi} = -\frac{g \sin \phi}{R}$$

and that

$$\frac{d\phi}{dt} = \left[\left(\frac{V}{R}\right)^2 - \frac{2g}{R}(1 - \cos \phi)\right]^{1/2}$$

Derive expressions for $\dot{\theta}$ and $\ddot{\theta}$ when the particle is at E and hence determine the velocity V that just allows the particle to pass over F without losing contact. (*Hint*: Write down component accelerations of the particle at C (either a_x and a_y or a_t and a_n) and relate to the free-body diagram, which should include the inertia force or its components; similarly at E.)

9.15 A particle, mass 2 kg, is travelling in a straight line on a horizontal table under the action of a horizontal force, magnitude 400 N. The resistance to motion due to the table and air resistance is proportional to (velocity)^2, and if the 400 N force is applied for a sufficient length of the time the mass finally attains a velocity of 20 m/s. Find the time taken and distance covered as the particle's velocity changes from 0 to 10 m/s. (*Hint*: Draw a free-body diagram and write down the equation of motion.)

9.16 A small particle is placed inside a smooth hole of radius R, the axis of which is horizontal. Show that if it is released from rest on the surface of the hole near the lowest point, its subsequent motion in an angular fashion will be simple harmonic if its displacement from the vertical through the hole centre is small. What is the periodic time? (*Hint*: Draw a free-body diagram, including inertia effects, for a symbolic displacement θ from the vertical and assume symbolic angular velocity and angular acceleration in the correct mathematical direction; compare the differential equation of motion with that for SHM (section 8.4) after making the required approximations.)

9.17 A particle B, mass 0.1 kg, is lying on a smooth horizontal surface and is connected by two elastic strings AB and BC to points A and C which are such a distance apart that the cords are just taut. The cords AB and BC have elastic constants 10 N/m and 2.5 N/m respectively. Show that if the mass is moved towards C such that AB is extended by 0.1 m and then released, the subsequent motion is periodic. Assume that if a cord is slack it has no effect.

Determine (a) the maximum velocity of the particle, (b) the maximum extension of cord BC and (c) the periodic time of oscillation. (*Hint*: Draw free-body diagrams of the particle on each side of the equilibrium position for a displacement x, inserting the corresponding acceleration $\ddot{x}$; hence show that the equations of motion are those of SHM in each case. Obtain solutions in both cases.)

9.18 A particle, mass 0.1 kg, slides on a wire that is bent in a curve lying in the horizontal plane. The particle has rectangular coordinates

$$x = 4(\beta + \sin\beta\cos\beta) \qquad y = 4\sin^2\beta$$

and the curve has a path length $s = 8\sin\beta$, all measured from the same origin; β is an independent variable.

If the particle is moving along the wire such that $s = 5t^2$, t being the time in seconds, find its total velocity and total acceleration when $t = 1$ s.

If the motion is caused by an applied force $\boldsymbol{P}$, that is always tangential to the path of the particle, find the magnitude of $\boldsymbol{P}$ and that of the reaction N of the wire on the particle when $t = 1$ s.

Does P vary along the path? (*Hint*: Write equations for $\dot{s}$ and $\ddot{s}$, and similarly for $\dot{x}$, $\ddot{x}$, $\dot{y}$, $\ddot{y}$, or find the radius of curvature and use a_t and a_n (see chapter 8). To find $\boldsymbol{P}$ and N draw a free-body diagram.)

9.19 Complete the solution of worked example 9.3.

10 Kinetics of Particles and Particle Systems II – Integrated Forms

In the previous chapter we considered applications of the vector equation $\Sigma F = ma$ as expressed in component form, both for a particle and for a particle system. Having expressed the acceleration components in terms of the external force components, we could obtain the velocity and position of a particle or of the mass-centre of a particle system at some specified instant – given the initial conditions. Conversely we were able to determine the resultant force required to bring about some specified motion. A review of the problems of the previous chapter will show that if the resultant force and its variation were known, then the complete solution for both velocity and position involved two successive integrations of the expression for the acceleration.

We shall now rewrite the basic equation of motion in two additional ways and then integrate these equations once in their general form before introducing the details of any specific problem. The equations we shall obtain are called *integrated forms* of the equation of motion and we shall find that these equations incorporate over-all changes in certain new quantities that are of fundamental importance.

Thus if we choose the x-component equation $\Sigma F_x = ma_x$, this can be manipulated in two ways.

(1)We write

$$\Sigma F_x = ma_x = m\frac{dv_x}{dt} = \frac{d}{dt}(mv_x)$$

and obtain

$$\Sigma(F_x dt) = d(mv_x)$$

an equation that introduces *linear impulse* of a force and change in *linear momentum* of a particle.

(2) We write

$$\Sigma F_x = ma_x = m\left(\frac{dv_x}{dx}\right)\left(\frac{dx}{dt}\right) = mv_x\frac{dv_x}{dx} = \frac{d}{dx}\left(\frac{1}{2}mv_x^2\right)$$

and obtain

$$\Sigma(F_x dx) = d(\tfrac{1}{2}mv_x^2)$$

an equation that introduces the *work* of a force and the change in *kinetic energy* of a particle.

We first formally define these new quantities and then proceed to consider their use in the analysis of the motion of particles and particle systems.

10.1 Linear Impulse and Linear Momentum

10.1.1 The Particle

Consider a particle mass m which, at time t, is at point P of its path (figure 10.1)

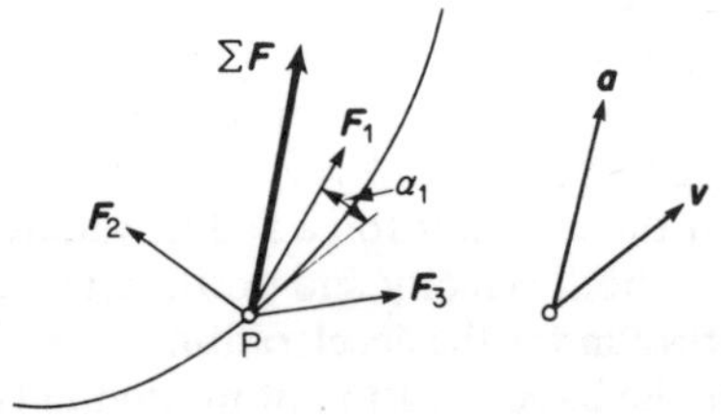

Figure 10.1

and moving at that instant with velocity $\boldsymbol{v}$ under the influence of a set of forces as indicated.

If the resultant force on the particle is ΣF (the direction being that of the acceleration $\boldsymbol{a}$) then we have

$$\Sigma F = ma = \frac{\mathrm{d}}{\mathrm{d}t}(mv) \tag{10.1}$$

The quantity $m\boldsymbol{v}$ is defined to be the linear momentum of the particle. This is a vector whose direction is that of $\boldsymbol{v}$. We shall use the symbol $\boldsymbol{G}$ for this quantity, the units of which will be kg m/s or N s. We have therefore

$$(\text{resultant force on a particle}) = \begin{pmatrix}\text{rate of change of linear momentum} \\ \text{of the particle with respect to time}\end{pmatrix}$$

Both the resultant force and the linear momentum can be resolved into rectangular components, namely ΣF_x, ΣF_y, and $mv_x = G_x$, $mv_y = G_y$. We can then write

$$\Sigma F_x = \frac{\mathrm{d}}{\mathrm{d}t}(mv_x) = \frac{\mathrm{d}}{\mathrm{d}t}(G_x)$$

$$\Sigma F_y = \frac{\mathrm{d}}{\mathrm{d}t}(mv_y) = \frac{\mathrm{d}}{\mathrm{d}t}(G_y)$$

Taking the first of these equations, multiplying through by dt and integrating over the time interval t_1 to t_2, the times at which the particle passes through points 1 and 2 of the path, we have

$$\Sigma \int_{t_1}^{t_2} F_x \, dt = (mv_x)_2 - (mv_x)_1 = G_{x2} - G_{x1} \qquad (10.2)$$

the expression on the left implying a term-by-term integration of the quantity $\Sigma(F_x \mathrm{d}t)$.

Each integral on the left-hand side of equation 10.2 is the x-component of a quantity that is defined to be the *linear impulse* of the corresponding force F. This quantity, which is symbolised as $\mathbf{Imp}_{1-2}$ is the vector sum of elementary quantities such as $F\delta t$ over the time interval t_1 to t_2; this vector will have components

$$\mathrm{Imp}_{x,1-2} = \int_{t_1}^{t_2} F_x \, dt$$

$$\mathrm{Imp}_{y,1-2} = \int_{t_1}^{t_2} F_y \, \mathrm{d}t$$

After rearranging, equation 10.2 and the corresponding y-component equation can be written in the form

$$G_{x1} + \Sigma \mathrm{Imp}_{x,1-2} = G_{x2} \qquad (10.3a)$$

$$G_{y1} + \Sigma \mathrm{Imp}_{y,1-2} = G_{y2} \qquad (10.3b)$$

or both equations can be combined into the vector form

$$\boldsymbol{G}_1 + \Sigma \mathbf{Imp}_{1-2} = \boldsymbol{G}_2 \qquad (10.3c)$$

We have thus arrived at our first integrated form, which can be referred to as the *impulse - momentum equation*, and which states that the initial linear momentum plus the sum of the linear impulses of the forces equals the final linear momentum. This is a vector equation and for calculation purposes is expressed in the component forms given in equations 10.3a and 10.3b. Since $\Sigma(F_x \, \mathrm{d}t) = \Sigma(F_x)\mathrm{d}t$ and similarly for the y components, then it follows that $\Sigma\mathbf{Imp}_{1-2}$, the sum of the linear impulses of the individual forces, is equal to the linear impulse of their resultant.

If the manner in which a particular force component (say the x-component) varies is shown on a force - time graph as in figure 10.2 then it is clear that the corresponding component of the linear impulse, namely $\int_{t_1}^{t_2} F_x \, \mathrm{d}t$, is represented by the area under the graph between the ordinates at $t = t_1$ and $t = t_2$.

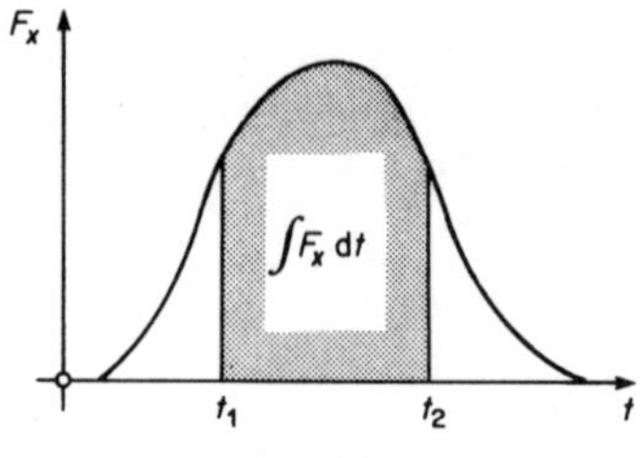

Figure 10.2

Worked Example 10.1

A plane having mass 14 000 kg is making its take-off run. The forward thrust of the engines less the drag on the plane is constant at 44.5 kN. If the required speed for lift-off is 200 km/h, determine the total time for take-off, starting from rest.

Solution

Since the linear impulse and the linear momentum are vectors we can represent the impulse–momentum equation by means of a diagram in which the vectors are shown. Figure 10.3 is such a diagram, in which only the x-components are

$$\boxed{}\!\!\longrightarrow \;+\; \boxed{}\!\!\longrightarrow \;=\; \boxed{}\!\!\longrightarrow$$

$$(mv_{x1}) \qquad\qquad\qquad \Sigma \int_1^2 F_x\, dt \qquad\qquad\qquad (mv_{x2})$$

$$G_{x1} \qquad + \qquad \Sigma\, Imp_{x,1-2} \qquad = \qquad G_{x2}$$

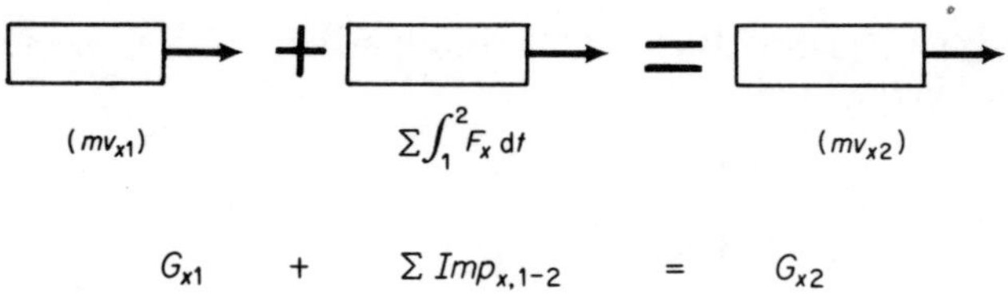

Figure 10.3

indicated for the purpose of this problem. The equation is applied between the instant at which the plane starts moving (t_1) and the time at lift-off (t_2).

If t is the total time for take-off

$$G_{x1} = 0$$

$$\Sigma\, Imp_{x1-2} = \Sigma \int_{t_1}^{t_2} F_x\, dt$$

$$= \int_{t_1}^{t_2} \Sigma F_x\, dt$$

$$= \int_{t_1}^{t_2} 44.5 \times 10^3\, dt$$

$$= (44.5 \times 10^3)t \text{ N s}$$

$$G_{x2} = 14\,000 \left(\frac{200 \times 10^3}{3600}\right)$$

$$= \frac{7 \times 10^6}{9} \text{ N s}$$

$$0 + 44.5 \times 10^3\, t = \frac{7 \times 10^6}{9}$$

and

$$t = 17.5 \text{ s}$$

This problem could also have been solved in two stages, by first calculating the acceleration a_x using $\Sigma F_x = ma_x$ and then deriving the time using the relation $v_{x2} = v_{x1} + a_x t$, a_x being constant in this case. The method using the impulse - momentum equation solves the problem in one stage. The time saved by using the method is trivial in this example but later we shall encounter problems for which the use of the impulse - momentum equation is more expeditious and also situations in which its use is essential.

10.1.2 Particle Systems

Having obtained the impulse - momentum equations for a single particle we can immediately extend our result to a many-particle system in which the particles may have differing momenta, by summing over the particles of the system. We now write

$$\Sigma \boldsymbol{G}_1 + \Sigma \mathbf{Imp}_{1-2} = \Sigma \boldsymbol{G}_2$$

summed over all particles and forces by vector addition. In carrying out the summation we find that the internal impulses occur in equal and opposite pairs and therefore reduce to zero: the second term is now the vector sum of the impulses of the external forces only, therefore we have

$$\Sigma \boldsymbol{G}_1 + \Sigma(\mathbf{Imp}_{1-2})_{\text{ext}} = \Sigma \boldsymbol{G}_2 \tag{10.4a}$$

In component form this is written

$$\Sigma G_{x1} + \Sigma(\text{Imp}_{x,1-2})_{\text{ext}} = \Sigma G_{x2} \tag{10.4b}$$

$$\Sigma G_{y1} + \Sigma(\text{Imp}_{y,\,1-2})_{\text{ext}} = \Sigma G_{y2} \tag{10.4c}$$

where typically

$$\Sigma G_{x1} = \Sigma(\delta m)v_{x1}$$

$(\delta m)v_{x1}$ being the x-component of the momentum of a typical particle.

Following the procedure we adopted for the basic form of the equation of motion we again introduce the mass-centre, the x-coordinate of which is given by

$$\Sigma(\delta m)x = m\bar{x}$$

On differentiating with respect to time

$$\Sigma(\delta m)\dot{x} = m\dot{\bar{x}} \text{ or } \Sigma(\delta m)v_x = m\bar{v}_x$$

where $\bar{v}_x$ is the x-component of the velocity of the mass-centre of the particle system.

Our equations now become

$$m\bar{v}_{x1} + \Sigma(\text{Imp}_{x,1-2})_{\text{ext}} = m\bar{v}_{x2} \qquad (10.5a)$$

$$m\bar{v}_{y1} + \Sigma(\text{Imp}_{y,1-2})_{\text{ext}} = m\bar{v}_{y2} \qquad (10.5b)$$

in which the second terms are the sums of the components of the linear impulses of the external forces acting on the system, or equivalently the components of the linear impulse of the resultant of the external forces.

The mass-centre moves as if it were a particle in which the whole mass of the system was concentrated, a result that is comparable to that of the previous chapter. In this case it is again emphasised that no statement has yet been made about the lines of action of the linear momentum and linear impulse vectors, but if the equation is applied to a rigid body, then, provided the body does not rotate, all points of the body have motions identical to that of the mass centre. Subject to this proviso, we are again justified in discussing the motion of rigid bodies as if they were particles, in which case equations 10.3 can be applied directly. Extending our discussion to a system of more than one rigid body: for each rigid body $\boldsymbol{G} = m\bar{\boldsymbol{v}}$ where $\bar{\boldsymbol{v}}$ is the velocity of the mass-centre of the body and it follows that equation 10.3c can be used for a *system of rigid bodies* provided $\boldsymbol{G}$ now means $\Sigma m\bar{\boldsymbol{v}}$. If a body is not rotating the quantity $\bar{\boldsymbol{v}}$ can be referred to as the velocity of the body.

10.2 Conservation of Linear Momentum

If the resultant of the external forces is zero and remains zero over some time interval, then since the total linear impulse is zero it follows from equation 10.3 that the total linear momentum of a system remains constant. Consequently the mass-centre of the system moves with constant velocity. The total linear momentum of the system is said to be *conserved.*

It is often one component of the resultant force that remains zero, in which case it is the corresponding component of linear momentum that is conserved. Thus if a body is moving in a horizontal plane the net vertical force on the body is zero and the vertical component of linear momentum remains zero.

Cases can arise in which the total linear impulse over a certain time interval is zero but the resultant force has non-zero values during the interval. In such cases the linear momenta are again the same at the beginning and end of the interval but may vary during the interval. Thus if the resultant force on a particle induces simple harmonic motion, then during one cycle the force takes zero values at two particular instants only (when the acceleration is zero), but the momenta are the same at *any* two instants separated by an interval equal to the periodic time.

Worked Example 10.2

A particle, mass 2 kg, has velocity components $v_x = -\ 10$ m/s and $v_y = +\ 5$ m/s when it is subjected to a force with components P_x and P_y that vary as indicated in figure 10.4. Determine the velocity components after (a) 2.5 s and (b) 5 s.

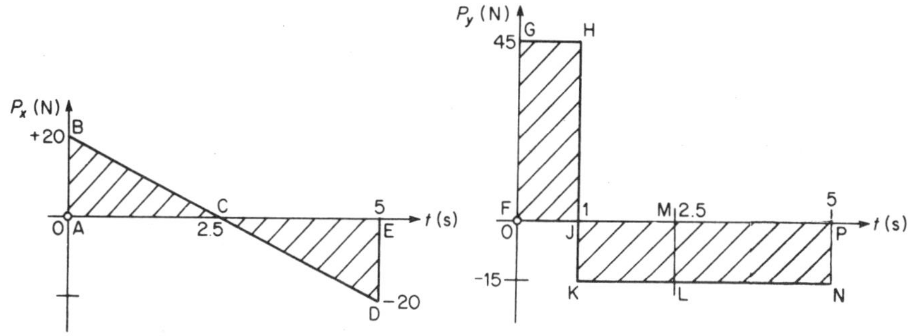

Figure 10.4

Solution

Applying the impulse - momentum equations 10.3

$$2 \times (-\ 10) + \Sigma \text{Imp}_{x,1-2} = 2\, v_{x2} \qquad (10.6)$$

$$2 \times (5) + \Sigma \text{Imp}_{y,1-2} = 2\, v_{y2} \qquad (10.7)$$

(a) For $t = 2.5$ s

$$\Sigma \text{Imp}_{x,1-2} = \int_0^{2.5} P_x \,\mathrm{d}t = \text{area ABC}$$

$$= \tfrac{1}{2} \times 20 \times 2.5 = +\ 25 \text{ N s}$$

$$\Sigma \text{Imp}_{y,1-2} = \int_0^{2.5} P_y \,\mathrm{d}t = \text{area FGHJKLM}$$

$$= 45 \times 1 - 15 \times 1.5 = +\ 22.5 \text{ N s}$$

Using these values in equations 10.6 and 10.7

$$-\ 20 + 25 = 2\, v_{x2}$$

and

$$v_{x2} = +2.5 \text{ m/s}$$

$$10 + 22.5 = 2 v_{y2}$$

and

$$v_{y2} = +16.25 \text{ m/s}$$

(b) For $t = 5$ s

$$\text{Imp}_{x,1-2} = \int_0^5 P_x \, dt = \text{area ABCDE} = 0$$

$$\text{Imp}_{y,1-2} = \int_0^5 P_y \, dt = \text{area FGHJKNP}$$

$$= 45 \times 1 - 15 \times 4 = -15 \text{ Ns}$$

Substituting in equations 10.6 and 10.7

$$-20 + 0 = 2 v_{x2}$$

and

$$v_{x2} = -10 \text{ m/s}$$

and

$$10 - 15 = 2 v_{y2}$$

$$v_{y2} = -2.5 \text{ m/s}$$

10.3 Impulsive Forces

In many cases of engineering interest a large force acts over a very short time interval; such a force is called an *impulsive force.* The linear impulse of the force is still $\int F \, dt$, but the manner in which the magnitude of the force varies is often difficult to measure directly, as also is the time for which it acts. The impulse of the force is therefore deduced indirectly by observing the change produced in the linear momentum of the system.

When impulsive forces act on a system those forces that are known to be non-impulsive can usually be neglected, since over the same short time interval their linear impulses are small in comparison with those of the impulsive forces. In cases of doubt they should be classed as impulsive until demonstrated to be otherwise.

10.4 Impact

If two bodies interact in such a way that the contact forces between them are impulsive forces, then the interaction is termed an *impact.*

Assuming the surfaces of the bodies are continuously curved, the common normal at the point of contact is termed the *line of contact.* If the mass-centres of the bodies both lie on this line, we have *central* impact; if otherwise the impact is *eccentric.* In the case of central impact, if the velocities of the mass-centres are both directed along the line of impact we have *direct* impact; but if inclined to the line of contact the impact is termed *oblique.* These terms are illustrated in figure 10.5.

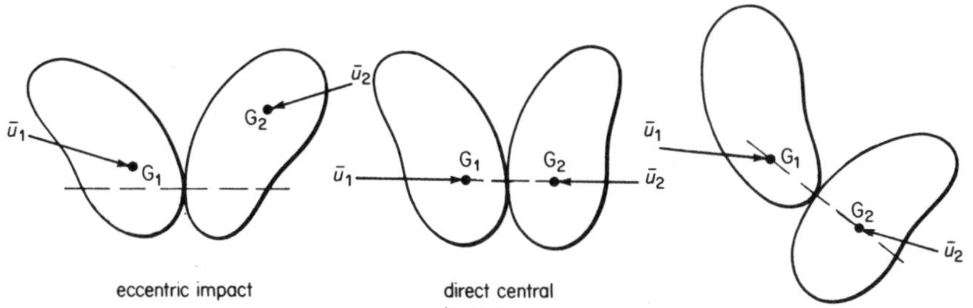

Figure 10.5

At this stage we shall discuss central impacts only and for discussion purposes we shall assume initially that the bodies are uniform spheres.

10.4.1 Direct Central Impact

Consider therefore the case of two spheres, A and B, having masses m_A and m_B and moving with speeds u_A and u_B in the same straight line. (figure 10.6). If $u_A > u_B$, they will collide and then separate. To justify this simple model we can assume the spheres are smooth, have equal diameters, and are sliding without rolling on a smooth horizontal table.

Without digressing into a discussion on the source of the mutual impulsive forces that are produced, we can accept that a condition for the forces to be generated is that real bodies, although nominally rigid, undergo deformation. Whether or not the original shape is restored after completion of the impact is dependent on the nature of the materials of the bodies, or in engineering terms, whether or not they are perfectly *elastic*. We shall use the terms *deformation* and *restitution* in our description of the motion of the spheres during the impact.

The behaviour of the spheres while they are in contact is complex, but we can assume a simplified picture of the sequence of events and separate the motion of the spheres into three phases, as explained in figure 10.6. Velocities to the right are assigned positive senses. Note in particular that there is some instant at which

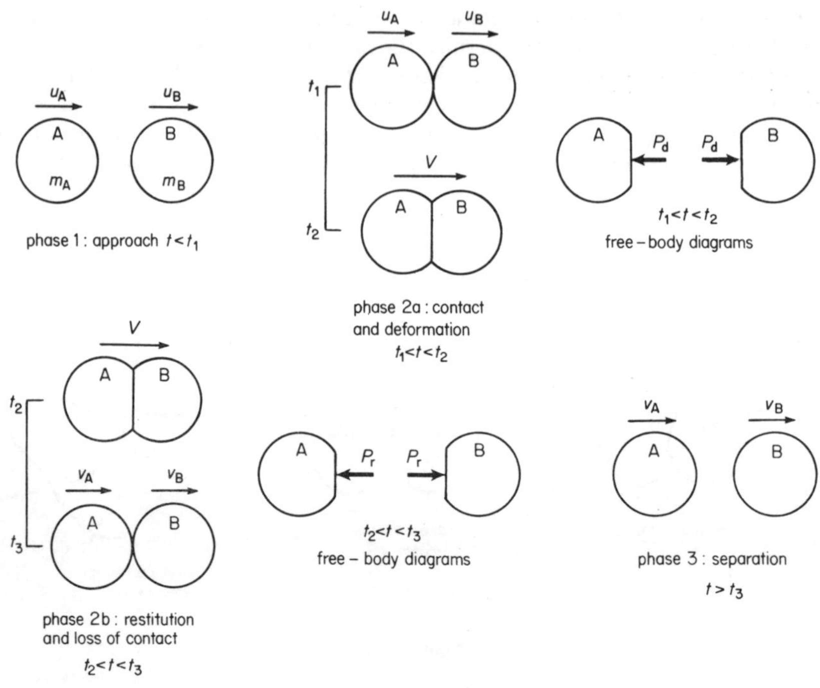

Figure 10.6

both spheres are moving at some common velocity; this will occur when the deformation of the spheres is greatest, since it is then that their relative velocities are zero.

We now apply the impulse - momentum equations 10.3, noting that $G = \Sigma m\bar{v}$, first to the whole system consisting of two spheres and then to each sphere in turn.

For the whole system

phases 1 and 2a; there is no external impulse

$$m_A u_A + m_B u_B = (m_A + m_B)V \tag{10.8a}$$

phases 2b and 3; there is no external impulse

$$(m_A + m_B)V = m_A v_A + m_B v_B \tag{10.8b}$$

phase 1 to phase 3; there is no external impulse

$$m_A u_A + m_B u_B = m_A v_A + m_B v_B \tag{10.8c}$$

For sphere A: phases 1 and 2a – sphere B is exerting a force $\boldsymbol{P}_d$ in the negative sense, varying from zero at $t = t_1$ to its value at $t = t_2$, during the deformation phase.

$$m_A u_A - \int_{t_1}^{t_2} P_d \, dt = m_A V \tag{10.9a}$$

Phases 2b and 3 – sphere B is exerting a force $\boldsymbol{P}_r$ in the negative sense, varying from its value at $t = t_2$ to zero at $t = t_3$, during the restitution phase

$$m_A V - \int_{t_2}^{t_3} P_r \, dt = m_A v_A \tag{10.9b}$$

Now the manner in which $\boldsymbol{P}_r$ diminishes to zero in the restitution phase will not, in general, be the same as the manner in which $\boldsymbol{P}_d$ increases from zero in the deformation phase and the impulse of the force $\boldsymbol{P}_r$ during restitution is not, in general, equal to the impulse of the force P_d during deformation. The ratio of the magnitudes of the two impulses is defined to be the *coefficient of restitution e*, and thus

$$e = \frac{\int_{t_2}^{t_3} P_r \, dt}{\int_{t_1}^{t_2} P_d \, dt} = \frac{\text{Impulse during restitution}}{\text{Impulse during deformation}} \tag{10.10}$$

An alternative way of expressing the coefficient of restitution is given by writing from equations 10.9

$$\frac{\int_{t_2}^{t_3} P_r \, dt}{\int_{t_1}^{t_2} P_d \, dt} = \frac{m_A V - m_A v_A}{m_A u_A - m_A V} = \frac{V - v_A}{u_A - V}$$

Similarly by considering sphere B, and noting that the mutual impact forces (and hence their impulses) are equal in magnitude but oppositely directed, we can obtain

$$\frac{\int_{t_2}^{t_3} P_r \, dt}{\int_{t_1}^{t_2} P_d \, dt} = \frac{m_B v_B - m_B V}{m_B V - m_B u_B} = \frac{v_B - V}{V - u_B}$$

Since these are both expressions for e, we can write

$$e = \frac{(V - v_A) + (v_B - V)}{(u_A - V) + (V - u_B)} = -\frac{v_B - v_A}{u_B - u_A}$$

that is

$$e = -\frac{(\text{relative velocity after impact})}{(\text{relative velocity before impact})} \tag{10.11}$$

or

$$v_B - v_A = -e(u_B - u_A)$$

The coefficient of restitution, therefore, can also be defined as the negative of the ratio of the relative velocity of separation and the relative velocity of approach. It is important to observe a uniform convention for the positive senses of the velocities. It is also advantageous to use consecutive symbols u, v, w, and so on for the successive velocities between multiple impacts, if these occur.

The definitions numbered 10.10 and 10.11 have been developed using the impulse - momentum equation but the value of the coefficient of restitution has to be determined experimentally. The second definition is useful for this purpose. The value of e is not a constant, and depends not only on the material of the spheres but also on other factors such as the relative velocity of approach. Its value is normally less than unity. If the spheres were perfectly elastic and the deformation was at all times uniquely related to the deforming force, then the value of e would be unity and the impact would be described as being *elastic*. At the other extreme, a zero value implies the absence of elasticity and the impact is then termed *plastic*. A value of e greater than unity can be envisaged, as for example, in the case of an impact that detonated a small explosive charge located at the point of contact.

Consider now the motion of the mass-centre. Since there is no external impulse on the whole system we anticipate that its velocity will be unchanged by the impact. We can verify this by assigning positions x_A and x_B to the spheres and $\bar{x}$ to the position of the mass-centre, measured from some origin. At any instant before the impact

$$(m_A + m_B)\bar{x} = m_A x_A + m_B x_B$$

and by differentiating with respect to time we obtain

$$(m_A + m_B)\bar{u} = m_A u_A + m_B u_B$$

After the impact

$$(m_A + m_B)\bar{v} = m_A v_A + m_B v_B$$

It follows, using equation 10.8c, that

$$\bar{u} = \frac{m_A u_A + m_B u_B}{m_A + m_B} = \frac{m_A v_A + m_B v_B}{m_A + m_B} = \bar{v} \tag{10.12}$$

and the velocity of the mass-centre is the same before and after the impact.

Worked Example 10.3

Two spheres, A with mass 5 kg, and B with mass 10 kg have initial velocities along the same path, + 10 m/s and – 2 m/s respectively.

(a) Assuming $e = 1$ determine (i) the common velocity of the two spheres when fully deformed, (ii) the final velocities of the two spheres and (iii) the maximum value of the impulsive force between the spheres if this is assumed to vary linearly with time and the total time of contact is 0.1 s.
(b) Repeat the problem for $e = 0.5$.

Solution

Let the common velocity be V and the final velocities be v_A and v_B:

(a) (i) Applying the impulse - momentum equation 10.3 to the whole system for the deformation phase, the external impulse is zero and

$$5(10) + 10(-2) = (5 + 10)V$$

$$V = +2 \text{ m/s}$$

(ii) Applying the impulse - momentum equation 10.3 to the whole system for the whole impact

$$5(10) + 10(-2) = 5v_A + 10v_B \qquad (10.13)$$

By the definition of coefficient of restitution, 10.11

$$1 = -\frac{(v_B - v_A)}{[(-2) - 10]} \qquad (10.14)$$

From equations 10.13 and 10.14

$$v_A = -6 \text{ m/s} \qquad v_B = +6 \text{ m/s}$$

(iii) Applying the impulse - momentum equations to sphere A, for the deformation phase

$$5(10) - \int_{t_1}^{t_2} P_d \, dt = 5(2)$$

and therefore

$$\int_{t_1}^{t_2} P_d \, dt = 40 \text{ N s}$$

Similarly for the restitution phase

$$5 \times 2 - \int_{t_2}^{t_3} P_r \, dt = 5(-6)$$

and therefore

$$\int_{t_2}^{t_3} P_r \, dt = 40 \text{ N s}$$

Both impulses are equal, as expected if $e = 1$.

Since both P_d and P_r are assumed to vary linearly with time, the plot of impulsive force against time must be as in figure 10.7a so that the integrals shall both be equal. It follows that $t_3 - t_2 = t_2 - t_1 = 0.05$ s and also

$$\int_{t_1}^{t_2} P_d \, dt = \tfrac{1}{2} P_{max} \, (t_2 - t_1) = 40 \text{ N s}$$

Thus

$$P_{max} = \frac{2 \times 40}{0.05} = 1600 \text{ N}$$

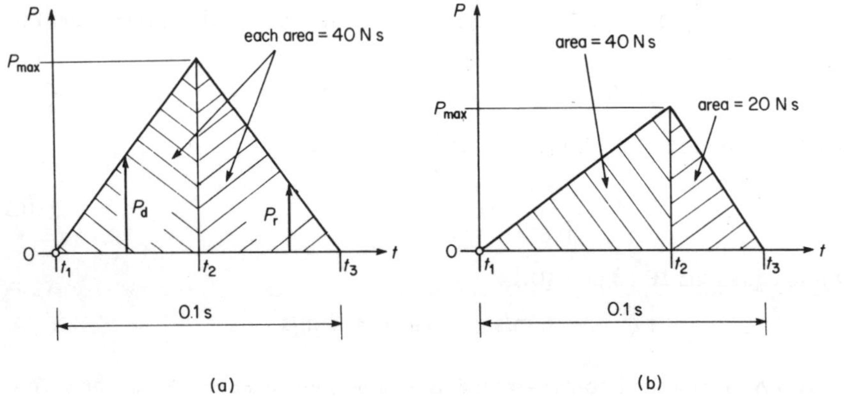

Figure 10.7

(b) (i) The value of e does not affect the result and $V = +\,2$ m/s.
(ii) Following the same procedure as before

$$5(10) + 10(-2) = 5\, v_A + 10\, v_B$$

$$0.5 = -\frac{(v_B - v_A)}{[(-2) - 10]}$$

giving

$$v_A = -\,2 \text{ m/s} \qquad v_B = +\,4 \text{ m/s}$$

(iii) By the same procedure as before

$$5(10) - \int_{t_1}^{t_2} P_d \, dt = 5(2)$$

and therefore

$$\int_{t_1}^{t_2} P_d \, dt = 40 \text{ N s}$$

as before. Also

$$5(2) - \int_{t_2}^{t_3} P_r \, dt = 5(-2)$$

and

$$\int_{t_2}^{t_3} P_r \, dt = +\ 20 \text{ N s}$$

(The last figure could have been found from the ratio of impulses – definition 10.10.)

It follows that the plot of impulsive force against time must now be as shown in figure 10.7b such that the areas representing the integrals are in the ratio of 2 : 1. From this we deduce

$$t_3 - t_2 = \frac{t_2 - t_1}{2}$$

and thus

$$t_2 - t_1 = 0.067 \text{ s} \qquad t_3 - t_2 = 0.033 \text{ s}$$

Hence

$$P_{max} = \frac{2 \times 40}{0.067} = 1200 \text{ N}$$

This type of problem is best solved by direct application of the impulse-momentum equation 10.3, supplemented by the defining relations for e, either 10.10 or 10.11.

10.4.2 Oblique Central Impact

Suppose the mass-centres of the spheres are now moving in the directions shown in figure 10.8 at the instant of contact. If we choose the line of contact to be

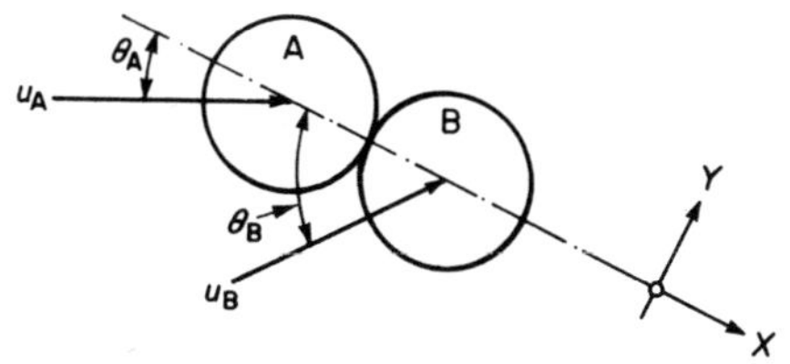

Figure 10.8

an x-direction then we can write the x-components of the velocities of approach as $(u_A)_x = u_A \cos\theta_A$ and $(u_B)_x = u_B \cos\theta_B$ respectively. Equation 10.8c, expressing the conservation of momentum, and equation 10.11, relating x-

components of velocities, can now be written for the x-direction, enabling the components of the velocities of separation $(v_A)_x$ and $(v_B)_x$ to be calculated. Now if we still assume that the spheres are smooth, then since for each sphere there is no y-direction frictional force, the y-components of the velocities are unchanged and we have

$$(v_A)_y = (u_A)_y = u_A \sin \theta_A$$

$$(v_B)_y = (u_B)_y = u_B \sin \theta_B$$

Having now ascertained the x- and y-components of the velocities of separation we can determine the values of v_A and v_B in magnitude and direction.

Worked Example 10.4

A small wooden block, mass 1.2 kg, is moving in a straight line on a smooth horizontal table at 100 m/s. It is struck by a bullet, mass 0.1 kg, moving in a horizontal plane along a line at 60° to the path of the block, its velocity component in the path of the block being opposed to the velocity of the block. If the bullet remains embedded in the block and the path of the latter is deflected through 30°, determine the velocity of the bullet and the impulse exerted on the block.

Rotations of the block and the bullet can be ignored.

Solution

The last sentence implies that only the motions of the mass-centres are to be considered and that the equations for a particle can be applied. x- and y-directions are chosen as in figure 10.9a and free-body diagrams, during impact, are drawn as in figure 10.9b for both the block and the bullet. Note the directions of the

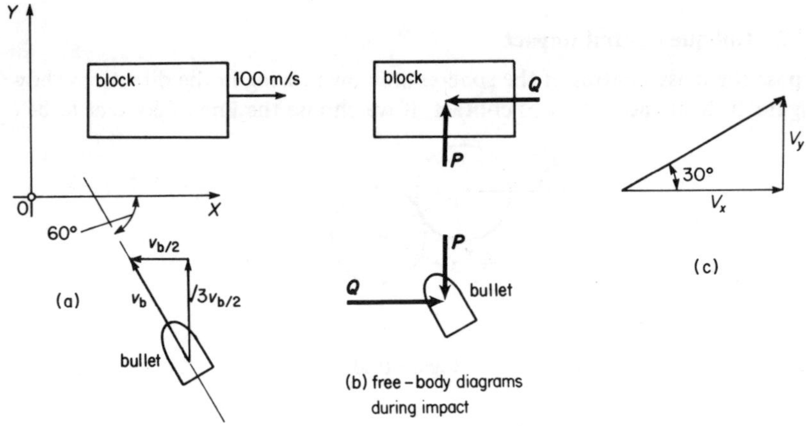

Figure 10.9

impulsive force components P and Q on the block; the same components, in reversed direction, act on the bullet.

Figure 10.9c shows the final velocity of the block (and bullet) with its components V_x and V_y.

The impulse - momentum equation 10.3 is applied in component form to both bodies.

For the block

$$G_{x1} = 1.2 \times 100 = 120 \text{ N s}$$

$$\Sigma\text{Imp}_{x,1-2} = \int_{t_1}^{t_2} (-Q)\,\mathrm{d}t$$

$$G_{x2} = 1.2\,V_x$$

and

$$120 - \int_{t_1}^{t_2} Q\,\mathrm{d}t = 1.2V_x \qquad (10.15)$$

Also

$$G_{y1} = 0$$

$$\Sigma\text{Imp}_{y,1-2} = + \int_{t_1}^{t_2} P\,\mathrm{d}t$$

$$G_{y2} = 1.2\,V_y$$

and

$$0 + \int_{t_1}^{t_2} P\,\mathrm{d}t = 1.2\,V_y \qquad (10.16)$$

For the bullet

$$G_{x1} = 0.1\left(\frac{-v_b}{2}\right) = -\,0.05v_b$$

$$\Sigma\text{Imp}_{x,1-2} = \int_{t_1}^{t_2} Q\,\mathrm{d}t$$

$$G_{x2} = 0.1\,V_x$$

and

$$-0.05v_b + \int_{t_1}^{t_2} Q\,\mathrm{d}t = 0.1\,V_x \qquad (10.17)$$

Also

$$G_{y1} = 0.1 \times (v_b\sqrt{3}/2) = 0.0866v_b$$

$$\Sigma\text{Imp}_{y,1-2} = \int_{t_1}^{t_2} -P\,\mathrm{d}t$$

$$G_{y2} = 0.1V_y$$

and

$$0.0866v_b - \int_{t_1}^{t_2} P\,dt = 0.1V_y \qquad (10.18)$$

There are five unknowns v_b, $\int P\,dt$, $\int Q\,dt$, V_x and V_y, and another equation is thus required; this is the relationship for V_x and V_y from figure 10.9c.

$$\frac{V_y}{V_x} = \frac{1}{\sqrt{3}} \qquad (10.19)$$

From equations 10.15 and 10.17

$$120 - 0.05v_b = 1.3V_x \qquad (10.20a)$$

From equations 10.16 and 10.18

$$0.0866v_b = 1.3V_y \qquad (10.20b)$$

Equations 10.19 and 10.20 give

$$v_b = 600 \text{ m/s}$$

$$V_y = 39.97 \text{ m/s}$$

$$V_x = 69.23 \text{ m/s}$$

Equation 10.16

$$\int_{t_1}^{t_2} P\,dt = 1.2 \times 39.97 = 47.96 \text{ N s}$$

Equation 10.15

$$\int_{t_1}^{t_2} Q\,dt = 120 - 1.2 \times 69.23 = 36.92 \text{ N s}$$

confirming that the directions of P and Q in figure 10.9b are physically correct.

The impulse exerted by the bullet on the block is the vector sum of

47.96 N s ↑ and 36.92 N s ←

= 60.53 N s ↖ 127.6°

If the impulse values had not been required the problem could be solved by expressing the conservation of linear momentum, since no external impulse acts on the system. Applying the impulse - momentum equation 10.3: In the x direction

$$1.2 \times 100 + 0.1\left(\frac{-v_b}{2}\right) = (1.2 + 0.1)V_x$$

in the y direction

$$1.2 \times 0 + 0.1 \left(\frac{v_b\sqrt{3}}{2}\right) = (1.2 + 0.1)\, V_y$$

These equations are exactly the same as equations 10.20.

10.5 Work and Kinetic Energy

10.5.1 The Particle

Consider a particle, mass m, which at time t is at point P of its path (figure 10.10) and moving at that instant with velocity $\boldsymbol{v}$ under the influence of a set of forces as

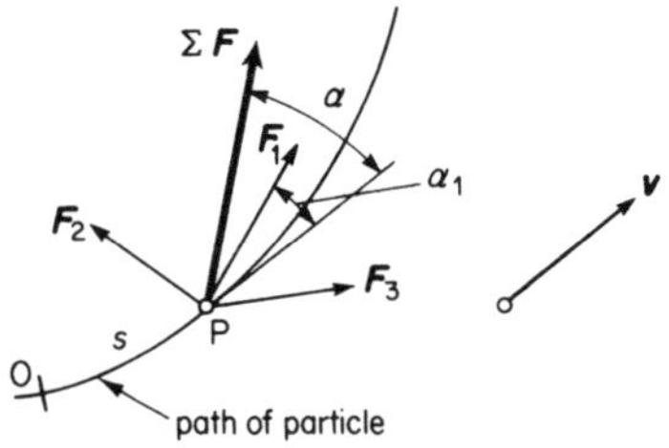

Figure 10.10

indicated. If the resultant force on the particle is $\Sigma \boldsymbol{F}$ then the tangential component in the direction of the velocity is $\Sigma F_s = |\Sigma \boldsymbol{F}| \cos \alpha$. We then have

$$\Sigma F_s = m a_s = \frac{\mathrm{d}}{\mathrm{d}t}(mv) = \frac{\mathrm{d}}{\mathrm{d}s}(mv)\frac{\mathrm{d}s}{\mathrm{d}t} = \frac{\mathrm{d}}{\mathrm{d}s}\left(\tfrac{1}{2}mv^2\right) \qquad (10.21)$$

The quantity $\frac{1}{2}mv^2$ is defined to be the *kinetic energy* of the particle. This is a scalar quantity for which we shall use the symbol T and which has units kg m^2/s^2 or N m. We have therefore

$$\left(\begin{array}{c}\text{component of the resultant force}\\ \text{on a particle in the direction of}\\ \text{motion of the particle}\end{array}\right) = \left(\begin{array}{c}\text{rate of change of the kinetic}\\ \text{energy of the particle with respect}\\ \text{to distance along the path}\end{array}\right)$$

Writing $\Sigma(F_s\, \mathrm{d}s) = \mathrm{d}\left(\frac{1}{2}mv^2\right)$ and integrating along the path between s_1 and s_2, the positions of the particle when at points 1 and 2 of the path

$$\Sigma \int_{s_1}^{s_2} F_s\, \mathrm{d}s = \tfrac{1}{2}m{v_2}^2 - \tfrac{1}{2}m{v_1}^2 = T_2 - T_1 \qquad (10.22)$$

Each integral on the left-hand side of equation 10.22 is a quantity that has already been defined in chapter 6 to be the *work* of the corresponding force F. This quantity, $\int F_s\, \mathrm{d}s$, is a scalar which we symbolise as U_{1-2} its unit being the joule (J) equal to 1 N m. After rearranging, equation 10.22 can be written in the form

$$T_1 + \Sigma U_{1-2} = T_2 \qquad (10.23)$$

This is our second integrated form which can be referred to as the *work - kinetic - energy equation,* and that states that the initial kinetic energy plus the work of the force equals the final kinetic energy. Since $\Sigma(F_s \, ds) = (\Sigma F_s) \, ds$ it follows that ΣU_{1-2}, the sum of the works of the individual forces, is equal to the work of the resultant.

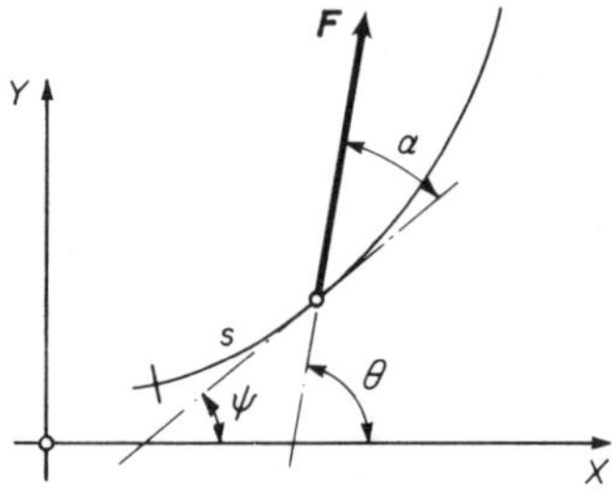

Figure 10.11

The quantity U_{1-2} is the summation of elementary quantities such as $F \cos \alpha \, \delta s$, that is, the product of the component of $\boldsymbol{F}$ in the direction of motion, and the actual elementary displacement of the particle. In order to calculate the work U_{1-2} in a particular case it is more convenient to carry out the summation using rectangular components of the force and the displacement. Referring to figure 10.11, we have

$$\begin{aligned}\delta U &= F \cos \alpha \, \delta s = F \, \delta s \cos (\theta - \psi) \\ &= F \, \delta s (\cos \theta \cos \psi + \sin \theta \sin \psi) \\ &= F \cos \theta (\delta s \cos \psi) + F \sin \theta (\delta s \sin \psi) \\ &= F_x \, \delta x + F_y \, \delta y\end{aligned}$$

The work integral now becomes

$$U_{1-2} = \int_{s_1}^{s_2} F_s \, ds = \int_{x_1}^{x_2} F_x \, dx + \int_{y_1}^{y_2} F_y \, dy \qquad (10.24)$$

This relationship expresses the fact that the work of a force acting on a particle undergoing a displacement is equal to the sum of the works of the components of the force in displacements whose magnitudes are the corresponding components of the particle displacement. In particular if a particle is moving in a fixed x-direction as in figure 10.12a.

$$U_{1-2} = \int_{x_1}^{x_2} F_x \, dx$$

since $dy = 0$, and if moving under the action of a force $\boldsymbol{F}$ having fixed x-direction, as in figure 10.12b.

$$U_{1-2} = \int_{x_1}^{x_2} F \, dx$$

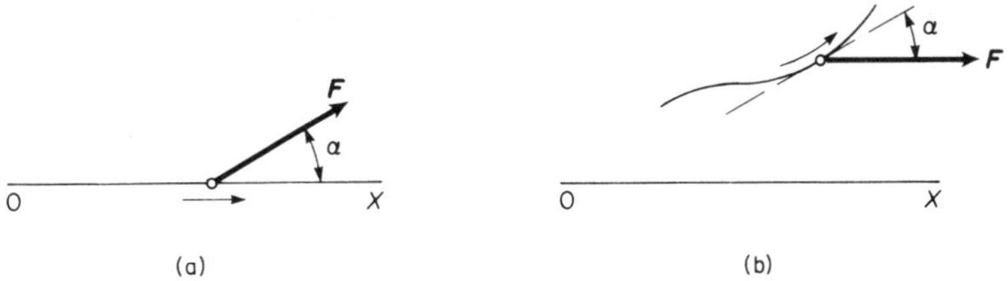

Figure 10.12

since $F_y = 0$.

If the component of the force in the direction tangential to the path has the same sense as the displacement then the element of work δU is positive, and if it has the opposite sense, δU is negative. Work can therefore be either a positive or a negative quantity, and this implies corresponding changes in the kinetic energy of the particle. We can also refer to positive work as work that is done by a force on a particle and negative work as work that is done against a force by a particle. We return to this point later, but mention at this stage that in applying the work-kinetic-energy relationship it is more satisfactory to speak only of work of a force on a particle and ascribe positive and negative signs to that work as the case may be.

A particular force component (say the x-component) will usually vary in a definite manner with the corresponding component of the position of the particle. If the variation is shown on a force-distance graph as in figure 10.13 then it is

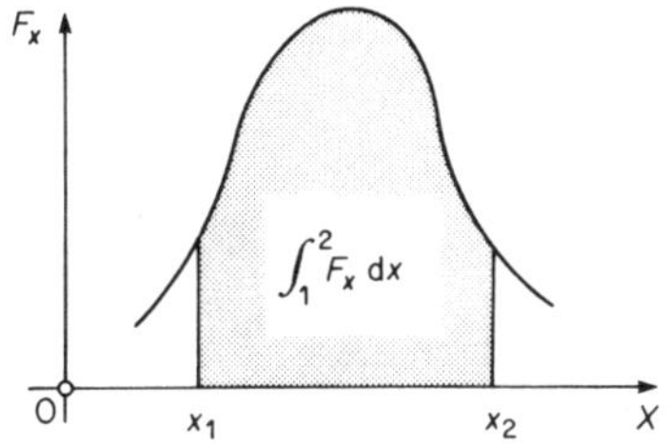

Figure 10.13

clear that the work of the force component is represented by the area under the graph between the ordinates at $x = x_1$ and $x = x_2$.

Power

Reverting to equation 10.21 on multiplying by ds/dt we obtain

$$(\Sigma F_s)\frac{ds}{dt} = \frac{d}{ds}(\tfrac{1}{2}mv^2)\frac{ds}{dt} = \frac{d}{dt}(\tfrac{1}{2}mv^2)$$

where the term on the left is the limiting value of $\Sigma(F\,\delta s)/\delta t$ at the point P. This represents the rate at which work is being done on the particle by the forces $\Sigma\boldsymbol{F}$

at time t. This quantity, which is equal to $(\Sigma F_s)v$, is given the name *power*. For the particle, power is seen to represent also the rate at which its kinetic energy is changing. If the numerical values of ΣF_s and v have opposing signs, then clearly the power is negative and the rate of change of kinetic energy is negative. For a system subject to external forces we can refer to net power of these forces, a quantity which can be positive or negative.

Power has particular significance in engineering applications where rates of performance of work by forces and rates of energy transfer are especially important. In such applications it should be made quite clear at what points the power is being evaluated.

It may be noted that since

$$\int_{s_1}^{s_2} F_s \, ds = \int_{t_1}^{t_2} \left(F_s \frac{ds}{dt}\right) dt$$

the work of a force on a particle as it moves between positions P_1 and P_2 can be expressed either as the space integral of the force or as the time integral of the power, the latter based if necessary on its corresponding graphic representation.

The unit of power is the watt (W) equal to 1 J/s or 1 N m/s. A more practical sized unit is the kilowatt (kW) which equals 1000 watts.

Worked Example 10.5

A small body, mass 1 kg, is moving in a horizontal plane along a smooth wire as shown in figure 10.14, under the influence of the constant applied force 25 N

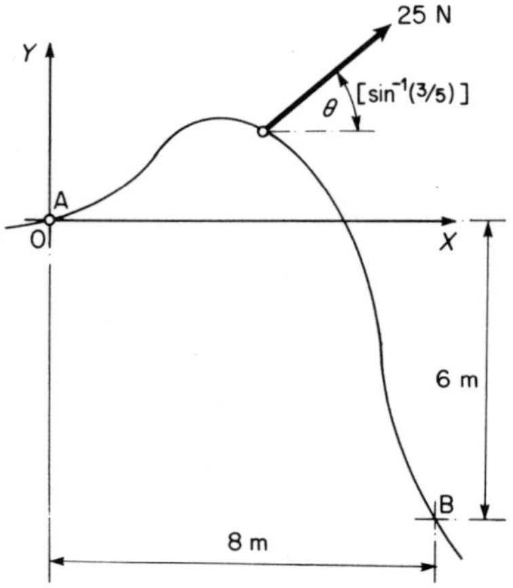

Figure 10.14

$\angle \sin^{-1} 3/5$. (a) Evaluate the work done on the particle as it moves from A to B. (b) If its speed at A is 6 m/s what is its speed at B?

Solution

There are two forces on the particle, the applied force and the reaction of the wire. Since the reaction is always normal to the path there is no work for this force.

(a) Since the shape of the path is not given we use equation 10.24 and evaluate the sum of the works of the components of the applied force.

$$F_x = 25 \cos \theta = 20 \text{ N}$$

$$F_y = 25 \sin \theta = 15 \text{ N}$$

$$U_{A-B} = \int_A^B F_x \, dx + \int_A^B F_y \, dy$$

$$= \int_0^8 20 \, dx + \int_0^{-6} 15 \, dy$$

$$= 160 - 90 = 70 \text{ N m}$$

Since the path was not specified and work is associated only with the applied force then it is clear that $\int F_s \, ds$ in equation 10.24 is the same for any path between A and B. However, if the wire were rough then this would not be so since the work of the friction force component would vary with the path.

(b) Applying the work-kinetic-energy equation 10.23

$$\tfrac{1}{2} m v_A^2 + U_{A-B} = \tfrac{1}{2} m v_B^2$$

$$\tfrac{1}{2} \times 1 \times 6^2 + 70 = \tfrac{1}{2} \times 1 \times v_B^2$$

and

$$v_B = 13.27 \text{ m/s}$$

Note how the use of the work-kinetic-energy equation enables us to avoid consideration of the speed variation between A and B. Is it possible to calculate the speed at B using the equation $\Sigma F = ma$? Remember that there are two forces to consider.

10.5.2 Particle Systems

Having obtained the work-kinetic-energy equation for a single particle we can immediately extend our result to a many-particle system by summing over the particles of the system. We now write

$$\Sigma T_1 + \Sigma U_{1-2} = \Sigma T_2 \tag{10.25}$$

summed over all particles and forces and taking due account of signs in the work term.

Considering first the kinetic energy summation, we again seek to introduce the mass-centre. If $\bar{v}_x$, $\bar{v}_y$ are the components of the velocity of the mass-centre, then those of a typical particle can be written $\bar{v}_x + v_x'$, $\bar{v}_y + v_y'$, where the dashed quantities are the components of the particle velocity relative to the mass-centre. For the particle

$$\begin{aligned} T &= \tfrac{1}{2}(\delta m)v^2 \\ &= \tfrac{1}{2}\delta m\,[(\bar{v}_x + v_x{}')^2 + (\bar{v}_y + v_y{}')^2] \\ &= \tfrac{1}{2}\delta m\,(\bar{v}_x{}^2 + \bar{v}_y{}^2 + 2\bar{v}_x v_x{}' + 2\bar{v}_y v_y{}' + v_x{}'^2 + v_y{}'^2) \end{aligned}$$

Summing over all the particles

$$\Sigma T = \tfrac{1}{2}m\,(\bar{v}_x{}^2 + \bar{v}_y{}^2) + \bar{v}_x\,\Sigma(\delta m)\,v_x{}' + \bar{v}_y\,\Sigma(\delta m)\,v_y{}' + \tfrac{1}{2}m\,(v_x{}'^2 + v_y{}'^2)$$

The second and third terms are zero, since

$$\Sigma(\delta m)v_x{}' = \frac{\mathrm{d}}{\mathrm{d}t}\Sigma(\delta m)x' \text{ and } \Sigma(\delta m)v_y{}' = \frac{\mathrm{d}}{\mathrm{d}t}\Sigma(\delta m)y'$$

where x', y' are the components of the position of the particle relative to the mass-centre, and by definition of the mass centre

$$\Sigma(\delta m)x' = \Sigma(\delta m)y' = 0$$

Therefore

$$T = \tfrac{1}{2}m\bar{v}^2 + \tfrac{1}{2}mv'^2$$

The kinetic energy of a particle system mass m can therefore be expressed as the sum of the kinetic energy of a particle mass m moving with the mass-centre and the kinetic energies of the separate particles in their motions relative to the mass-centre.

If this equation is applied to a rigid body, then, provided the body does not rotate, all points of the body have motion identical to that of the mass-centre, and the relative velocity v' for each particle is zero. Subject to this proviso the kinetic energy of a rigid body can be calculated as if the mass of the body were concentrated into a particle having the motion of the mass-centre and

$$\Sigma T = \tfrac{1}{2}m\bar{v}^2 \tag{10.26}$$

When we examine the work of the forces, both internal and external, we find that we can no longer disregard the work of the internal forces. Figure 10.15

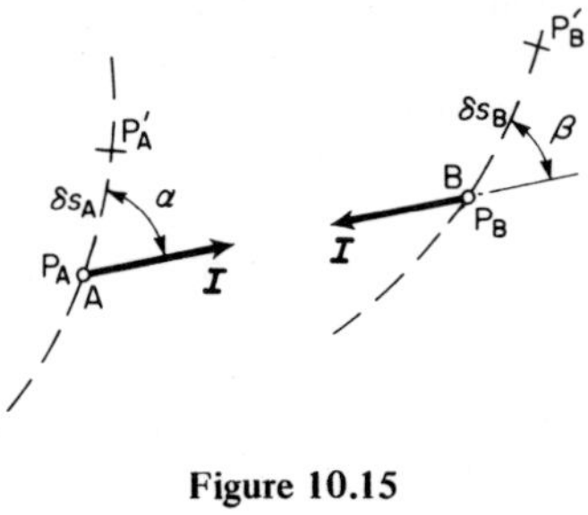

Figure 10.15

shows two typical particles A and B of a particle system and a pair of equal and opposite internal forces each having magnitude I. In a small displacement of the system the two particles move from positions P_A and P_B to P_A' and P_B' respectively, the displacements being δs_A and δs_B. The work of the pair of forces shown in the figure

$$\begin{aligned}\delta U &= I \cos\alpha\, \delta s_A - I \cos\beta\, \delta s_B \\ &= -\, I\,(\delta s_B \cos\beta - \delta s_A \cos\alpha) \\ &= -\, I \times \text{(change in length of } P_A P_B)\end{aligned}$$

a quantity which will be positive or negative depending on the displacements of the particles and the senses of the internal forces. In general therefore, there is the work of the internal forces to consider. However, in the case of rigid bodies, or systems made up of rigid bodies connected by inextensible links, the work of internal forces is zero, and only the work of external forces enters into the work-kinetic-energy equation.

We can state this result in more general terms by recognising that external forces are those that are exerted on a system by the surroundings. By virtue of Newton's third law, an equal and opposite set of forces is exerted by the system on the surroundings. In the former case we speak of the work of the surroundings on the system, a quantity that may be positive or negative in a particular case; in the latter case we can refer to the work of the system on the surroundings. In our discussions we shall invariably be concerned with the work of the surroundings on the system – with appropriate signs – and again it becomes essential to identify the system under consideration, and use a free-body diagram to display that identification.

10.6 Evaluation of Work in Standard Cases

The forces that the engineer has to take account of in his analyses are given by the conditions of the problem in hand, and in applying the work-kinetic-energy relation work can be calculated without discriminating between the source or nature of one force and another. It is soon found that certain forces are associated with the position of the particle or particle system, in the sense that associated with each point of a region of space is a definite magnitude and direction of force, the force being that which would be experienced by a particle if it were placed at the point in question. We call such a region a field of force. Knowing the characteristics of the field of force and given the position of the particle, we can immediately state the magnitude and direction of the force acting on the particle, in so far as the force is of a type which can be described in this manner.

We select two classes of this kind of force for consideration.

10.6.1 Gravitational Forces

The phenomenon of gravitation has already been brought into our discussion since it gives rise to a force that we cannot avoid in motions occurring on or near the surface of the Earth. The Earth's gravitational force is a particular example of the mutual forces that exist between any two particles having masses m_1, m_2 respectively and distance r apart. These forces are the subject of Newton's law of gravitation, according to which two such particles attract each other with forces $\boldsymbol{F}$ and $\boldsymbol{F}'$, directed along the line joining the particles, and with magnitudes given by

$$F = F' = G\,\frac{m_1 m_2}{r^2}$$

where G is a universal constant having the value 6.67×10^{-11} N m^2/kg^2.

An interesting feature of this equation is that the property mass, which measures the resistance to change of motion of a particle, also enters into a law of universal gravitation.

The total resultant gravitational force between two particle systems is obtained by summing the mutual particle attractions. This could be a formidable task for large numbers of particles, but fortunately the summation can be simplified in certain cases. Omitting formal proof we state the results for three cases.

(1) If two bodies are a very large distance apart then they can be treated as particles and no summation is required.
(2) If the two bodies are uniform solid spheres or uniform spherical shells then the particle attractions are equivalent to those of particles at the centres of the spheres, with masses respectively equal to the masses of the spheres.
(3) If one body is a uniform sphere attracting a single particle we have the case which is of immediate interest to us in which the Earth, having mass M_e and assumed to be uniform and spherical, is attracting a particle mass m, the distance of the particle from the centre of the Earth being r. The gravitational force on the particle now has magnitude

$$F_g = G\,\frac{M_e m}{r^2} \tag{10.27}$$

and is directed towards the Earth's centre. To avoid having to specify a particular mass we choose a particle having unit mass and then characterise the gravitational force field by stating that the force per unit mass which would be experienced at any point is given by

$$F_g = G\,\frac{M_e}{r^2} \tag{10.28}$$

directed towards the Earth's centre.

Suppose the motion of our particle is confined to small elevations z above the Earth's surface. Then if R_e is the radius of the Earth, $r^2 = (R_e + z)^2$, which is negligibly different from $R_e{}^2$, and the gravitational field has a magnitude that is more or less constant and given by

$$F_g = \frac{GM_e}{R_e{}^2} \text{ per unit mass} \tag{10.29}$$

Furthermore, over a small region of the Earth's surface the gravitational force field, although strictly radial is unidirectional, and we say that the field is uniform in both vertical and horizontal directions.

For a particle falling freely vertically downwards the acceleration a_z in the vertical z-direction is given by

$$-\frac{GM_e m}{R_e{}^2} = ma_z \text{ and } a_z = -\frac{GM_e}{R_e^2} \tag{10.30}$$

therefore for a particle a_z has a magnitude that is independent of the mass of the particle. We signify this magnitude by the symbol g, now referred to as the *acceleration due to gravity.*

The magnitude and direction of the experimentally observed acceleration vector $\boldsymbol{g}$ will vary with locality since the Earth is not a uniform sphere, and furthermore, our observations are made in a frame of reference that is fixed to the Earth's surface and rotates with it. However, in our work we shall neglect variations of this kind and adopt the value $g = 9.81$ m/s^2 that we have already used.

We now consider the work of the gravitational force on a particle as the particle moves radially outwards in the Earth's gravitational field from position 1 to position 2, at distances r_1, r_2 respectively from the centre of the Earth (figure 10.16a).

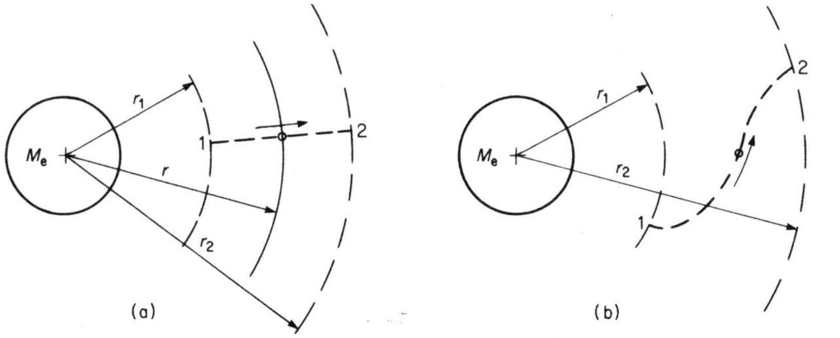

Figure 10.16

At any distance r, $F_g = -GM_e m/r^2$, therefore

$$U_{1-2} = -\int_{r_1}^{r_2} \frac{GM_e m}{r^2}\,dr = -\left[-\frac{GM_e m}{r}\right]_{r_1}^{r_2} \tag{10.31}$$

$$= -\left[\left(-\frac{GM_e m}{r_2}\right) - \left(-\frac{GM_e m}{r_1}\right)\right]$$

It is important to note that the same result would have been obtained for any other path such as that shown in figure 10.16b, provided the end points were also at radii r_1, r_2 from the Earth's centre, since such a path could be considered as made up of small radial displacements together with displacements along spherical surfaces for which the work of the gravitational force is zero. We shall return to this point.

If the motion of the particle were confined to regions near the Earth's surface for which the changes in elevation were small, then the gravitational field would

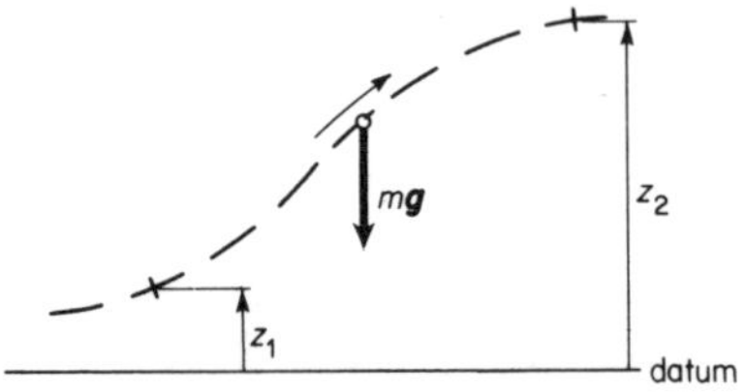

Figure 10.17

be sensibly uniform and the gravity force on the particle would be *mg* vertically downwards (figure 10.17) and

$$U_{1-2} = -\int_{z_1}^{z_2} mg\,\mathrm{d}z = -mg(z_2 - z_1) \tag{10.32}$$

We note that the work of the force *mg* depends only on the difference of elevations. The datum for measurement of *z* is now arbitrary.

10.6.2 Spring Forces

A close-coiled helical spring serves as an elastic deformable body with some well-defined relationship between axial deformation and applied axial force. If the deformation δ is proportional to the magnitude F of the applied force then $F = k\delta$. The spring is now referred to as a linear spring, and k is termed the spring constant, with units N/m. The spring can also serve as a means of applying a force to a body, the magnitude of the force, by Newton's third law, being related through $F = k\delta$ to the deformation of the spring.

In figure 10.18a a particle is moving in the x-direction when under the influence of the force exerted by a linear spring. The position of the particle is measured relative to its rest position, that is, when the spring is undeformed. When the particle is at the position x then the force on the particle in the x-direction is $-k\delta$, δ now being equal to x. As the particle moves from position 1, $x = x_1$, to position 2, $x = x_2$, the work of the spring force

$$U_{1-2} = \int_{x_1}^{x_2} -kx\,\mathrm{d}x = -\left[\frac{kx_2^2}{2} - \frac{kx_1^2}{2}\right] = -\left[\frac{k\delta_2^2}{2} - \frac{k\delta_1^2}{2}\right]$$

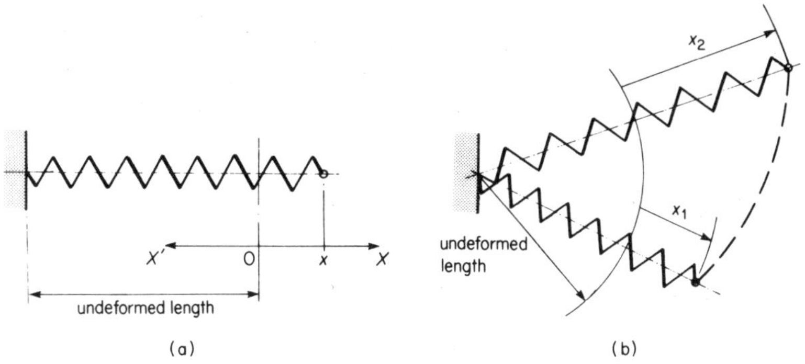

Figure 10.18

If the spring had been in compression during the motion of the particle in the x'-direction shown, then the force in that direction would have been $-k\delta = -kx'$ and in moving from position 1, $x' = x_1'$ to position 2, $x' = x_2'$, the work of the spring force

$$U_{1-2} = \int_{x_1'}^{x_2'} -kx' \, dx' = -\left[\frac{kx_2'^2}{2} - \frac{kx_1'^2}{2}\right] = -\left[\frac{k\delta_2^2}{2} - \frac{k\delta_1^2}{2}\right]$$

The same result would have been obtained if the particle had followed a path such as that shown in figure 10.18b and the orientation of the spring had changed during the motion. The force of the spring is now radially inwards; only the radial displacements of the particle enter into the calculation of U_{1-2}, and δ_1 and δ_2 are still the deformations of the spring at positions 1 and 2.

It can be verified that in all cases, including those in which the particle passes through its rest position, the work of the spring force on the particle is given by

$$U_{1-2} = -\left[\frac{k\delta_2^2}{2} - \frac{k\delta_1^2}{2}\right] \tag{10.33}$$

where δ_1, δ_2 are the initial and final deformations of the spring, or equivalently, the corresponding positions of the particle relative to the rest position, measured along the axis of the spring.

10.7 Potential Energy and Conservative Forces

If we examine the expressions we have obtained in equations 10.31, 10.32 and 10.33 for the work of gravitational forces and spring forces, we see that U_{1-2} is the negative of the change in value of some quantity that depends only on the position of the particle. In the cases considered the quantities encountered were respectively $-GM_em/r$, mgz and $k\delta^2/2$.

We now define each of these quantities to be the *potential energy* of the particle with respect to the particular force field in which it is situated. This is a scalar symbolised by V, with units N m. We use a subscript to indicate the nature of the field, whether gravitational force or spring force. We have therefore

potential energy due to a non-uniform gravitational force field $V_g = -GM_e m/r$

potential energy due to a uniform gravitational force field $V_g = mgz$

potential energy due to a spring force field $V_s = \frac{1}{2}k\delta^2$

and we can write

(1) for motion in a gravitational force field

$$U_{1-2} = -(V_{g2} - V_{g1}) \tag{10.34a}$$

(2) for motion in a spring force field

$$U_{1-2} = -(V_{s2} - V_{s1}) \tag{10.34b}$$

Both of the forces that we have discussed are characterised by the following properties

(a) the magnitude and direction of the force is a function only of the position of the particle on which it acts;
(b) the work of the force, U_{1-2}, if the particle moves between initial and final positions is independent of the path followed by the particle.

We call such forces *conservative forces*, for reasons that follow directly. If at every point of a region the force acting on a particle is a conservative force we say we have a *conservative field of force*.

It is a consequence of the definition of a conservative force that to each point of the field there can be ascribed a potential energy V, and the force field can then be described fully in terms of the scalar function $V = V(x, y, z)$. The component of the force on a particle at any given point in some specified direction can be determined by evaluating the negative of the rate of change of V in the chosen direction. Thus the x-component of the conservative force at the point x_1, y_1, z_1 in a force field described by the potential function $V(x, y, z)$ is given by $[-\partial V/\partial x]_{x_1, y_1, z_1}$. For the cases discussed

$$V_g = -\frac{GM_e m}{r} \qquad -\frac{\partial V_g}{\partial r} = -\frac{GM_e m}{r^2}$$

the r-directed gravitational force

$$V_g = mgz \qquad -\frac{\partial V_g}{\partial z} = -mg$$

the z-directed gravitational force

$$V_s = \frac{1}{2}kx^2 \qquad -\frac{\partial V_s}{\partial x} = -kx$$

the x-directed spring force.

10.7.1 Total Mechanical Energy

In general for a particle moving under the influence of conservative forces only, from equations 10.34 $\Sigma U_{1-2} = -(V_2 - V_1)$. If we insert this expression for ΣU_{1-2} into the work - kinetic-energy relationship, equation 10.23 we have

$$T_1 - (V_2 - V_1) = T_2$$

and therefore

$$T_1 + V_1 = T_2 + V_2 \tag{10.35}$$

where V implies the summation of quantities such as V_g and V_s.

We shall refer to the sum of the kinetic energy and the total potential energy as the *total mechanical energy* and we can then state that, for a particle moving in a conservative force field, the total mechanical energy is *conserved.*

10.7.2 Application to Rigid Bodies

Having introduced the potential energy of a particle, it is now necessary to extend the discussion to a rigid body. We have found (sections 7.2 and 7.3) that if the gravitational force field is uniform the resultant gravitational force on a rigid body acts at the centre of gravity, or equivalently, for a uniform field, at the centre of mass. The potential energy, as already defined, is based on the evaluation of work, and for a rigid body it can be determined by the summation of works over all the particles, or equivalently, the work of the resultant at the mass-centre. It follows that the potential energy of a rigid body in a uniform gravitational field is equal to that of a particle at the mass-centre having mass equal to that of the body.

As far as the spring force is concerned the definition of potential energy cannot be usefully extended to rigid bodies except in a limited number of cases, such as, for example, motion without rotation in a straight path along a line lying on the axis of the spring. This is because the potential energy cannot be defined unambiguously by the position of the body since it will vary according to the point of attachment of the spring on the body and the orientation of the body. Problems involving spring forces where a potential energy cannot be defined unambiguously are best treated by use of the work - kinetic-energy equation 10.23, the work of the spring forces being evaluated directly.

It is emphasised that the spring force is an external force as far as the rigid body is concerned. In chapter 13 it will be shown that a potential energy can be associated with the internal forces in the spring; this will enable the internal forces to be included in an energy accounting and thus the difficulty we have noted can be circumvented.

For the simpler cases where the total potential energy can be defined equation 10.35 can be used. In the next section a further equation is introduced, of which equation 10.35 is a special case.

10.8 Extraneous Forces

Usually a particle or a rigid body, while moving in a conservative force field, is subject to the action of other, non-conservative forces. Such forces are arbitrarily applied forces and forces depending on the direction of motion or the magnitude of the particle or body velocities. These are referred to as extraneous forces. We can now write two equations for a particle or a rigid body moving without rotation, from position 1 to position 2: from equation 10.23

work of conservative forces and extraneous forces

$$= \Sigma(U_{1-2})_{\text{cons}} + \Sigma(U_{1-2})_{\text{extr}} = T_2 - T_1$$

from equations 10.34

work of conservative forces only

$$= \Sigma(U_{1-2})_{\text{cons}} = -(V_2 - V_1)$$

Subtracting the second equation from the first

work of extraneous forces

$$= \Sigma(U_{1-2})_{\text{extr}} = (T_2 + V_2) - (T_1 + V_1)$$

$$= \text{change in total mechanical energy } (T + V) \qquad (10.36)$$

The work of the extraneous forces can be positive or negative as calculated in any particular case. In particular the friction force of a stationary surface on a rigid body is always in the opposite direction to that of the body's motion; the work of this extraneous friction force is therefore always negative and if acting alone, friction forces bring about a reduction in the total mechanical energy.

Equation 10.36 in the rearranged form

$$(T_1 + V_1) + \Sigma(U_{1-2})_{\text{extr}} = (T_2 + V_2) \qquad (10.37)$$

$$\begin{pmatrix}\text{initial total}\\ \text{mechanical energy}\end{pmatrix} + \begin{pmatrix}\text{work of}\\ \text{extraneous forces}\end{pmatrix} = \begin{pmatrix}\text{final total}\\ \text{mechanical energy}\end{pmatrix}$$

indicates the accounting of energy in this situation and is a useful form – referred to as the *work - energy equation.*

If any forces, although conservative, are not treated as such, then they must be included in equation 10.37 and dealt with as extraneous forces.

Worked Example 10.6

A body, mass 0.5 kg, drops vertically through a distance of 2 m before striking a vertical spring, constant k = 100 N/m, and after the impact rebounds in the same vertical path. Determine its velocity just before it strikes the spring. Deduce an equation relating its velocity while in contact with the spring with the spring deflection during deformation and obtain the maximum deflection and maximum force in the spring. What impulse does the spring impose on the body during the deformation stage?

If e for the impact (between the body and the Earth with the spring regarded as the deforming part of the latter) is 0.5 find the height of the rebound and the change in total mechanical energy of the body as a result of the impact.

Assume (1) that the spring behaves linearly during deformation, (2) that the spring has regained its original length when the body leaves the spring and (3) that the total mechanical energy is conserved during deformation.

Solution

Assume that the undeformed length of the spring is L and that the datum for gravitational potential energy is at the base of the spring.

Several different conditions relevant to the problem should be noted as follows, they are also illustrated in figure 10.19 – the velocity directions shown in this figure are the physical directions.

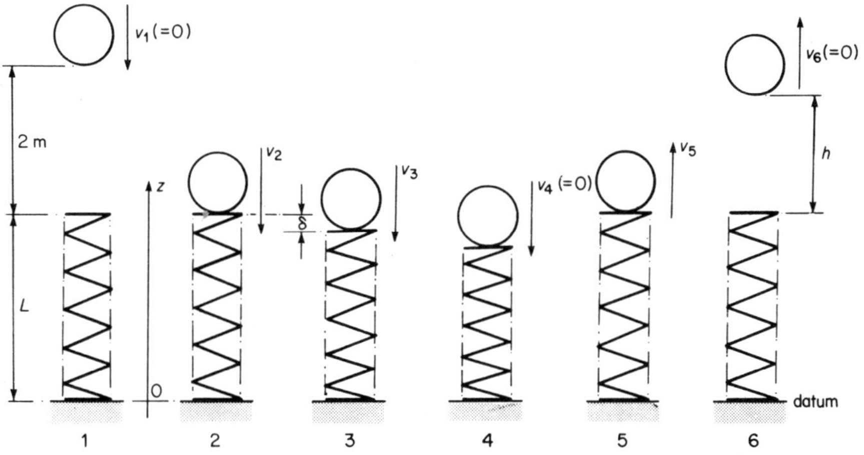

Figure 10.19

(1) The body is at rest 2 m above the spring.
(2) The body is just about to strike the spring.
(3) The body is in contact with the spring during the deformation stage with a spring deformation δ.
(4) The spring has a maximum deflection and it follows that the body is instantaneously stationary.
(5) The body just loses contact with the spring.
(6) The body becomes instantaneously stationary at the top of its rebound at a distance h above the top of the spring.

(a) Applying the work - energy equation 10.37 to the body between conditions 1 and 2 in order to obtain v_2

$$T_1 = 0$$

$$V_1 = V_{g1} = mgz_1 = 0.5 \times 9.81 \times (2 + L) \text{ N m}$$

$$T_2 = \tfrac{1}{2}mv_2^2 = \tfrac{1}{2} \times 0.5 \times v_2^2 \text{ N m}$$

$$V_2 = V_{g2} = 0.5 \times 9.81 \times L \text{ N m}$$

$$\Sigma(U_{1-2})_{\text{extr}} = 0$$

since there are no extraneous forces. Thus

$$0 + 0.5 \times 9.81 \times (2 + L) = \tfrac{1}{2} \times 0.5 \times v_2^2 + 0.5 \times 9.81 \times L$$

and

$$v_2 = \sqrt{(2 \times 9.81 \times 2)} = 6.27 \text{ m/s downwards}$$

(b) Applying the work - energy equation 10.37 between conditions 1 and 3

$$T_3 = \tfrac{1}{2}mv_3^2 = \tfrac{1}{2} \times 0.5 \times v_3^2 = 0.25\, v_3^2$$

$$V_3 = V_{g3} + V_{s3} = mgz_3 + \tfrac{1}{2}k\delta^2$$

$$= 0.5 \times 9.81 \times (L - \delta) + \tfrac{1}{2} \times 100 \times \delta^2$$

$$\Sigma(U_{1-3})_{\text{ext}} = 0$$

because there are no extraneous forces. Thus

$$0 + 0.5 \times 9.81\,(2 + L) = 0.25\, v_3^2 + 0.5 \times 9.81\,(L - \delta) + 50\delta^2$$

giving

$$v_3^2 = 2 \times 9.81\,(2 + \delta) - 200\,\delta^2$$

(c) The spring deflection will be a maximum when v_3 is zero (the latter being its value at condition 4). Thus

$$0 = 2 \times 9.81\,(2 + \delta_{\text{max}}) - 200\,\delta_{\text{max}}^2$$

$$200\,\delta_{\text{max}}^2 - 19.62\,\delta_{\text{max}} - 39.24 = 0$$

giving

$$\delta_{\text{max}} = +\,0.495 \text{ or } -\,0.396$$

(The negative value is not applicable in this case. It is applicable if the spring remains adhered to the body on the rebound.)

The maximum force in the spring is

$$k\delta_{\text{max}} = 100 \times 0.495 = 49.5 \text{ N}$$

Note that this is much greater than the weight of the particle – the further the mass is dropped the larger the force it will cause in the spring.

(d) In order to find the impulse during the deformation stage we apply the impulse - momentum equation 10.3 in the z-direction between conditions 2 and 4.

$$G_{z2} + \Sigma \mathrm{Imp}_{z,2-4} = G_{z4}$$

$$\Sigma \mathrm{Imp}_{z,2-4} = G_{z4} - G_{z2}$$

Now $G_{z4} = 0$ since the mass is instantaneously at rest and $G_{z2} = m \times (-6.27)$ since the direction of the velocity in position 2 is in the negative z-direction; therefore

$$\Sigma \mathrm{Imp}_{z,2-4} = 0 - 0.5 \times (-6.27)$$

$$= 3.135 \text{ kg m/s or } 3.135 \text{ N s upwards}$$

(e) In order to find the height of rebound we shall equate the total mechanical energy of the body in condition 6 to that in condition 5 (since there are no extraneous forces acting between these conditions); to find the total mechanical energy at 5 we require the body's velocity, which we determine using equation 10.11 for e, applied between conditions 2 and 5.

$$e = -\frac{(\text{relative velocity after impact})}{(\text{relative velocity before impact})}$$

Since the other body in the impact is the Earth which has zero velocity, this equation becomes

$$e = -\frac{(v_5 - 0)}{(-6.27 - 0)}$$

hence

$$v_5 = 0.5 \times 6.27 = 3.135 \text{ m/s}$$

The definition 10.10 for e, together with the impulse - momentum equation could alternatively be used to determine v_5.

(f) Applying the work - energy equation 10.37 between conditions 5 and 6

$$T_5 = \tfrac{1}{2} \times 0.5 \times (3.135)^2 \text{ N m}$$

$$V_5 = V_{g5} = 0.5 \times 9.81 \times L \text{ N m}$$

$$T_6 = 0$$

$$V_6 = V_{g6} = 0.5 \times 9.81 \times (L + h) \text{ N m}$$

therefore

$$\tfrac{1}{2} \times 0.5\,(3.135)^2 + 0.5 \times 9.81 \times L = 0 + 0.5 \times 9.81\,(L + h)$$

and

$$h = \frac{(3.135)^2}{2 \times 9.81} = 0.5 \text{ m}$$

(g) The change in the total mechanical energy of the body due to the impact can be computed by evaluating the quantity $(T + V)$ at any two conditions separated by the impact; in this case it is simpler to use conditions 1 and 6.

At 1 $$T_1 + V_1 = 0 + 0.5 \times 9.81 \times (L + 2)$$

At 6 $$T_6 + V_6 = 0 + 0.5 \times 9.81 \times (L + 0.5)$$

The change being

$$(T_6 + V_6) - (T_1 + V_1) = -0.5 \times 9.81 \times 1.5 = -7.35 \text{ N m}$$

the negative sign indicating, of course, a decrease in total mechanical energy.

10.9 Summary

(1) Linear momentum of a particle $G = m\boldsymbol{v}$.

(2) Linear impulse of a force

$$\mathbf{Imp}_{1-2} = \int_{t_1}^{t_2} \boldsymbol{F} \, \mathrm{d}t$$

(3) Impulse - momentum equation for a particle

$$G_1 + \Sigma \, \mathbf{Imp}_{1-2} = G_2 \qquad (10.3)$$

for a particle system

$$m\bar{\boldsymbol{v}}_1 + \Sigma(\mathbf{Imp}_{1-2})_{\text{ext}} = m\bar{\boldsymbol{v}}_2 \qquad (10.5)$$

for a set of rigid bodies use equation 10.3 with $G = \Sigma \, m\bar{\boldsymbol{v}}$.

(4) Conservation of linear momentum: If $\Sigma\mathbf{Imp}_{1-2}$ is zero: for a particle

$$G_1 = G_2$$

for a particle system

$$m\bar{\boldsymbol{v}}_1 = m\bar{\boldsymbol{v}}_2$$

for a set of rigid bodies

$$\Sigma m\bar{\boldsymbol{v}}_1 = \Sigma m\bar{\boldsymbol{v}}_2$$

(5) Impulsive force is the name given to $\boldsymbol{F}$ when it is large and acts for a very short time.

(6) Impact of moving bodies: use equation 10.3 for each body, relating the impulsive forces on each to the momentum change. Coefficient of restitution is

$$e = \frac{\text{impulse during restitution}}{\text{impulse during deformation}} \qquad (10.10)$$

$$= -\frac{\text{(relative velocity of separation)}}{\text{(relative velocity of approach)}} \qquad (10.11)$$

(7) Kinetic energy of a particle $T = \frac{1}{2}mv^2$
(8) Work of a force

$$U_{1-2} = \int_{s_1}^{s_2} F \cos \alpha \, ds$$

where α is the angle between F and ds. Also

$$U_{1-2} = \int_{x_1}^{x_2} F_x \, dx + \int_{y_1}^{y_2} F_y \, dy \qquad (10.24)$$

(9) The sum of the works of the individual forces is equal to the work of the resultant.
(10) Work - kinetic-energy equation: for a particle

$$T_1 + \Sigma U_{1-2} = T_2 \qquad (10.23)$$

for a rigid body in translation

$$\tfrac{1}{2}m\bar{v}_1^{\,2} + \Sigma(U_{1-2})_{\text{ext}} = \tfrac{1}{2}m\bar{v}_2^{\,2}$$

(11) For a particle, power = $(\Sigma F_s) \times v$, the rate of performance of work, which can be positive or negative.
(12) Potential energy of a particle: due to a uniform gravitational force field

$$V_g = mgz$$

due to a spring force field

$$V_s = \tfrac{1}{2}k\delta^2$$

(13) Total mechanical energy = $T + V$, V being equal to $V_g + V_s$; the total mechanical energy is conserved during motion in a conservative force field.
(14) Potential energy of a rigid body: due to a uniform gravitational force field $V_g = mg\bar{z}$; due to a spring force field $V_s = \frac{1}{2}k\delta^2$ for the simple case where motion is in a straight line along the axis of the spring.
(15) Work - energy equation for a particle or rigid body moving without rotation

$$T_1 + V_1 + \Sigma(U_{1-2})_{\text{extr}} = T_2 + V_2 \qquad (10.37)$$

$\Sigma(U_{1-2})_{\text{extr}}$ is the work of those forces that are treated as being extraneous;

Problems

10.1 A body of mass 10 kg is moving to the right in a straight line on a smooth horizontal table at 10 m/s when a force of 20 N, opposed to its motion, is applied for 8 s. What is its final velocity? (*Hint*: Draw a free-body diagram and apply equation 10.3.)

10.2 A body of mass 10 kg is initially moving at 10 m/s to the left on a rough ($\mu = 0.5$) horizontal table when a force $P = (100 + 20t)$ N is applied to the right, t being the time in seconds from the instant of application. What is velocity of the body after 5 s? (*Hint*: Write impulse - momentum equations for horizontal and vertical directions; split problem into (a) motion to left and (b) motion to right.)

10.3 A body of mass 5 kg is initially at rest when $t = 0$ on a rough ($\mu = 0.2$) inclined plane making an angle of $\sin^{-1}$ (5/13) with the horizontal. A force $P = (5t^2)$ N is applied to the body and makes an angle of $\sin^{-1}$ (3/5) to the plane in an upward direction. Find (a) the time t_2 when the body first moves up the plane, (b) the time t_3 when it loses contact with the plane and (c) its velocity at this instant. (*Hint*: Use a free-body diagram and hence write down impulse - momentum equations for motion parallel and perpendicular to the plane; decide which way the body will first move; will the body remain stationary for any interval of time? Note that both the normal and friction components will vary.)

10.4 A body of mass 100 kg is dropped on to a pile (a civil engineering term meaning a column that is driven into the ground) of mass 50 kg in order to drive it into the ground. If the body drops 5 m before striking the pile and does not rebound from the latter, find their common velocity just after completion of the impact (ignore ground resistance during the impact). If the pile and body together move 0.2 m after the impact what is the ground resistance if this is assumed constant? (*Hint*: Use work - energy to deduce velocity just before impact, impact - momentum during impact and work - energy for ground resistance.)

10.5 A body of mass 2 kg moving with a velocity of 20 m/s along a smooth horizontal surface is opposed by a force of $10V$ N, V being its velocity in m/s. Use the impulse - momentum equation to find the distance required before the velocity drops to 10 m/s. What work is done by the body against the force during this distance? (*Hint*: Note relationship of v, s and t; use work - energy for last part.)

10.6 A particle, mass m, rests at the bottom of a smooth cylindrical hole of radius R, the axis of which is horizontal. Find the impulse needed to be applied horizontally to the particle so that it will just lose contact with the surface of the hole at the uppermost point. (*Hint*: Draw a free-body diagram at the uppermost point to relate forces and accelerations, hence find velocity; use work - energy after and impulse - momentum during the impact.

10.7 If the impulse in problem 10.6 is to be delivered by allowing another particle of mass $5m$ to slide downwards on the surface of the hole to strike the original mass, find the vertical distance through which it must slide if $e = 1$. Decide whether the $5m$ mass rebounds or carries on in the same direction, and with what velocity. (*Hint*: Use impulse - momentum during the two stages of impact and use the equation for e.)

10.8 Body A of mass 1 kg hangs vertically by a cord of length 2 m. Body B of mass 2 kg is on a cord of length 1 m and, when hanging freely, just touches A. They are both dropped from the positions with the cords horizontal and with such coordination that they strike each other when the cords are vertical. If $e = 0.8$ find (a) the velocities of the bodies at the end of the impact (b) the impulse exerted by A on B and (c) the energy lost during impact. (*Hint*: Use impulse - momentum and e during impact; work energy before and during impact — the latter to determine energy loss.)

10.9 Two trucks A and B, masses 1000 kg and 2000 kg respectively, are connected by a rope 1000 m long and are on the same horizontal rail, which is directed perpendicular to the edge of a cliff. The trucks are being used to tip refuse over the cliff with A standing on the edge and B 200 m from A with the rope slack. By some mishap A falls over the cliff. Find (a) the velocity of A just before the rope tightens (b) the common velocity of the trucks just after the rope becomes taut (c) their velocity when B falls over the cliff if $e = 0$ for the impulse action.

Ignore the weight of A as an impulsive force and the frictional effect of the cliff edge on the rope. [*Hint*: Draw a free-body diagram for each truck; use work-energy for (a); impulse - momentum (for each truck) for (b) and work - energy (or equation of motion) for (c)]

10.10 A small wooden block of mass 1.2 kg is moving in a straight line on a smooth horizontal table at 100 m/s. It is struck by a bullet whose path is in a plane normal to the path of the block and inclined at 40° to the horizontal. If the bullet (mass 0.1 kg) remains fixed in the block and the new path of the latter makes an angle of 30° to its original path, find the velocity of the bullet.

What impulse is exerted by the block on the table? [*Hint*: Consider first the horizontal component of the bullet velocity, apply the impulse - momentum equation in component form in the horizontal plane to both bodies separately (drawing free-body diagrams of external forces for each body); secondly consider the velocities and impulses in a vertical plane.]

10.11 Two bodies A and B, of mass 5 kg and 3 mg respectively are connected by a spring of stiffness 200 N/m. They are stationary on a smooth horizontal table with the spring undeflected when body B is given an impulse of 30 N s in the direction B to A. Assuming that the impulse is of such short duration that the movement of B is negligible during the impulse find

(a) the velocity of B at the completion of the impulse
(b) the velocity of A when B has a velocity of 5 m/s
(c) the instantaneous common velocity of the two masses and the spring deflection at that time (assume no energy losses) and
(d) the velocities of the bodies at the instant when the spring is next undeflected.
(e) What are the ranges of velocity of each body?

[*Hint*: (a) If B does not move what of the force in the spring, hence impulse - momentum; for (b) - (e), write impulse - momentum equations (after drawing free-body diagrams) for both bodies; use energy considerations where necessary.]

10.12 Solve problem 9.2 by work - energy methods.

10.13 Check the velocity in problem 9.7 by work - energy methods.

10.14 Show when $\mu = 0$ and ω is constant that the power required to drive the tube in problem 9.12 is $m\omega^3 r_0{}^2 (e^{2\omega t} - e^{-2\omega t})/2$.

10.15 In problem 9.14 check the equation for $d\phi/dt$ by work - energy methods.

10.16 Refer to problem 8.10. Show that $v_y = V \tan \theta$. Hence find the work done by the external force as the particle moves along the arc subtending an angle ϕ at the centre of the arc.

10.17 A body of mass m is released on a rough inclined plane that makes an angle θ with the horizontal. It slides a distance L down the plane and then strikes an ideally elastic spring. Ignoring the friction of the plane during contact with the spring show, by using work - energy considerations, that the distance it rebounds up the plane (measured from the undeflected end of the spring) is given by

$$x = \frac{L(\sin \theta - \mu \cos \theta)}{(\sin \theta + \mu \cos \theta)}$$

where μ is the coefficient of friction. $(\mu < \tan \theta)$

If $\mu = 0.25$ and $\theta = \sin^{-1} 0.6$ find the number of impacts required with the spring before $x < L/6$.

10.18 Two particles A and B, of mass m_1 and m_2 respectively, are connected by a string length L of elastic constant K and lie on a smooth horizontal surface. The particles are ini ially at rest with the string just taut when A is moved a distance L_1 and B a distance L_2 such that the string now has length $L + L_1 + L_2$. If the particles are now released find expressions for (a) their velocities when the string is again just unstretched, (b) their positions at this instant and (c) the positions at which the particles meet.

Positions should be stated with reference to initial positions. (*Hint*: (a) write down impulse - momentum equations for eacl particle to relate their velocities; use work - energy to relate velocities and strain energy. (b) Note relationship of displacements and velocities. (c) string is slack in this situation.)

10.19 Refer to problem 9.16. Since no external work is done the work energy equation can be written $T + V_g$ = constant. After substitution, differentiate this equation and apply to the problem to obtain the required differential equation of motion.

10.20 Check parts (a) and (b) of problem 9.17 by work - energy methods.

10.21 For problem 9.18 find the work done by the applied force P (= 1 N) during the interval $0 < t < 1$ s. Use work - energy considerations to check the absolute velocity of the particle when $t = 1$ s. What is the power of the force at this instant?

11 Kinematics of Rigid Bodies

In the previous chapters we discussed the kinematics and kinetics of a single particle, and we were able to extend the discussion to particle systems and obtain significant equations relating to the kinetics of such systems. In this and the following chapters we apply the equations to rigid bodies, since engineering interest centres on particle systems of this particular kind. We first discuss the kinematics of rigid bodies, recalling that a rigid body is, by definition, a particle system for which the configuration of the particles is unchanging.

11.1 Types of Rigid Body Motion

We shall limit the discussion at this stage to so-called *plane motion*, If a body is undergoing plane motion the distance of each particle from some reference plane remains the same, and all particles therefore move in parallel planes. The motion of a cross-section of the body in any one of these planes, chosen as the reference plane, can be used to describe the motion, and the motion of the body can be represented by the motion of the so-called *representative slab* or *lamina* of negligible thickness moving in the plane of the lamina.

We recognise two basic types of rigid body motion, namely *translation* and *fixed-axis rotation.*

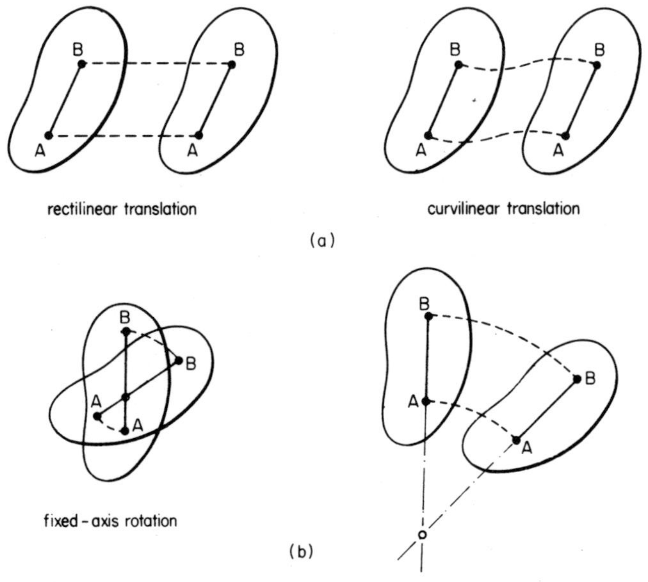

Figure 11.1

Translation

A rigid body is said to be in translation if the straight line joining any two particles of the body, such as A and B in figure 11.1, maintains the same orientation throughout the motion. This implies that all particles of the representative lamina travel along parallel paths. If these paths are straight lines we have *rectilinear* translation and if curved, *curvilinear* translation (figure 11.1a).

Fixed-axis Rotation

A rigid body is said to be in fixed-axis rotation if all particles of the body move in concentric circles about some fixed axis, which, in plane motion, is perpendicular to the reference plane. This implies that all particles of the representative lamina move in concentric circles, the centre of the circles not necessarily being on the representative lamina (figure 11.1b).

Although these are the basic types of motion it is evident that a rigid body can have some arbitrary motion that is neither basic translation nor fixed-axis rotation. We shall refer to such motion as *general plane motion*, but we shall find that this too can be analysed in terms of the basic types described.

We now proceed to establish a means of describing the motion of rigid bodies using the definitions already introduced in chapter 8 in connection with particle motion and angular motion of a line.

11.1.1 Translation

For a given orientation of the body the position of the representative lamina at time t can be described by specifying the position of any point of the lamina.

In a time interval δt all particles of the lamina have displacements that are equal in magnitude and direction (figure 11.2a) and therefore $\delta s' = \delta s''$. It follows

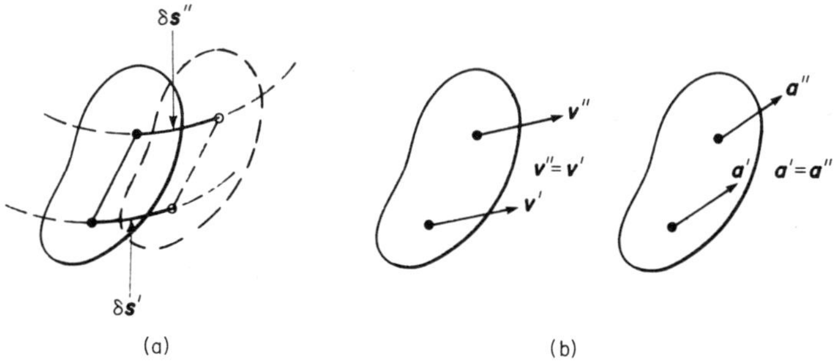

Figure 11.2

that in the limit all particles have the same velocities and accelerations, as indicated in figure 11.2b, and the velocity and acceleration of the body can be described by reference to the velocity and acceleration of any selected particle.

11.1.2 Fixed-axis Rotation

If O is the intersection of the axis of rotation with the reference plane (figure 11.3a) OQ is chosen as a fixed direction through O. Select any particle or point P on

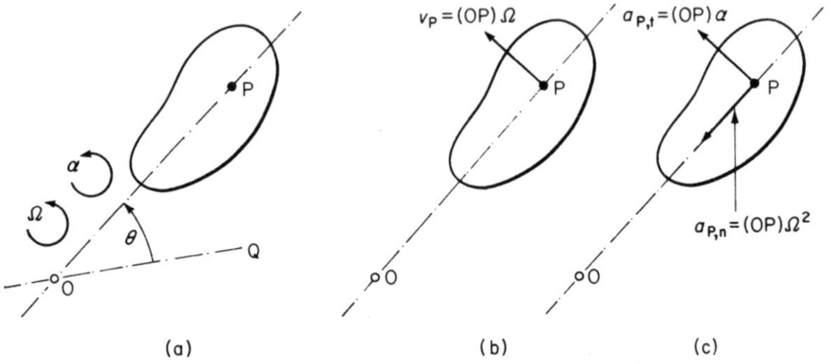

Figure 11.3

the representative lamina. The angular position of the body is then described by reference to the angular position θ of a line such as OP. The angular velocity, $\dot{\theta}$ or Ω, and angular acceleration, $\ddot{\theta}$ or α, of the line are as defined in chapter 8. From that discussion we see that Ω and α do not depend upon the particular line on the representative lamina that is selected and therefore we can now properly refer to the angular velocity Ω and angular acceleration α of the body.

A typical point P on the lamina moves in a circular path radius OP; the velocity of P is therefore tangential to the path (or perpendicular to OP) in a direction decided by the sense of Ω, and has magnitude $v_P = (OP)\,\Omega$. The acceleration of P has two components, namely, the tangential component $a_{P,t} = (OP)\,\alpha$,perpendicular to OP in a direction decided by the sense of α, and the normal or centripetal component $a_{P,n} = (OP)\,\Omega^2$ directed towards O. These are shown in figures 11.3b and 11.3c. Note that P is any selected point of the lamina and that the magnitudes of v_P, $a_{P,t}$ and $a_{P,n}$ for all points are, at any instant, each proportional to the distances of the points from the axis of rotation.

11.1.3 General Plane Motion

The representative lamina is now assumed to be moving in any arbitrary manner which is neither translation nor fixed-axis rotation. To examine its motion it is useful to have available the services of three observers, each of whom views the motion that is taking place and describes that motion by reference to his own set of axes attached to him. Referring to figure 11.4, observer 1 with his attached axes remains fixed in an inertial frame; observer 2 can translate with some velocity $\boldsymbol{v}_0$ and acceleration $\boldsymbol{a}_0$ but is not allowed to rotate; observer 3 if called upon, can translate with velocity $\boldsymbol{v}_0$ and acceleration $\boldsymbol{a}_0$ and in addition, may rotate with angular velocity Ω_0 and angular acceleration α_0.

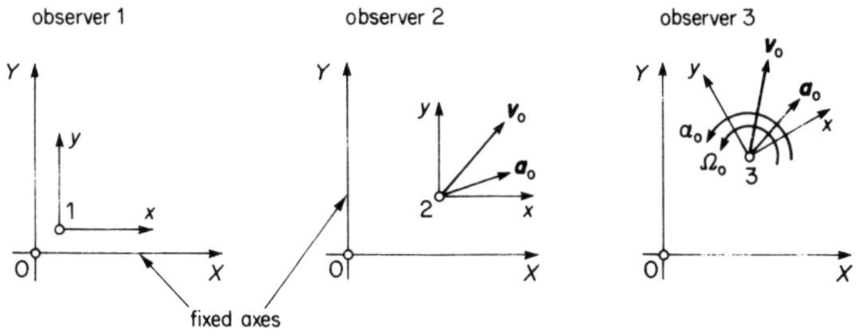

Figure 11.4

Consider a representative lamina, identified by three of its points A, B and C arbitrarily chosen (figure 11.5), undergoing general plane motion and displaced in some time interval to a new location represented by A′, B′, C′, the particles being joined by full lines in both locations.

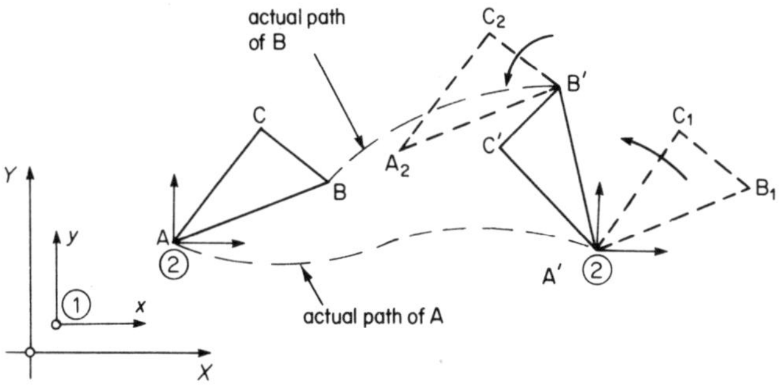

Figure 11.5

Utilising the services of observers 1 and 2: observer 1 sees the complete displacement; observer 2 if translating with A, sees a net rotation from A′ B_1 C_1 to A′ B′ C′. Observer 1 can interpret the complete displacement as a curvilinear translation with A together with an anticlockwise rotation about A′.

Alternatively: observer 2, if translating with B, sees a net rotation from B′ C_2 A_2 to B′C′A′; observer 1 can interpret the complete displacement as a curvilinear translation with B together with an anticlockwise rotation about B′.

Comparing the two interpretations made by observer 1, we note that the path length AA′ $\neq$ the path length BB′, but that the rotation about A′ = the rotation about B′. The intermediate positions through which the lamina translates between the initial and final positions of the complete displacement are therefore different. We can generalise this result and describe the displacement of the lamina as being made up of a curvilinear translation with some point X of the lamina, together with a fixed-axis rotation about that point X: the rotation is the same whichever point is selected but the amount of translation will depend on the choice of the point X.

Consider now the displacement of a lamina in a time interval δt, as shown in figure 11.6. The displacement of any line AB to the position A′B′ as δt becomes

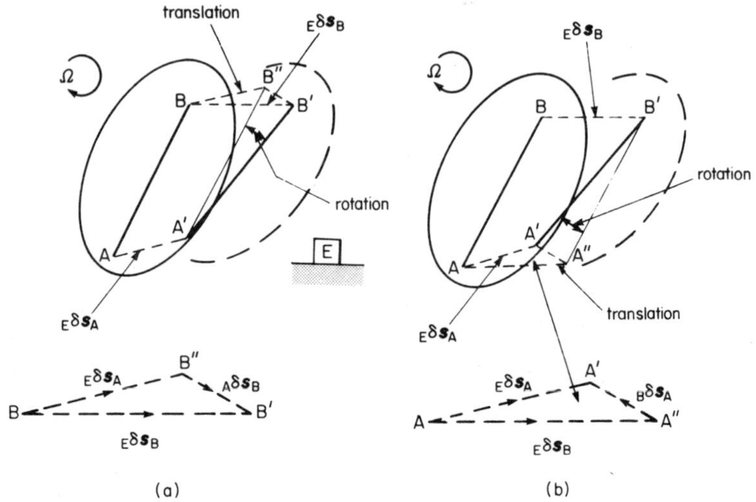

Figure 11.6

very small is made up of a rectilinear translation from AB to A′B″ together with a rotation about A′ from A′B″ to A′B′ (figure 11.6a). The displacement of the point B is given by

$$\overrightarrow{BB'} = \overrightarrow{BB''} + \overrightarrow{B''B'}$$
$$= \overrightarrow{AA'} + \overrightarrow{B''B'}$$

(since $\overrightarrow{BB''} = \overrightarrow{AA'}$ for translation). If we refer the motion to the earth as our fixed inertial reference frame we can write

$$_E\delta s_B = \underset{\left(\begin{array}{c}\text{translation}\\ \text{with A}\end{array}\right)}{_E\delta s_A} + \underset{\left(\begin{array}{c}\text{movement of B due to}\\ \text{rotation about A}\end{array}\right)}{_A\delta s_B}$$

where $_E\delta s_B$ is the displacement of B as seen from, or with respect to, a fixed point E on the Earth.

After division by δt, then in the limit as δt tends to zero

$$_E\nu_B = \underset{\text{(trans)}}{_E\nu_A} + \underset{\text{(rot)}}{_A\nu_B} \tag{11.1}$$

Thus the velocity of B is the sum of (1) the velocity of B arising from translation with some other point A and (2) the velocity of B relative to A arising from the rotation of AB about A.

From a similar consideration of figure 11.6b we can write alternatively for the same motion

$$ {}_E v_A = \underset{\text{(trans)}}{{}_E v_B} + \underset{\text{(rot)}}{{}_B v_A} \tag{11.2} $$

an equation that can be obtained directly from equation 11.1 by noting that ${}_B v_A = -\,{}_A v_B$. Note that ${}_A v_B$ arising from the rotation is given by (AB) Ω similarly ${}_B v_A$ is also given by (AB)Ω, in which Ω is the angular velocity of the lamina. It follows in general that, for a lamina undergoing plane motion, if the velocity of any point is known together with the angular velocity of the lamina, then the velocity of any other point can be determined.

Differentiation of equation 11.1 with respect to t yields an expression for the acceleration of B

$$ {}_E a_B = {}_E a_A + {}_A a_B \tag{11.3} $$

in which ${}_A a_B$ is the acceleration of B relative to A arising out of the rotation of AB about A. It must be noted that ${}_A a_B$ has two components, namely ${}_A a_{B,t} = (\mathrm{AB})\alpha$ perpendicular to AB (in the sense determined by the sense of α) and ${}_A a_{B,n} = (\mathrm{AB})\Omega^2$ directed from B towards A. It follows that, for a lamina undergoing plane motion, if the acceleration of any point is known together with the angular velocity and angular acceleration of the lamina, then the acceleration of any other point can be determined.

Worked Example 11.1

The rod AB (2 m long) in figure 11.7a has, at some instant, the angular velocity and the angular acceleration indicated; the point A has the linear velocity and the linear

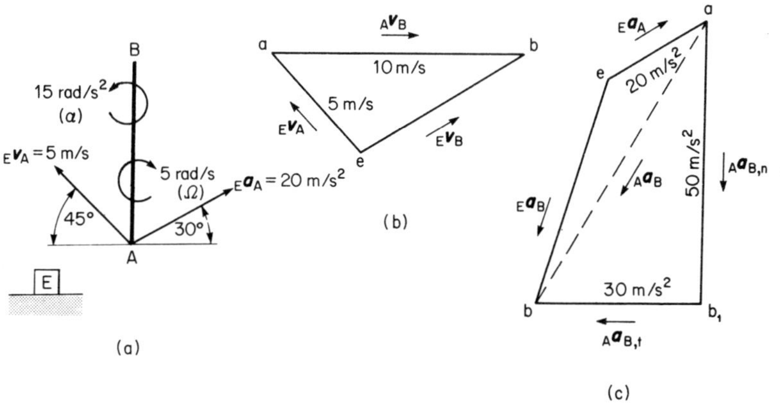

Figure 11.7

acceleration indicated. Determine the absolute velocity and absolute acceleration of B when in the position shown. (Note that absolute velocities or accelerations are with respect to the Earth's surface which is usually denoted by E.)

Solution

Using equation 11.1

$$_E v_B = {}_E v_A + {}_A v_B$$

in which

$$_A v_B = (AB)\,\Omega \text{ perpendicular to AB}$$
$$= 2 \times 5 = 10 \text{ m/s} \angle 0^\circ$$

the sense being consistent with that of Ω.

The vector addition is carried out using a vector diagram, figure 11.7b, the solution being given by the directed line $\overrightarrow{eb}$, and

$$_E v_B = 7.37 \text{ m/s} \angle 28.7^\circ$$

Using equation 11.3

$$_E a_B = {}_E a_A + {}_A a_B$$

in which

$$_A a_B = {}_A a_{B,n} + {}_A a_{B,t}$$
$$_A a_{B,n} = (AB)\,\Omega^2 \text{ in direction } \overrightarrow{BA}$$
$$= 2(5)^2 \text{ m/s}^2 \angle -90^\circ$$
$$_A a_{B,t} = (AB)\,\alpha \text{ perpendicular to AB}$$
$$= 2(15) \text{ m/s}^2 \angle 180^\circ$$

the sense being consistent with that of α.

The vector addition is carried out in figure 11.7c, the solution being given by the line $\overrightarrow{eb}$ and

$$_E a_B = 42 \text{ m/s}^2 \angle -107.6^\circ$$

11.2 Instantaneous Centre of Rotation I

Since we can now relate the velocities and accelerations of any two points on a representative lamina we can say that we are able to describe the motion of that lamina at any instant. That motion is the result of a translation with some arbitrary point together with a fixed-axis rotation about that point. It may be surmised that there is some point for which the instantaneous velocities and accelerations of the particles of the lamina are the result of rotation alone about that point. As far as velocities are concerned such a point – the instantaneous centre of rotation – can indeed be found.

Consider the lamina in figure 11.8a. If the velocity of A is ${}_E v_A$ (E being a fixed point on the Earth) and the angular velocity of the lamina is Ω, then the velocity

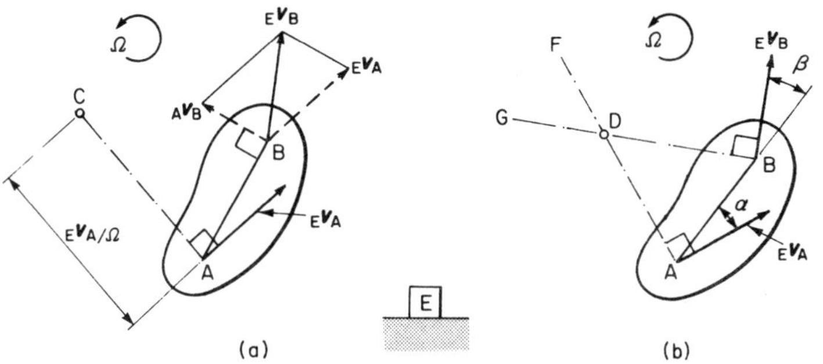

Figure 11.8

of A is consistent with rotation of the lamina about a point C such that AC is perpendicular to ${}_E v_A$ and AC = ${}_E v_A/\Omega$. Since ${}_E v_B$ is determined from ${}_E v_A$ and Ω, the velocity of B is given by ${}_E v_B$ = CB × Ω in the direction perpendicular to CB and consistent with the sense of Ω.

Alternatively it may be that the directions of ${}_E v_A$ and ${}_E v_B$ are known at some instant if the lamina is constrained to move in a certain manner, but that the angular velocity is not known directly. If in figure 11.8b we take lines AF and BG perpendicular to ${}_E v_A$ and ${}_E v_B$ respectively, then the directions of ${}_E v_A$ and ${}_E v_B$ are consistent with rotation about the intersection D of AF and BG. Now the length AB is constant being the distance between two points of a rigid body, therefore

$$ {}_E v_A \cos\alpha = {}_E v_B \cos\beta $$

$$ \frac{{}_E v_A}{{}_E v_B} = \frac{\cos\beta}{\cos\alpha} = \frac{\sin(\pi/2 - \beta)}{\sin(\pi/2 - \alpha)} = \frac{AD}{BD} $$

(by the sine rule for triangle ABD). Therefore

$$ \frac{{}_E v_A}{{}_E v_B} = \frac{\Omega \times AD}{\Omega \times BD} $$

It follows that the ratio of the magnitudes of ${}_E v_A$ and ${}_E v_B$ is also consistent with rotation about D.

The point C in figure 11.8a or D in figure 11.8b therefore represent a point on the lamina (or fixed relative to the lamina), which, at the given instant, is a fixed axis of rotation. This point, which we now label I, and whose location relative to the lamina is found by either of the above methods, is instantaneously at rest and is therefore referred to as the *instantaneous centre of rotation* of the lamina.

Note carefully that at a subsequent instant the instantaneous centre is in general a different point relative to the lamina. It is therefore important to appreciate that the point I, although instantaneously at rest, is in general accelerating and cannot be used directly to compare the accelerations of points of a lamina.

Worked Example 11.2

In the mechanism in figure 11.9a the arm DA rotates at 10 rad/s clockwise. Find, for the configuration shown, the instantaneous centres of rotation of the links ACB

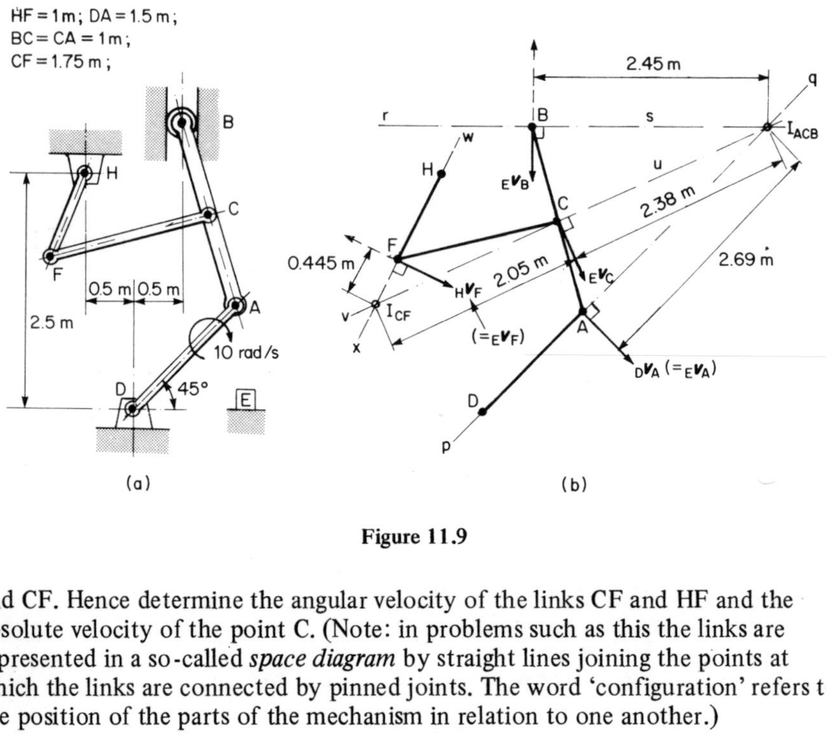

Figure 11.9

and CF. Hence determine the angular velocity of the links CF and HF and the absolute velocity of the point C. (Note: in problems such as this the links are represented in a so-called *space diagram* by straight lines joining the points at which the links are connected by pinned joints. The word 'configuration' refers to the position of the parts of the mechanism in relation to one another.)

Solution

The construction is laid out in figure 11.9b. At the outset our knowledge consists of the following facts.

(1) $_D v_A = {_E v_A}$ and is in the direction given in figure 11.9b

$$_E v_A = (DA)\,\Omega = 1.5 \times 10 = 15 \text{ m/s}$$

(2) $_E v_B$ is constrained to be vertical
(3) $_H v_F = {_E v_F}$ and is perpendicular to HF.

This is sufficient information to locate the instantaneous centre of ACB, I_{ACB}, since the directions of the velocities of two points on ACB are now known. I_{ACB} must lie on DA extended (line pq) in the direction perpendicular to ${}_D\nu_A$; it must also lie on a line rs passing through B and perpendicular to ${}_E\nu_B$. It is thus at the intersection of lines pq and rs. From the sense of ${}_D\nu_A$ it follows that Ω_{ACB} is anticlockwise and that ${}_E\nu_B$ is directed downwards.

Since

$$ {}_D\nu_A = \Omega_{ACB}\,(I_{ACB}A) $$

$$ \Omega_{ACB} = \frac{15}{2.69} = 5.58 \text{ rad/s anticlockwise} $$

Thus the absolute velocity of C

$$ {}_E\nu_C = \Omega_{ACB}\,(I_{ACB}C) = 5.58 \times 2.38 $$

$$ = 13.3 \text{ m/s} $$

in the direction shown, perpendicular to $I_{ACB}C$.

Considering now the link FC, the velocity of C, ${}_E\nu_C$ is known in direction and sense (and magnitude) and the velocity of F, ${}_E\nu_F$ is also known in direction. I_{CF} must lie on a line $I_{ACB}C$ extended (line uv) in the direction perpendicular to ${}_E\nu_C$; it must also lie on the line wx perpendicular to ${}_H\nu_F$. I_{CF} is at the intersection of lines uv and wx.

Since the velocity of C as a point on CF is ${}_E\nu_C$

$$ {}_E\nu_C = \Omega_{CF}\,(I_{CF}C) $$

and thus

$$ \Omega_{CF} = \frac{13.3}{2.05} = 6.47 \text{ rad/s} $$

Its direction is clockwise, consistent with the direction of ${}_E\nu_C$.

Now ${}_H\nu_F$ must, from the sense of Ω_{CF}, be in the direction shown by the full line in figure 11.9b and is given by

$$ {}_H\nu_F = \Omega_{CF}\,(I_{CF}F) = 6.47 \times 0.445 $$

$$ = 2.88 \text{ m/s} $$

For the link HF

$$ {}_H\nu_F = \Omega_{HF}\,HF $$

and

$$ \Omega_{HF} = \frac{2.88}{1} = 2.88 \text{ rad/s} $$

and its direction, consistent with that of ${}_H\nu_F$, is anticlockwise.

11.3 Engineering Applications

The rigid bodies that are the subject of engineering applications either move or are required to move in a definite, predictable manner. In such applications, therefore, the ability to analyse the motion of rigid bodies that make up mechanical systems is essential. Such systems are usually composed of a number of rigid bodies or elements that are connected to one another, thereby enabling the motion of one body to influence the motion of another. A mechanism is a system of this kind. At this stage we are considering only the kinematics of rigid bodies and some simple mechanisms in general plane motion; in chapter 13 we shall consider the forces involved in the motion of rigid bodies, again confining ourselves to plane motion. The proportioning of machine elements and the design of mechanisms to bring about a specified motion is beyond the scope of this text, but the ability to analyse rigid body motion and the associated forces is a necessary prerequisite for this design process.

The velocities and accelerations of points in bodies that are undergoing simple translation or fixed-axis rotation can be found directly, as we have seen, by applying the principles relating to the motion of a particle and the angular motion of a line. In discussing general plane motion for engineering applications we shall first consider two basic cases of constrained motion, namely the rolling disc and the constrained link, before discussing the analysis of the motion of connected bodies for application to simple mechanisms.

11.3.1 The Rolling Disc

The thin circular disc or lamina radius R in figure 11.10a is rotating in a clockwise

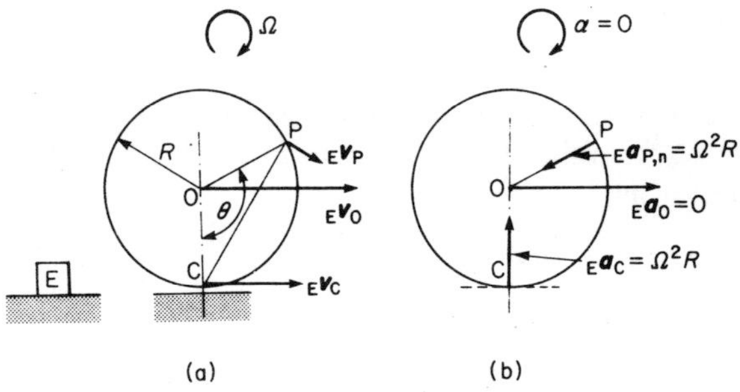

Figure 11.10

direction while its centre is moving to the right parallel to the horizontal plane shown. The disc is not in contact with the plane. This is a case of general plane motion since the orientation of any line on the disc is changing and there is no fixed axis about which the particles of the disc are moving in concentric circular paths.

(1) If the velocity of the centre O of the disc is ${}_{\rm E}v_{\rm O}$ and the angular velocity of the disc is Ω, then the velocity ${}_{\rm E}v_{\rm C}$ of the point C to the right is the result of translation with O and rotation about O, and is given generally from equation 11.1 by

$${}_{\rm E}v_{\rm C} = {}_{\rm E}v_{\rm O} + {}_{\rm O}v_{\rm C}$$

and

$${}_{\rm E}v_{\rm C} = {}_{\rm E}v_{\rm O} - R\Omega$$

since ${}_{\rm O}v_{\rm C} = -R\Omega$). If the disc is in contact with the plane and is rolling without slipping then at any instant the point C in contact with the plane has zero absolute velocity, therefore

$$0 = {}_{\rm E}v_{\rm O} - R\Omega$$

and

$${}_{\rm E}v_{\rm O} = R\Omega$$

This condition must be satisfied if rolling is taking place. Alternatively, since C is at rest at the instant under consideration, it is the instantaneous centre of rotation and again it follows that ${}_{\rm E}v_{\rm O} = R\Omega$. It also follows that the velocity of any point P on the rim is perpendicular to CP and has magnitude CP $\times$ $\Omega = 2R\Omega \sin(\theta/2)$ where θ is the angle subtended at the centre by the line CP.

(2) If the acceleration of the centre O of the disc is ${}_{\rm E}a_{\rm O}$ to the right and the angular acceleration of the disc is α clockwise, then ${}_{\rm E}a_{\rm C}$ of the point C is given generally from equation 11.3 by

$${}_{\rm E}a_{\rm C} = {}_{\rm E}a_{\rm O} + {}_{\rm O}a_{\rm C}$$

where ${}_{\rm O}a_{\rm C}$ has two components, ${}_{\rm O}a_{\rm C,t} = R\alpha$ to the left and ${}_{\rm O}a_{\rm C,n} = R\Omega^2$ directed towards O.

If the disc is in contact with the plane and is rolling without slipping then at any instant the point C in contact with the plane has zero *tangential* acceleration, therefore in the tangential direction

$$0 = {}_{\rm E}a_{\rm O} - R\alpha$$

and

$${}_{\rm E}a_{\rm O} = R\alpha$$

This condition must be satisfied if rolling is taking place.

Now if the disc is rolling, the acceleration of the centre cannot have a downwards component, therefore in the upward direction

$${}_{\rm E}a_{\rm C} = 0 + R\Omega^2$$

and

$${}_{\rm E}a_{\rm C} = R\Omega^2$$

towards O.

The point C has therefore an absolute acceleration even though it is instantaneously at rest.

Suppose ${}_{\mathrm{E}}a_{\mathrm{O}} = 0$ as in figure 11.10b: then the acceleration of P is also directed towards O and has magnitude $R\Omega^2$. This result demonstrates that the acceleration of P cannot be calculated by reference to an instantaneous fixed-axis rotation about the instantaneous centre C. However, we can again show that the acceleration of P is the result of translation with C and rotation about C, given by

$$ {}_{\mathrm{E}}a_{\mathrm{P}} = {}_{\mathrm{E}}a_{\mathrm{C}} + {}_{\mathrm{C}}a_{\mathrm{P}} $$

where

$$ {}_{\mathrm{E}}a_{\mathrm{C}} = R\Omega^2 $$

directed towards O, and

$$ {}_{\mathrm{C}}a_{\mathrm{P}} = \left(2R \sin\frac{\theta}{2}\right)\Omega^2 $$

directed towards C.

These two accelerations can be combined to give ${}_{\mathrm{E}}a_{\mathrm{P}}$. The component of ${}_{\mathrm{E}}a_{\mathrm{P}}$ in the direction PO

$$ = \left[2R \sin\left(\frac{\theta}{2}\right)\Omega^2\right] \sin\frac{\theta}{2} - R\Omega^2 \cos(\pi - \theta) $$

$$ = R\Omega^2 \left[2 \sin^2\left(\frac{\theta}{2}\right) + \cos\theta\right] $$

$$ = R\Omega^2 $$

The component of ${}_{\mathrm{E}}a_{\mathrm{P}}$ perpendicular to PO

$$ = -\left[2R \sin\left(\frac{\theta}{2}\right)\Omega^2\right] \cos\frac{\theta}{2} + R\Omega^2 \sin(\pi - \theta) $$

$$ = R\Omega^2 \left(-2 \sin\frac{\theta}{2} \cos\frac{\theta}{2} + \sin\theta\right) $$

$$ = 0 $$

The previous result is therefore confirmed.

If slipping is taking place then the velocity and acceleration of C tangential to the plane are no longer zero and the conditions for rolling are no longer applicable; ${}_{\mathrm{E}}v_{\mathrm{O}}$ and Ω are not dependent on each other, and ${}_{\mathrm{E}}a_{\mathrm{O}}$ and α are also independent of each other. The general forms of equations 11.1 and 11.3 must therefore be retained. Alternatively, to determine velocities of points on the disc the position of the instantaneous centre I can be found since $\mathrm{OI} = {}_{\mathrm{E}}v_{\mathrm{O}}/\Omega$ on a line perpendicular to ${}_{\mathrm{E}}v_{\mathrm{O}}$ and the velocity of any point can be determined on the basis of an instantaneous fixed-axis rotation about I.

11.3.2 The Constrained Link

It is useful in this context to reiterate and expand on some of the work covered in section 11.1.3.

Notation

The symbol ${}_{Y}v_{X}$ means the velocity of X with respect to Y, or the velocity of X as seen by a non-rotating observer travelling with Y. Its representation by a directed line is shown in figure 11.11, its sense being from y to x, corresponding

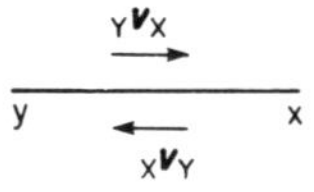

Figure 11.11

to the sense in which X is moving as seen from Y. The velocity ${}_{X}v_{Y}$ is also indicated by the same line, its sense now being from x to y. An arrow may be placed on the line to indicate which interpretation is intended.

Now considering equation 11.1 we should make special note of the juxtaposition of the subscripts. This equation may be written generally as

$$ {}_{X}v_{Z} = {}_{X}v_{Y} + {}_{Y}v_{Z} $$

It is the position of the letter Y that is important; the two terms on the right must be connected by the same letter (Y in this case) and the other two letters (X and Z) must be in the same order on both sides of the equation. Thus, given ${}_{Q}v_{P}$ and ${}_{Q}v_{R}$, we may state directly that ${}_{R}v_{P}$ is given by

$$ {}_{Q}v_{P} = {}_{Q}v_{R} + {}_{R}v_{P} $$

It also follows that

$$ {}_{R}v_{P} = {}_{Q}v_{P} - {}_{Q}v_{R} $$

or

$$ {}_{R}v_{P} = {}_{R}v_{Q} + {}_{Q}v_{P} $$

These remarks apply equally well to accelerations, in which case a replaces v.

The rigid body shown in figure 11.12a is typical of a link in a mechanism. At a certain instant, point A on the body is moving to the right with absolute velocity v, or more precisely ${}_{E}v_{A}$, meaning the velocity of A with respect to the Earth E. Point B is constrained to move in a vertical direction and we need to determine its absolute velocity ${}_{E}v_{B}$.

Applying equation 11.1

$$ {}_{E}v_{B} = {}_{E}v_{A} + {}_{A}v_{B} \qquad (11.4) $$

In this equation ${}_{E}v_{A}$ is known completely, the direction of ${}_{A}v_{B}$ is known to be perpendicular to AB (B can only have a tangential velocity when viewed from A)

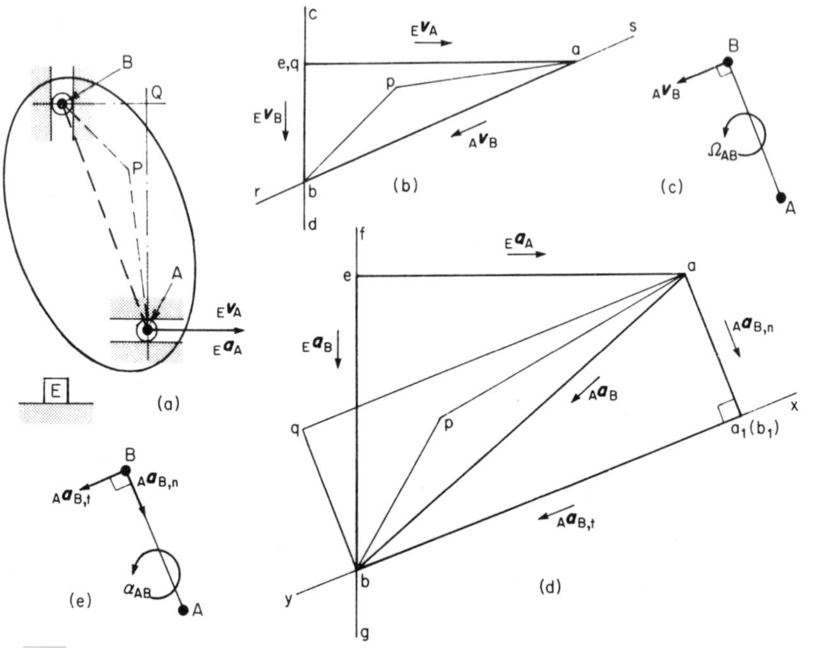

Figure 11.12

and the direction of $_E v_B$ is known to be vertical (from the constraint on B).

Equation 11.4 is best interpreted graphically as in figure 11.12b. ea is drawn to a suitable scale to represent $_E v_A$, a being to the right of e since $_E v_A$ is to the right; we now have to add a vector $_A v_B$ to $_E v_A$. $_A v_B$ is known in direction only (perpendicular to AB) and thus a line rs drawn in this direction and passing through a and the point b must be on this line. We now have to make use of the known direction of $_E v_B$, which is vertical; a vertical line cd is thus drawn to pass through e; the point b must also lie on this line. The point b must therefore be at the intersection of rs and cd in which case $\vec{eb} = \vec{ea} + \vec{ab}$ and thus equation 11.4 is represented. Thus the correct senses of $_A v_B = \vec{ab}$ and $_E v_B = \vec{eb}$ can now be put in the figure and their magnitudes scaled off.

The angular velocity of AB can also be determined since $_A v_B = (AB)\Omega_{AB}$; consequently $\Omega_{AB} = ab/AB$ and is anticlockwise. This is consistent with the direction of $_A v_B$ (see figure 11.12c). The same vector diagram, figure 11.12b, referred to as a *velocity diagram,* can be used to determine the velocity $_E v_P$ of any other point on the body such as P. To locate p in the velocity diagram note that $_A v_P$ is perpendicular to AP; if a line is drawn in this direction to pass through a, then p must be on this line. By the same argument $_B v_P$ is perpendicular to BP, and if a line is drawn in this direction to pass through b then p must also be on this line. The point p is thus located at the intersection of the two lines. It should be noted that the triangle apb is similar to APB and is the so-called *velocity image* of the link as defined by APB. If P is on the line AB it follows that p is on the line ab and the ratios ap/ab and AP/AB must be equal; this also follows from ${}_{A}v_{P}/{}_{A}v_{B}$ = (AP)Ω/(AB)Ω. The absolute velocity of P, ${}_{E}v_{P}$ is the vector $\overrightarrow{ep}$. If this general procedure is carried out for the point Q, for which AQ and BQ are perpendicular to ${}_{E}v_{A}$ and ${}_{E}v_{B}$ respectively, it will be found that q coincides with e. This means that ${}_{E}v_{Q} = 0$; Q is instantaneously at rest and is the instantaneous centre of rotation of the body.

Consider now the solution for the acceleration of B if A has an absolute acceleration ${}_{E}a_{A}$ to the right. Equation 11.3 is now written

$$ {}_{E}a_{B} = {}_{E}a_{A} + {}_{A}a_{B} \tag{11.5} $$

We must particularly note that ${}_{A}a_{B}$ will have two components, namely a tangential component ${}_{A}a_{B,t} = (AB)\alpha_{AB}$ in direction perpendicular to AB and consistent with the sense of α_{AB}, and a normal component ${}_{A}a_{B,n} = (AB)\Omega^{2}{}_{AB}$ directed towards A; that is in direction and sense $\overrightarrow{BA}$ (see figure 11.12e). The angular velocity Ω_{AB} has already been determined but α_{AB} is unknown. The graphical representation of equation 11.5 is given in figure 11.12d. The line ea is first drawn to represent ${}_{E}a_{A}$ thus fixing point a; aa_1 is then drawn in direction $\overrightarrow{BA}$ to represent ${}_{A}a_{B,n}$ with magnitude $(AB)\Omega^{2}{}_{AB}$ thus locating the point a_1. The component ${}_{A}a_{B,t}$ is perpendicular to AB and a line xy is drawn in this direction to pass through a_1; the point b representing the total acceleration of B must, by equation 11.5, lie on this line. We now make use again of the known direction of ${}_{E}a_{B}$, which, from the constraint on B, must be vertical. A line fg is drawn in this direction to pass through e; b must also be on this line. The intersection of lines xy and fg thus locate b such that

$$ \overrightarrow{eb} = \overrightarrow{ea} + (\overrightarrow{aa_1} + \overrightarrow{a_1b}) = \overrightarrow{ea} + \overrightarrow{ab} $$

and equation 11.5 is represented.

The acceleration of B, ${}_{E}a_{B}$ is thus represented to scale by $\overrightarrow{eb}$.

The vector diagram so constructed is referred to as an *acceleration diagram.*

The angular acceleration of the link can now be determined in magnitude since ${}_{A}a_{B,t} = (AB)\alpha_{AB}$, consequently $\alpha_{AB} = a_1b/AB$. Its direction is decided by the direction of ${}_{A}a_{B,t}$ and recalling that this means the tangential acceleration of B as seen from A, it follows (see figure 11.12e) that α_{AB} is anticlockwise.

The acceleration ${}_{E}a_{P}$ of any other point, such as P, can be determined by locating P in the acceleration diagram. Because AP and BP have the same angular velocity and angular acceleration as AB, the triangles abp (in the acceleration diagram) and ABP are again similar and abp is described as the *acceleration image* of ABP. If P is on the line AB it again follows that p is on the line ab such that ap/ab = AP/AB. If q is located by this general process it will be found not to coincide with e thus demonstrating that the instantaneous centre is not a point of zero acceleration.

In figure 11.12d an additional letter b_1 has been indicated adjacent to the point a_1. This notation will be found useful in the construction of acceleration diagrams. It implies that not only is $\overrightarrow{ab}$ the sum of two components $\overrightarrow{aa_1}$ and $\overrightarrow{a_1b}$ but that $\overrightarrow{ba}$ is the sum of the components $\overrightarrow{bb_1}$ and $\overrightarrow{b_1a}$.

It should be noted that we have shown that by graphical interpretation of equations 11.1 and 11.3 we can determine the velocity and acceleration of any point on a constrained link; we must now accept that our notation always ensures that these equations are satisfied. In both the velocity and acceleration diagrams, once the points e and a were located, there was only one logical position for b and it was after locating this that vector quantities were identified and fully determined in magnitude and direction. For the same points a, b and e the following relations could also be written

(1) $\overrightarrow{ea} = \overrightarrow{eb} + \overrightarrow{ba}$, or
(2) $\overrightarrow{ab} = \overrightarrow{ae} + \overrightarrow{eb}$, or
(3) $\overrightarrow{ba} = \overrightarrow{be} + \overrightarrow{ea}$

which all represent the equations if the correct notation is used.

11.4 Velocity and Acceleration Diagrams

11.4.1 Simple Mechanisms

The graphical method described is now extended to determine the velocities and accelerations of points in simple mechanisms. Following from the preceding paragraphs the method used is to locate the representative points in a logical step-by-step manner, our notation ensuring that equations 11.1 and 11.3 are represented. The method is best illustrated by the following worked example.

Worked Example 11.3

In the mechanism whose space diagram for a particular instant is shown in figure 11.13a, the link PR is rotating with angular velocity 2 rad/s clockwise and angular acceleration 2 rad/s^2 anticlockwise. Determine the angular velocity and angular acceleration of the link QS and also the linear acceleration of C, the mid-point of RS.

Solution

Note that in both velocity and acceleration diagrams all points such as p and q corresponding to fixed points in the mechanism must coincide with e since fixed points have zero absolute velocity and acceleration. Starred values in these diagrams indicate that these are ascertained from the diagrams.

The mechanism is first drawn to scale as in figure 11.13a. For the velocity diagram, figure 11.13b, the point r is first located (with respect to p) by drawing $pr = {}_P v_R = (PR)\Omega_{PR}$ in the direction perpendicular to PR and taking account of

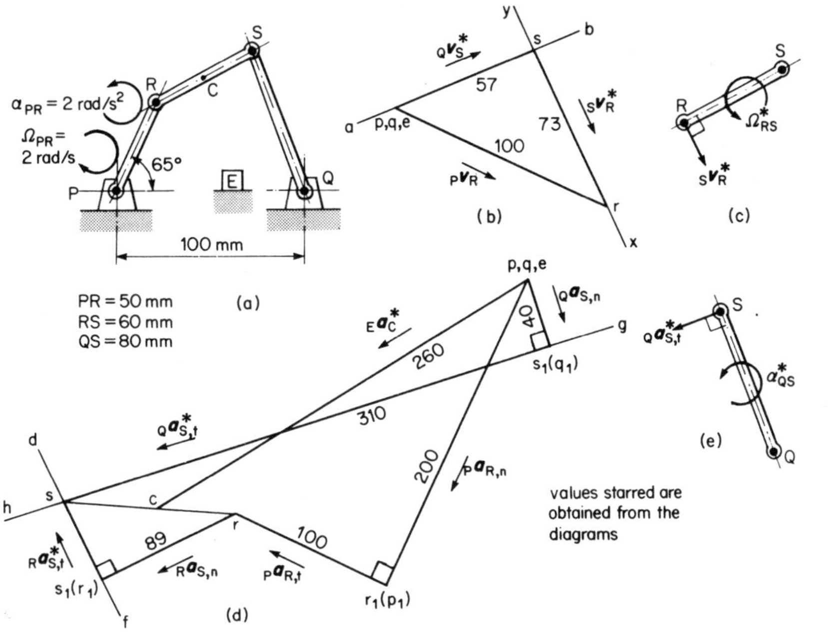

Figure 11.13

the sense of Ω_{PR}. Thus pr = 100 mm/s $\angle -25°$. Now ${}_{R}v_{S}$ is perpendicular to RS; a line xy is therefore drawn through r (since r must be at one end of the vector $\vec{rs}$) in this direction and s must lie on this line. We now approach the problem from the other fixed point Q; the velocity ${}_{Q}v_{S}$ is perpendicular to QS and a line ab is thus drawn in this direction to pass through q; s must also lie on this line. s is thus located at the intersection of lines xy and ab and the velocity diagram prsq is now completed. There is no need to locate the point c in the velocity diagram unless its velocity is required. The unknown velocities may now be determined

$$
\begin{aligned}
{}_{S}v_{R} &= 73 \text{ mm/s in the direction of } \vec{sr} \\
{}_{Q}v_{S} &= 57 \text{ mm/s in the direction of } \vec{qs}
\end{aligned}
$$

Note that we could also define ${}_{R}v_{S}$ in the direction of $\vec{rs}$ or ${}_{S}v_{Q}$ in the direction $\vec{sq}$ if we so desired.

The angular velocity of links QS and RS may now be calculated.

$$\Omega_{RS} = \frac{{}_{S}v_{R}}{RS} = \frac{rs}{RS} = \frac{73}{60} = 1.22 \text{ rad/s}$$

and from the direction of ${}_{S}v_{R}$ is anticlockwise (see figure 11.13c). Similarly

$$\Omega_{QS} = \frac{qs}{QS} = \frac{57}{80} = 0.71 \text{ rad/s clockwise}$$

In order to draw the acceleration diagram the normal components of relative accelerations must first be calculated for every link. Thus

$$_{P}a_{R,n} = (PR)\Omega^2{}_{PR} = 200 \text{ mm/s}^2 \quad \text{in direction } \overrightarrow{RP}$$

$$_{R}a_{S,n} = (RS)\Omega^2{}_{RS} = 89 \text{ mm/s}^2 \quad \text{in direction } \overrightarrow{SR}$$

$$_{Q}a_{S,n} = (QS)\Omega^2{}_{QS} = 40 \text{ mm/s}^2 \quad \text{in direction } \overrightarrow{SQ}$$

If it is possible to calculate any tangential components, as for example $_{P}a_{R,t}$ this should now be done.

$$_{P}a_{R,t} = (PR)\alpha_{PR} = 100 \text{ mm/s}^2$$

perpendicular to PR taking account of the sense of α_{PR}, that is $\angle 155°$.

We can only write for the remaining components

$_{R}a_{S,t}$ is perpendicular to RS

$_{Q}a_{S,t}$ is perpendicular to QS

since α_{RS} and α_{QS} are not known.

Starting at p in the acceleration diagram, figure 11.13d, we first locate r. This is found by drawing $\overrightarrow{pr_1} = {}_{P}a_{R,n}$ in direction $\overrightarrow{RP}$ to fix $r_1(p_1)$ and then adding $\overrightarrow{r_1r} = {}_{P}a_{R,t}$ perpendicular to RP (that is, at $\angle 155°$); this locates r. We now proceed to locate s. We can locate $s_1(r_1)$ by drawing $\overrightarrow{rs_1} = {}_{R}a_{S,n}$ in direction $\overrightarrow{SR}$. Now $_{R}a_{S,t}$ is perpendicular to RS and thus a line df is drawn in this direction through $s_1(r_1)$; s must be on this line. Working now from q we draw in $\overrightarrow{qs_1} = {}_{Q}a_{S,n}$ in direction $\overrightarrow{SQ}$ to fix $s_1(q_1)$. $_{Q}a_{S,t}$ is perpendicular to this and a line gh is drawn through $s_1(q_1)$ in this direction; since s must also be on this line, its intersection with line df locates s and all necessary points have been found. We may now put in the line $rs = {}_{R}a_{S}$ and hence locate the point c; this is on rs such that rc/rs = RC/RS = 1/2. The absolute acceleration of C, $_{E}a_{C} = \overrightarrow{ec}$ and by measurement is found to be 260 mm/s² $\angle 212°$.

The angular acceleration of link QS is determined by $_{Q}a_{S,t} = (QS)\alpha_{QS}$, thus $\alpha_{QS} = {}_{Q}a_{S,t}/QS = 310/80 = 3.88 \text{ rad/s}^2$. From figure 11.13e it is seen to be anticlockwise, the sense of $_{Q}a_{S,t}$ being deduced from the line joining the points $s_1(q_1)$ and s. Note that Ω_{QS} is clockwise whereas α_{QS} is anticlockwise; this simply means that QS is slowing down.

In certain problems it may be that only accelerations are required; even so the velocity diagram must still be drawn first in order to determine the angular velocities required for the calculation of the normal components of acceleration.

11.4.2 Mechanisms in which Coriolis Accelerations Occur

From the discussion in section 8.5.3 (2) it may be stated that if the path of a particle B (moving along the path with velocity $_{P}v_{B}$ with respect to its coincident point P) is rotated at angular velocity Ω then a Coriolis component of acceleration is observed. In general terms the total acceleration $_{P}a_{B}$ of B with respect to P is given (as in equation 8.25) by

$$_{P}a_{B} = {}_{P}a'_{B} + {}_{P}a''_{B}$$

where ${}_{\mathrm{P}}a'_{\mathrm{B}}$ has a normal (centripetal) component, ${}_{\mathrm{P}}v_{\mathrm{B}}^2/\rho$, and a tangential component $\mathrm{d}({}_{\mathrm{P}}v_{\mathrm{B}})/\mathrm{d}t$, and ${}_{\mathrm{P}}a''_{\mathrm{B}}$ is the Coriolis component of acceleration with magnitude $2\,{}_{\mathrm{P}}v_{\mathrm{B}}\Omega$ and in the direction of ${}_{\mathrm{P}}v_{\mathrm{B}}$ rotated through 90° in the same sense as Ω.

If we now confine ourselves to the simplified (and in mechanisms the more usual) case where the rotating path is a straight line, then ${}_{\mathrm{P}}a'_{\mathrm{B}}$ reduces to the one component $\mathrm{d}({}_{\mathrm{P}}v_{\mathrm{B}})/\mathrm{d}t$ along the path. In this case

$$ {}_{\mathrm{P}}a_{\mathrm{B}} = \underset{\text{(along the path)}}{{}_{\mathrm{P}}a'_{\mathrm{B}}} + \underset{\text{(perpendicular to the path)}}{{}_{\mathrm{P}}a''_{\mathrm{B}}} \qquad (11.6) $$

We may, for example, apply this equation to the small block sliding on a rod as in figure 11.14a; the rod itself is assumed to be part of a mechanism.

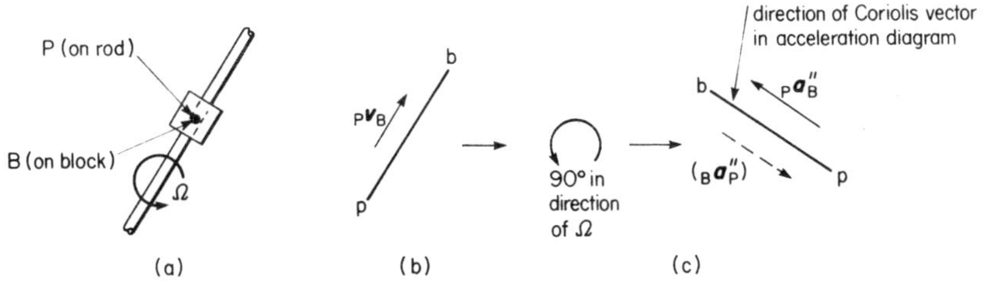

Figure 11.14

The velocity ${}_{\mathrm{P}}v_{\mathrm{B}}$ can usually be determined in magnitude, sense and direction (its direction is known since it is constrained to be parallel to the rod) by drawing a velocity diagram for the mechanism. The velocity diagram will yield a representation pb such as that indicated in figure 11.14b.

With reference to equation 11.6, ${}_{\mathrm{P}}a'_{\mathrm{B}}$ is directed along the rod as stated; ${}_{\mathrm{P}}a''_{\mathrm{B}}$ is in the direction of ${}_{\mathrm{P}}v_{\mathrm{B}}$ rotated through 90° in the same sense of Ω as shown in figure 11.14c. In this context it is useful to describe ${}_{\mathrm{P}}v_{\mathrm{B}}$ by its vector $\overrightarrow{\mathrm{pb}}$ (figure 11.14b) and to retain the letters (p and b in this case) on the vector as it is turned (see figure 11.14c). The final vector in figure 11.14c then indicates either the direction of ${}_{\mathrm{P}}a''_{\mathrm{B}}$ or that of ${}_{\mathrm{B}}a''_{\mathrm{P}}$ as required in the acceleration diagram. This has certain advantages since in many problems we will know the total acceleration of B and will be endeavouring to determine the acceleration of P.

In so far as the velocity and acceleration diagrams are concerned, the important facts relating to a block sliding on a rotating rod or a roller moving in a rotating slot are

(1) the velocity of B relative to P, ${}_{\mathrm{P}}v_{\mathrm{B}}$, has direction parallel to the rod or slot; it can be referred to as the *sliding velocity*,
(2) the acceleration of B relative to P has two components, namely,
(a) a component ${}_{\mathrm{P}}a'_{\mathrm{B}}$ with direction parallel to the rod but unknown in magnitude; it can be referred to as the *sliding acceleration*, and
(b) the Coriolis component ${}_{\mathrm{P}}a''_{\mathrm{B}}$ which, after drawing the velocity diagram, is known in magnitude, sense and direction.

The method of constructing velocity and acceleration diagrams for mechanisms incorporating sliding connections is illustrated in the following worked example.

Worked Example 11.4

In the mechanism shown in figure 11.15a it is required to find the angular velocity and angular acceleration of the slotted member QDF when $\theta = 120^\circ$ and the member OA is rotating at 120 rev/min clockwise.

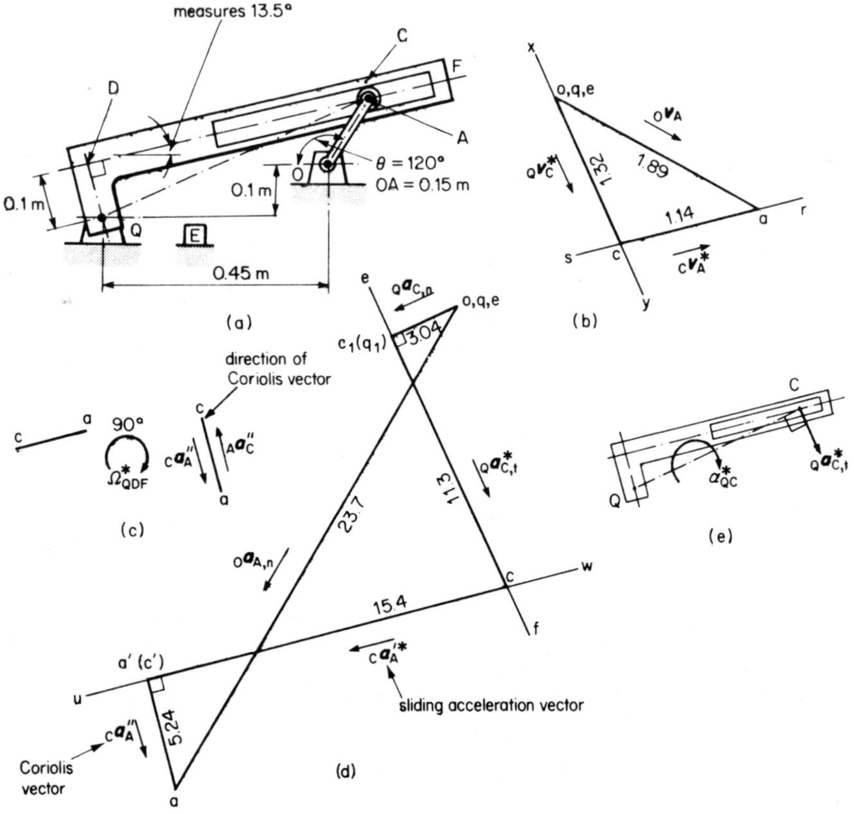

Figure 11.15

Solution

The mechanism is drawn to scale with $\theta = 120^\circ$ (figure 11.15a). The small roller A is on the link OA but is sliding in the rotating link QDF. A has a coincident point on the member QDF; this is the point C which is on the line DF, but to emphasise that C is on the link QDF, C is shown in the diagram on the link QDF adjacent to A.

For the velocity diagram, figure 11.15b, the point a is first located with respect to o by drawing oa = ${}_{O}v_{A}$ = (OA)Ω_{OA} perpendicular to OA, taking account of the sense of Ω_{OA}.

$$\Omega_{OA} = \frac{2\pi\ 120}{60} = 4\pi \text{ rad/sec}$$

and

$$\text{oa} = 4\pi \times 0.15 = 1.89 \text{ m/s} \angle -30^\circ$$

Now ${}_{C}v_{A}$ is along the slot, so a line rs is drawn through a in this direction; c must lie on this line. With respect to the fixed point Q, ${}_{Q}v_{C}$ is perpendicular to the line QC (remember that C is actually coincident with A) and a line xy in this direction is drawn through q; c must also lie on this line. The intersection of lines rs and xy locates c and enables ${}_{C}v_{A}$ (the sliding velocity) and ${}_{Q}v_{C}$ to be determined.

Then

$$\Omega_{QDF}\ (= \Omega_{QC}) = \frac{{}_{Q}v_{C}}{QC} = \frac{1.32}{0.574} = 2.3 \text{ rad/s}$$

and, from the direction of ${}_{Q}v_{C}$, is clockwise (QC is scaled from figure 11.15a).

To draw the acceleration diagram we first calculate the normal components of the accelerations and the Coriolis acceleration component

$$\begin{aligned} {}_{O}a_{A,n} &= (OA)\Omega^2{}_{OA} = 0.15 \times (4\pi)^2 \\ &= 23.7 \text{ m/s}^2\text{, in direction } \overrightarrow{AO} \end{aligned}$$

$${}_{Q}a_{C,n} = \frac{{}_{Q}v_{C}{}^2}{QC} = \frac{1.32^2}{0.574} = 3.04 \text{ m/s}^2\text{, in direction } \overrightarrow{CQ}$$

The Coriolis acceleration component

$${}_{C}a''_{A} = 2{}_{C}v_{A}\ \Omega_{QDF} = 2 \times 1.14 \times 2.3 = 5.24 \text{ m/s}^2$$

its direction being given by rotating the sliding velocity vector 90° in the sense of Ω_{QDF} as indicated in figure 11.15c.

We know that ${}_{O}a_{A,t} = 0$ since $\alpha_{OA} = 0$ but we can only write for the remaining components

${}_{Q}a_{C,t}$ is perpendicular to QC
${}_{C}a'_{A}$ (the sliding acceleration) is parallel to the slot.

In the acceleration diagram, figure 11.15d, we first locate the point a by drawing oa = ${}_{O}a_{A,n}$ = 23.7 m/s^2 in direction $\overrightarrow{AO}$ which fixes a since ${}_{O}a_{A,t}$ = 0. The vector oa represents the absolute acceleration of A with respect to O; this acceleration should correspond with the acceleration of A with respect to Q. Starting from q the point $c_1(q_1)$ can be plotted since qc_1 = ${}_{Q}a_{C,n}$ = 3.04 m/s^2 in direction $\overrightarrow{CQ}$. A line ef is drawn through $c_1(q_1)$ in the direction of ${}_{Q}a_{C,t}$ (perpendicular to QC) and the point c must lie on this line. We cannot proceed further since ${}_{Q}a_{C,t}$ and ${}_{C}a'_{A}$ are not known in magnitude or sense. However, the Coriolis component ${}_{C}a''_{A}$ is known and can be represented by a′a terminating at a; with this drawn in position the point a′(c′) is fixed. Through a′(c′) a line uw is drawn in the direction ${}_{C}a'_{A}$ (parallel to the slot); c must lie on this line in order that ca′ may represent ${}_{C}a'_{A}$, the sliding acceleration. The point c is therefore at the intersection of ef and uw.

The sliding acceleration ${}_{C}a'_{A}$ can now be obtained and has magnitude 15.4 m/s^2 and direction as indicated in figure 11.15d. Also from this diagram ${}_{Q}a_{C,t}$ = 11.3 m/s^2 and therefore

$$\alpha_{QDF} = \alpha_{QC} = \frac{{}_{Q}a_{C,t}}{QC} = \frac{11.3}{0.574} = 19.7 \text{ rad/s}^2$$

and, from figure 11.15e, is clockwise.

11.4.3 Rules of Thumb

In the preceding section a rule of thumb was given to determine the direction of the Coriolis component in the acceleration diagram ('retain the symbols on the velocity vector during the 90° rotation'). Another rule of thumb that may be of help to some students is one to decide the direction of the normal components of acceleration. This is as follows.

'If a link **AB** rotates with angular velocity Ω (say clockwise) its velocity image ab is the direction of the link turned 90° in the direction of Ω. The normal component of acceleration is this velocity image turned another 90° in the direction of Ω'.

This is illustrated in figure 11.16.

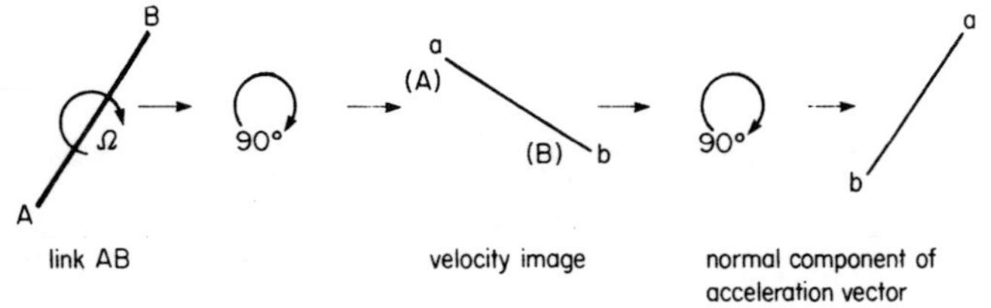

Figure 11.16

11.5 Analytical Methods

The graphical methods described in the previous section for determining velocities and accelerations of points in mechanisms have one serious disadvantage – the solutions are true for one particular instant only. For instance, if in worked example 11.3 (figure 11.13) the variation of the linear acceleration of point C was required for a complete revolution of PR, we should be faced with the tedious job of drawing velocity and acceleration diagrams for a multitude of positions of RS. A better approach when we require a continuous solution is to use analytical methods. Unfortunately these do not always yield simply equations but the methods have the advantage that they lend themselves to solution by digital computation.

11.5.1 Useful Relationships

The equations deduced in chapter 8 relating to the motion of a particle are relevant and are reiterated in the form most useful for our purposes in figure 11.17. Note

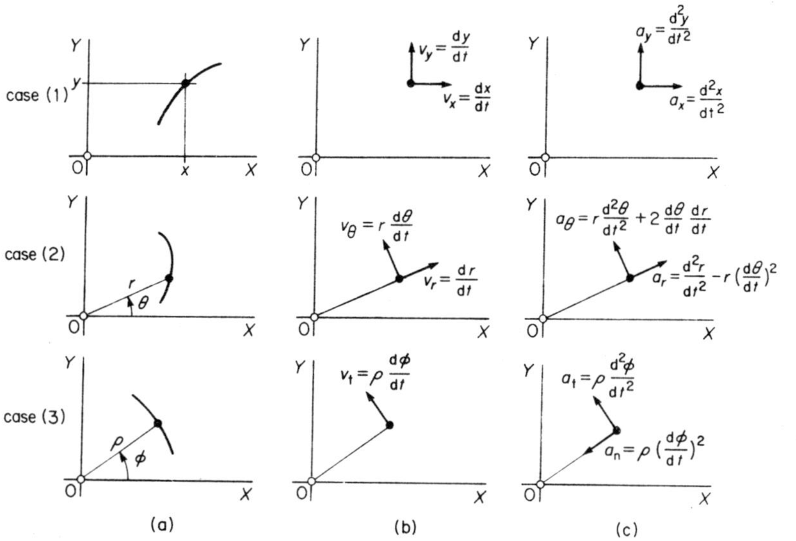

Figure 11.17

that cases (1) and (2) apply when the path is completely arbitrary whereas case (3) only applies where the path is an arc of a circle with its centre at the origin.

Column (a) shows three ways of defining the position of a point in a plane, columns (b) and (c) show the component velocities and accelerations according to which set of coordinates are used; note that the positive senses are those indicated. It is emphasised that case (3) is only applicable when ρ is constant; otherwise the components shown for cases (1) and (2) must be used.

11.5.2 Direct Method

This consists of choosing some point on a link of the mechanism, writing its coordinates and calculating its absolute velocity and absolute acceleration by application of one of the cases illustrated in the previous section. For connected bodies the chosen point will usually be at a connection and its absolute velocity and absolute acceleration can be determined with respect to different fixed points; the equality of the two velocities and the two accelerations so determined then lead to a solution. The method is best illustrated by the following worked example.

Worked Example 11.5

In the mechanism shown in figure 11.18 the link OA rotates about the fixed axis with the indicated angular velocity and acceleration. The rod ACD slides through a block B, which is free to rotate in a fixed pivot at F. If, when $\phi = 20°$, $\Omega = -40$

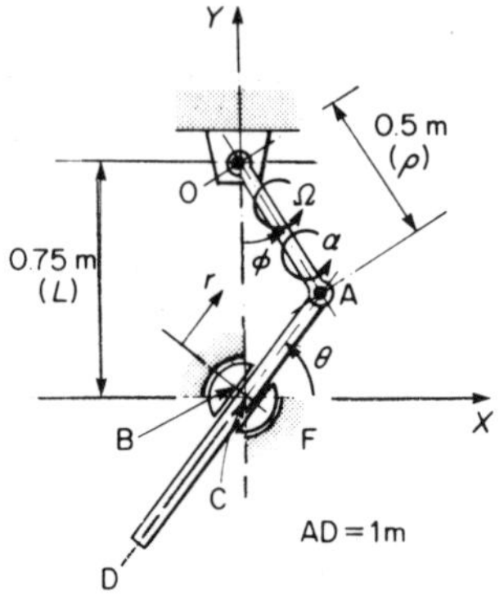

Figure 11.18

rad/s and $\alpha = +600$ rad/s^2, determine (a) the angular velocity and angular acceleration of ACD and (b) the linear acceleration of D at this instant.

Solution

A general solution will be outlined for (a), into which desired numerical values can be substituted. Polar coordinates will be used, the y-axis shown being adopted as a reference axis for the angle ϕ and the x-axis for the angle θ.

Put OF = L, OA = ρ and FA = r_A. Relative to O, A is at the point (ρ, ϕ) where ρ is constant, and relative to F, A is at the point (r_A, θ) where r_A is varying.

Using the expressions of figure 11.17, case (3), for the motion of A relative to O

$$v_n = 0 \qquad v_t = \rho\dot{\phi}$$

$$a_n = \rho\dot{\phi}^2 \qquad a_t = \rho\ddot{\phi}$$

Using the expressions of figure 11.17, case (2) for the motion of A relative to F

$$v_r = \dot{r}_A \qquad v_\theta = r_A\dot{\theta}$$

$$a_r = \ddot{r}_A - r_A\dot{\theta}^2 \quad a_\theta = r_A\ddot{\theta} + 2\dot{r}_A\dot{\theta}$$

(a) Since A is common to both links and O and F are both fixed points: ${}_Ov_A = {}_Fv_A$, and resolving in the x- and y-directions shown

$$v_t \cos\phi = v_r \cos\theta - v_\theta \sin\theta$$

$$v_t \sin\phi = v_r \sin\theta + v_\theta \cos\theta$$

Substituting for v_t, v_r and v_θ

$$\rho\dot{\phi}\cos\phi = \dot{r}_A\cos\theta - r_A\dot{\theta}\sin\theta \qquad (11.7)$$

$$\rho\dot{\phi}\sin\phi = \dot{r}_A\sin\theta + r_A\dot{\theta}\cos\theta \qquad (11.8)$$

The angle θ is related to the angle ϕ since

$$\tan\theta = \frac{L - \rho\cos\phi}{\rho\sin\phi}$$

also

$$r_A = \frac{\rho\sin\phi}{\cos\theta}$$

For given values of ρ and L, and for particular values of ϕ and $\dot{\phi}$ we therefore have two simultaneous equations (11.7) and (11.8) for $\dot{r}_A$ and $\dot{\theta}$. For the values $\rho = 0.5$ m, $L = 0.75$ m, $\phi = 20°$, $\dot{\phi} = -40$ rad/s it is found that $\dot{r}_A = -15.63$ m/s and $\dot{\theta} = 38.02$ rad/s.

(b) Again since A is common to both links ${}_Oa_A = {}_Fa_A$ and resolving in the x- and y-directions

$$-a_n\sin\phi + a_t\cos\phi = a_r\cos\theta - a_\theta\sin\theta$$

$$a_n\cos\phi + a_t\sin\phi = a_r\sin\theta + a_\theta\cos\theta$$

Substituting for a_n, a_t, a_r and a_θ

$$-\rho\dot{\phi}^2\sin\phi + \rho\ddot{\phi}\cos\phi = (\ddot{r}_A - r_A\dot{\theta}^2)\cos\theta - (r_A\ddot{\theta} + 2\dot{r}_A\dot{\theta})\sin\theta$$

$$\rho\dot{\phi}^2\cos\phi + \rho\ddot{\phi}\sin\phi = (\ddot{r}_A - r_A\dot{\theta}^2)\sin\theta + (r_A\ddot{\theta} + 2\dot{r}_A\dot{\theta})\cos\theta$$

Again we have

$$\tan\theta = \frac{L - \rho\cos\phi}{\rho\sin\phi}$$

and

$$r_A = \frac{\rho\sin\phi}{\cos\theta}$$

For given values of ρ and L, and for particular values of ϕ, $\dot{\phi}$ and $\ddot{\phi}$, together with the values of $\dot{r}_A$ and $\dot{\theta}$ previously determined, we therefore have two simultaneous equations for $\ddot{r}_A$ and $\ddot{\theta}$.

For the values $\rho = 0.5$ m, $L = 0.75$ m, $\phi = 20°$, $\dot{\phi} = -40$ rad/s, $\ddot{\phi} = 600$ rad/s^2; also $\dot{r}_A = -15.63$ m/s, $\dot{\theta} = 38.02$ rad/s as above it is found that $\ddot{r}_A = 1207$ m/s and $\ddot{\theta} = 4957$ rad/s^2.

As an alternative the rectangular coordinates of A with respect to both O and F can be written, namely ${}_Ox_A = \rho\sin\phi$, ${}_Oy_A = -\rho\cos\phi$ and ${}_Fx_A = r\cos\theta$,

${}_{F}y_A = r\sin\theta$. By repeated differentiation ${}_{O}\dot{x}_A$, ${}_{O}\ddot{x}_A$, ${}_{O}\dot{y}_A$ and ${}_{O}\ddot{y}_A$ can be found, also ${}_{F}\dot{x}_A$, ${}_{F}\ddot{x}_A$, ${}_{F}\dot{y}_A$ and ${}_{F}\ddot{y}_A$. The following pairs of equations can now be written

$$ {}_{O}\dot{x}_A = {}_{F}\dot{x}_A \qquad {}_{O}\dot{y}_A = {}_{F}\dot{y}_A $$

and

$$ {}_{O}\ddot{x}_A = {}_{F}\ddot{x}_A \qquad {}_{O}\ddot{y}_A = {}_{F}\ddot{y}_A $$

These are essentially the same pairs of simultaneous equations that were encountered above, and for particular values of ϕ, $\dot{\phi}$ and $\ddot{\phi}$ can be solved as before for $\dot{r}_A$ and $\dot{\theta}$, and for $\ddot{r}_A$ and $\ddot{\theta}$.

(c) The linear acceleration of D can be found directly using case (2) now that $\dot{r}_A$, $\ddot{r}_A$ and $\dot{\theta}$ and $\ddot{\theta}$ are known.

The absolute acceleration of D has two components

$$ a_r = \ddot{r}_D - r_D\dot{\theta}^2 $$

and

$$ a_\theta = r_D\ddot{\theta} + 2\dot{r}_D\dot{\theta} $$

Note that (1) because ACD is rigid $\dot{r}_A = \dot{r}_D$ and $\ddot{r}_A = \ddot{r}_D$, and (2) $r_D = -(1 - r_A)$ being negative because it is in the opposite sense to the positive direction for r.

With these relationships

$$ a_r = 2179\ \text{m/s}^2 \qquad a_\theta = -4517\ \text{m/s}^2 $$

which combine to give

$$ {}_{F}a_D = 5015\ \text{m/s}^2 \angle -5.6^\circ $$

This solution has been outlined in order to indicate one method of analytical solution. It is not to be implied that this is the only analytical method or that analytical methods are suitable for all problems. Clearly the arithmetical work involved in the analysis of a complete cycle of the motion of the mechanism is best entrusted to a digital computer.

11.6 Summary

(1) If A and B are any two points on a rigid body in general plane motion

$$ {}_{E}v_B = {}_{E}v_A + {}_{A}v_B \tag{11.1} $$

$$ {}_{E}a_B = {}_{E}a_A + {}_{A}a_B \tag{11.3} $$

(2) The instantaneous centre of rotation I is located at the intersection of the lines drawn through any two points on the body in directions perpendicular to the absolute velocities at those points. The point I is instantaneously stationary; it usually does not have zero acceleration.

(3) The notation ${}_{Y}v_{X} = \vec{yx}$ means the velocity of X as seen from Y, and similarly for acceleration. Equations 11.1 and 11.3 are automatically valid if the ordering of the subscripts is strictly adhered to.
(4) Velocity diagrams: locate the unknown points logically, making use of the constraints and noting that, for a rigid link AB, ${}_{A}v_{B} = (AB)\Omega_{AB}$ is perpendicular to AB with sense dependent upon the sense of Ω_{AB}.
(5) Acceleration diagrams; locate the unknown points logically, making use of constraints and noting that for a rigid link AB

$$ {}_{A}a_{B} = {}_{A}a_{B,n} + {}_{A}a_{B,t} $$

${}_{A}a_{B,n} = (AB)\Omega_{AB}{}^{2}$ can always be calculated and its sense is always $\vec{BA}$; ${}_{A}a_{B,t} = (AB)\alpha_{AB}$ which generally cannot be calculated (unless α_{AB} is stated) but is perpendicular to AB with sense dependent on the sense of α_{AB}.
(6) If the mechanism contains a slotted member, or a block sliding on a member, and the member is rotating, a Coriolis acceleration component will be present.
(7) The Coriolis acceleration component is equal to $2v\Omega$ where v is the sliding velocity; the direction is decided by rotating the sliding velocity 90° in the sense of Ω.
(8) For analytical methods make use of the equations relating to motion of a particle; express the position of a point in a mechanism by use of a particular set of coordinates then either express its velocity and accelerations (as in case (2) of figure 11.17) or deduce these by repeated differentation with respect to time of, for example, the rectangular coordinates.

Problems

Many of the following problems have been solved graphically and it must therefore be expected that the answers obtained may differ by a small amount from the quoted answer due to variation in graphical accuracy.

11.1 A link AB, length 1.5 m, has a constrained motion such that its end A has velocity 2 m/s $\angle 0°$ and acceleration 5 m/s^2 $\angle 180°$. If AB at this instant is in the angular position $\angle 150°$ and has angular velocity 2 rad/s clockwise and angular acceleration 5 rad/s^2 anticlockwise, determine the absolute velocity and acceleration of B. (*Hint*: Use vector diagrams to solve equations 11.1 and 11.3).

11.2 A uniform rod AB, length 5 m, is moving at a particular instant with $\vec{AB}$ in the angular position $\angle 90°$; the absolute acceleration of A is 20 m/s^2 $\angle 0°$ and of B is 40 m/s^2 $\angle -45°$. Obtain the angular velocity and angular acceleration of the rod and the absolute acceleration of its mass-centre. (*Hint*: Use a vector diagram; note components of acceleration of one end of the rod with respect to the other.)

11.3 For the mechanism shown in figure 11.19 determine for the configuration shown, the velocity of H and the angular velocity of BDF. (*Hint*: The method of

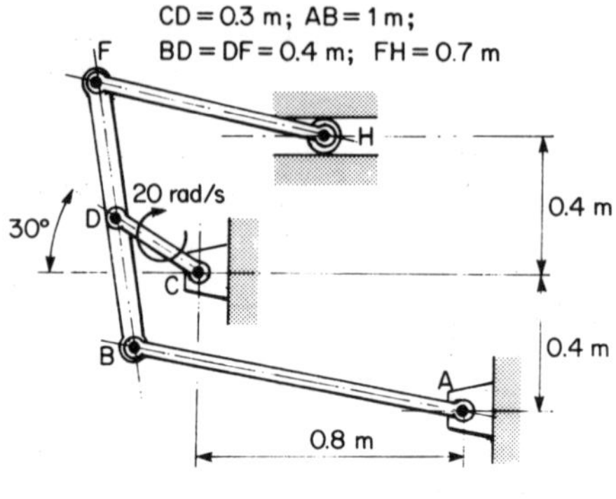

Figure 11.19

instantaneous centres is suggested; the answers should then be checked using a velocity diagram.)

11.4 Use the method of instantaneous centres to determine (a) the linear velocity of C and (b) the angular velocity of HJ for the configuration of the mechanism in figure 11.20. Check the answers using a velocity diagram. (*Hint*: Consider absolute velocities of B and D to find I_{BCD}.)

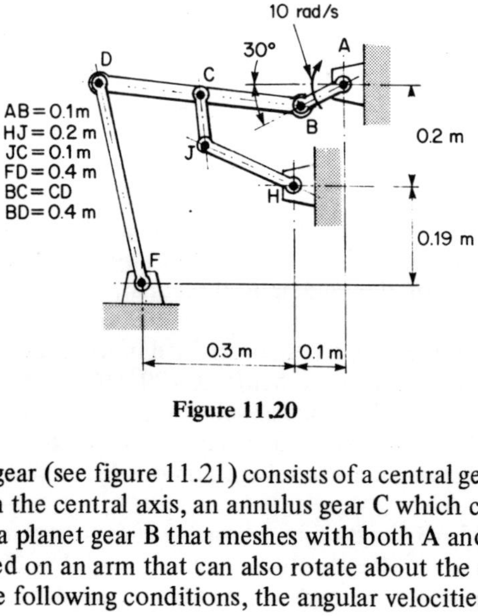

Figure 11.20

11.5 An epicyclic gear (see figure 11.21) consists of a central gear A (called the sun gear) rotating on the central axis, an annulus gear C which can also rotate on the central axis and a planet gear B that meshes with both A and C. B rotates on a pin S, which is carried on an arm that can also rotate about the central axis.

Determine for the following conditions, the angular velocities of (a) B and (b) the arm

(i) $\Omega_A = +10$ rad/s; $\Omega_C = +5$ rad/s
(ii) $\Omega_A = +10$ rad/s; $\Omega_C = -5$ rad/s
(iii) $\Omega_A = +10$ rad/s; $\Omega_C = 0$

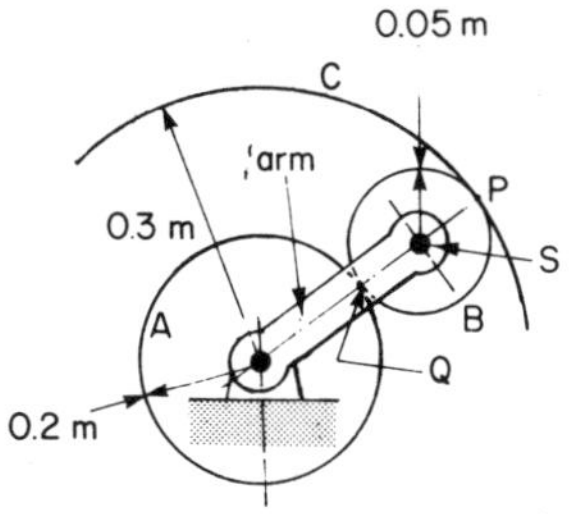

Figure 11.21

(*Hint*: Use the absolute velocities of points P and Q to find the instantaneous centre of B; hence find Ω_B and then the velocity of S; Ω_{arm} may then be found.)

11.6 Figure 11.22 shows the ouline of a mechanical digger. Link BCD is rigid and link AC is hinged to the body at A and to BCD at C. The point B is moved

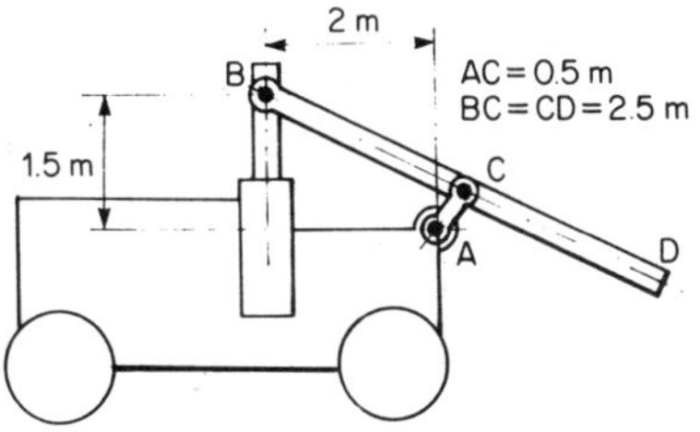

Figure 11.22

vertically, with respect to the digger, by being pinned to the end of a piston moving in a vertical hydraulic cylinder. If, at the instant shown, the digger is moving to the right at 4 m/s and B is moving upwards with respect ot the digger at 2 m/s, find the angular velocity of AC and the velocity of D. (*Hint*: A velocity diagram is suggested.)

11.7 For the given configuration of the mechanism shown in figure 11.23 draw

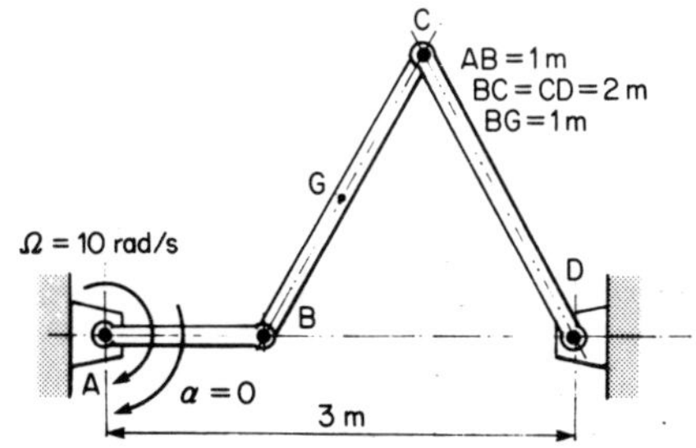

Figure 11.23

the velocity and acceleration diagrams. What are (a) the angular velocity and angular acceleration of BC and (b) the acceleration of G?

11.8 In the slider - crank mechanism shown in figure 11.24 the end P of the link

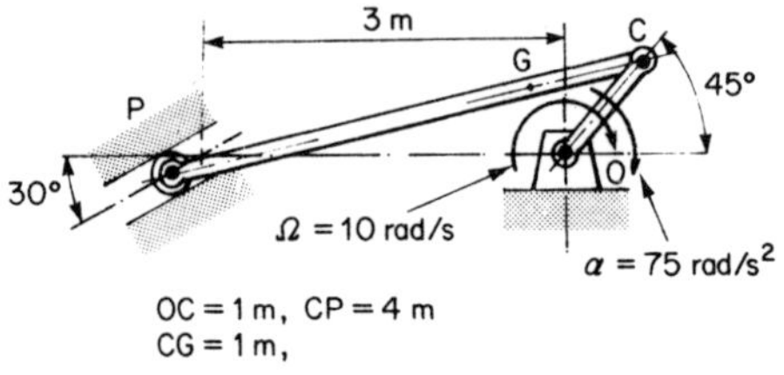

Figure 11.24

CP is constrained to slide in a slot. For the given configuration draw velocity and acceleration diagrams and find (a) the absolute velocity of P, (b) its absolute acceleration, (c) the absolute acceleration of G, (d) the angular velocity of CP and (e) its angular acceleration; use instantaneous centres to check (a) and (d); (f) is P accelerating or retarding in its guide?

11.9 Figure 11.25 shows a mechanism (a four-bar chain) in which DA rotates

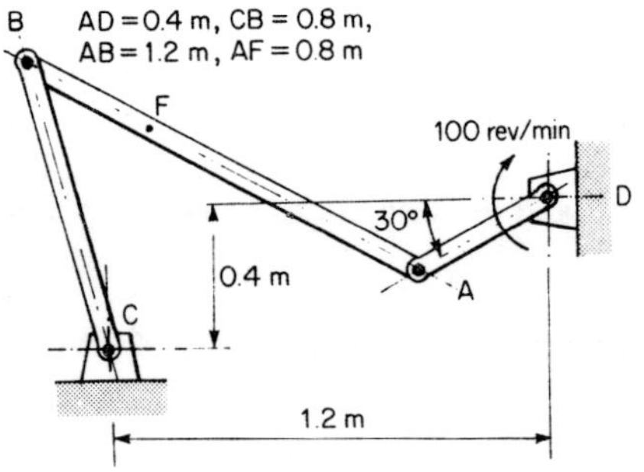

Figure 11.25

at a constant speed. For the configuration shown find (a) the velocity and acceleration of F and (b) the angular velocity and angular acceleration of AB.

11.10 For the mechanism shown in figure 11.26 find (a) the velocity of F, (b) the angular velocity of AB and (c) the angular acceleration of BD. (*Hint*: Having located b recall that bcd is exactly similar to BCD in both velocity and acceleration diagrams).

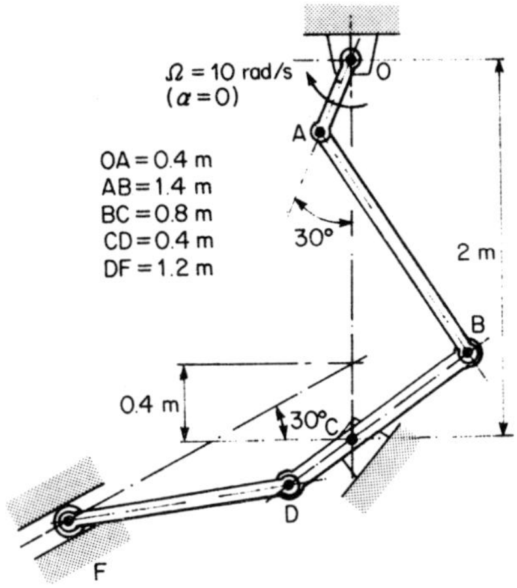

Figure 11.26

11.11 Figure 11.27 shows a quick-return mechanism in which QC rotates clockwise at a constant speed 150 rev/min. Determine for the configuration shown the acceleration of D and the sliding acceleration of C in the slotted link. (*Hint*: A sliding velocity is being rotated.)

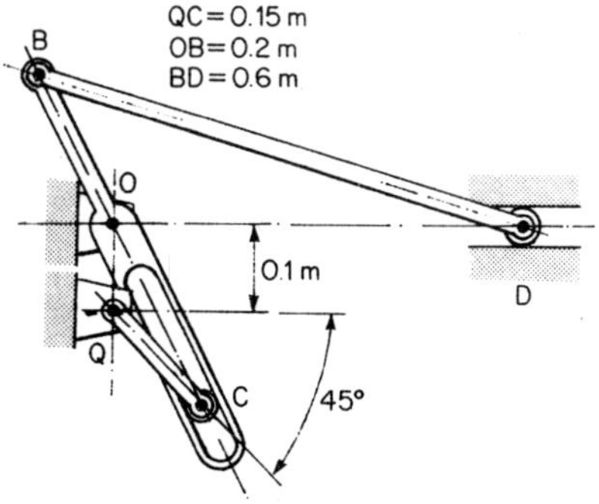

Figure 11.27

11.12 In figure 11.28, C is a roller fixed to the link OB and sliding in the slot in QD. Determine the velocity and acceleration of A when $\theta = 30^\circ$ if QD rotates at constant speed 10 rad/s clockwise.

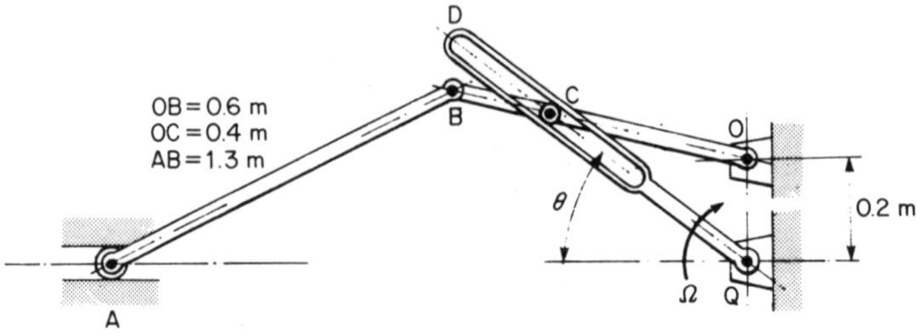

Figure 11.28

11.13 In the mechanism illustrated in figure 11.29 the block J is free to rotate in a housing at the end of the rod FG and J itself slides on the rod BC. For the configuration shown find the acceleration of the rod FG and the angular acceleration of BC.

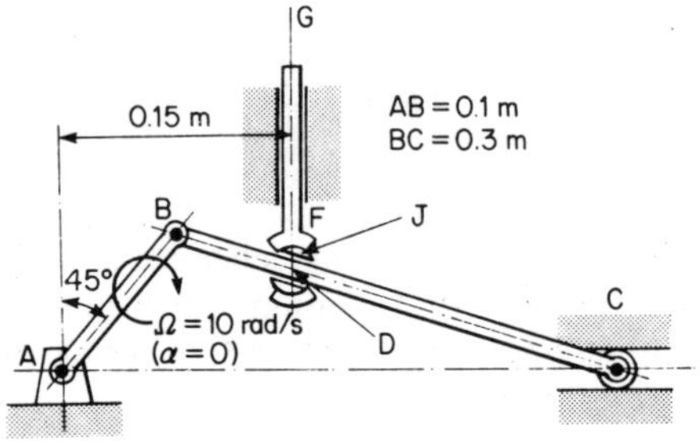

Figure 11.29

11.14 For the mechanism in figure 11.30 write down x as a function of θ. Hence derive expressions for the velocity and acceleration of the roller C. If when $\theta = 60°$ C has velocity 5 m/s to the right and acceleration 2 m/s^2 to the left, find the instantaneous values of the angular velocity and angular acceleration of AB.

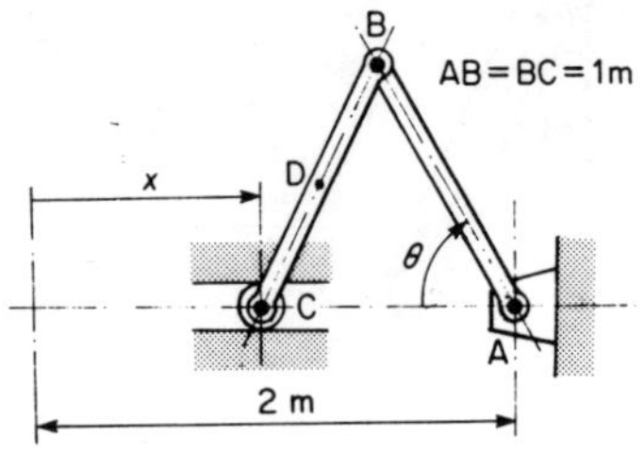

Figure 11.30

Determine also the velocity and acceleration of D the mid-point of BC. Confirm your values by drawing velocity and acceleration diagrams. (*Hint*: To find analytical values for D write expressions for x_D and y_D and differentiate them.)

11.15 In figure 11.31 the ends A and B of the rigid link AB, length 2 m, are constrained to move in mutually perpendicular directions. If when $\theta = 30°$ A has velocity 5 m/s downwards and acceleration 60 m/s^2 upwards find, analytically, the velocity and acceleration of B at this instant.

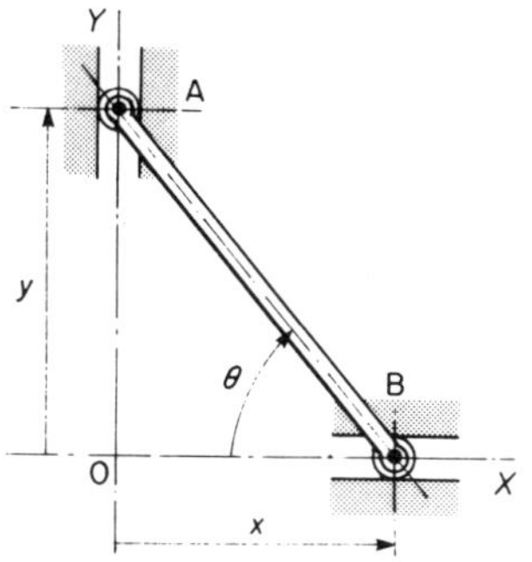

Figure 11.31

Check your answers by drawing velocity and acceleration diagrams. (*Hint*: Write expressions for x and y in terms of θ and differentiate them.)

11.16 The rod AB in figure 11.32 is hinged at A and slides in a block F which is free to rotate in a carrier on the rod DC. The latter is constrained to move horizontally.

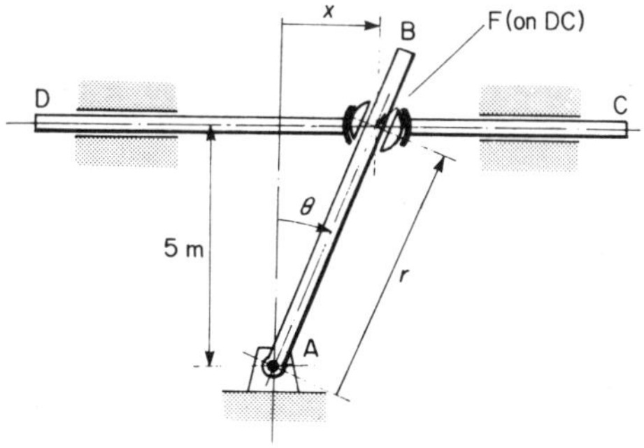

Figure 11.32

If $x = 3t^2$, t being the time in seconds, find analytically the angular velocity and angular acceleration of AB when $t = 1$.

What then is the sliding velocity and sliding acceleration of the rod in the block?

Check your answers by drawing velocity and acceleration diagrams (*Hint*: Equate the absolute velocity and acceleration of point F (on DC), defined by x, to its absolute velocity and acceleration when sliding on AB; polar coordinates probably offer a quicker solution than rectangular coordinates; for the diagrams first denote the point adjacent to F on the link AB as J (F is actually the coincident point of J), use the calculated values of the velocity and acceleration of F as a starting point for the diagrams.)

11.17 A link PQ (see figure 11.33) is arranged so that the end P moves along a horizontal path with simple harmonic motion. PQ is 1.0 m long, the frequency

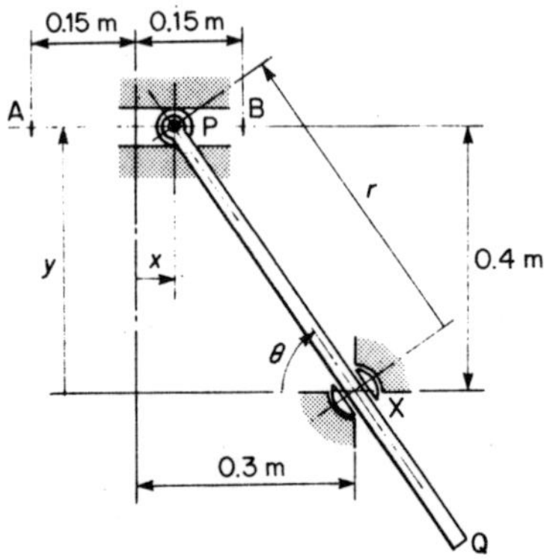

Figure 11.33

of oscillation is 20 Hz and the travel of P between extreme positions A and B is 0.3 m. The link PQ slides through a block, which is free to rotate in a fixed pivot at X.

Determine analytically, for the situation when P is 0.05 m from A and travelling to the right, (a) the angular velocity and angular acceleration of PQ, (b) the velocity and acceleration of Z, a point on PQ instantaneously at the centre of the pivot and (c) the velocity and acceleration of Q.

Check your answers by drawing velocity and acceleration diagrams. (*Hint*: See previous problem.)

11.18 The link AB in figure 11.34 is fitted with smooth rollers at each end and

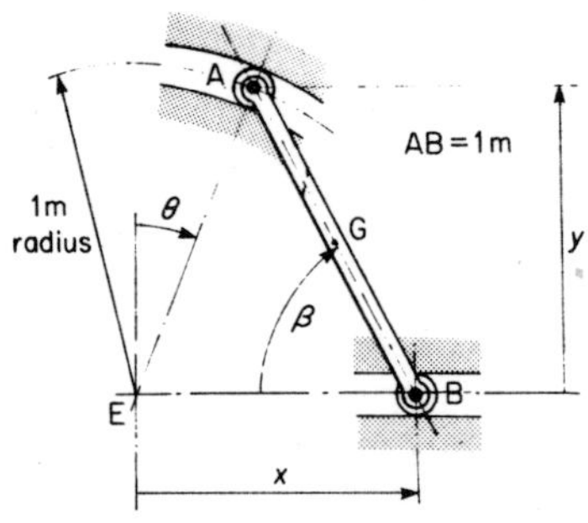

Figure 11.34

is constrained to move with these rollers in the guides. If the movement is such that at the instant when $\beta = 60°$ the velocity of B is constant at 5 m/s to the left find the angular velocity and angular acceleration of AB and the acceleration of G the mid point of AB. (*Hint*: For the analytical method write x in terms of θ and β and differentiate for velocities and acceleration; relate θ and β from y. For the graphical method note that A has a sliding velocity along its guide and also that it has normal and tangential components of acceleration with respect to E.)

12 Moments of Inertia

12.1 Moment of Inertia and Radius of Gyration

In considering the kinetics of a rigid body we shall meet the quantity $\Sigma(\delta m)r^2$ where δm is the mass of some particle at a perpendicular distance r from a particular axis. The summation symbol implies the addition of all the products indicated over the complete rigid body, and is termed the *moment of inertia I* with respect to the particular axis, which is then signified by a double subscript. For example I_{xx} is used to signify the moment of inertia about the x-axis through the origin (which is labelled $X'X$), $I_{z_1 z_1}$ signifies the moment of inertia about an axis $Z_1'Z_1$ which is parallel to, but displaced from, the z-axis ($Z'Z$) at the origin; for other special axes the moment of inertia may be signified by the axis itself and thus $I_{B'B}$ is the moment of inertia about the axis $B'B$.

The moment of inertia of a body, as will be seen in the next chapter, is a measure of the resistance of the body to angular acceleration in the same way as the mass of a body is a measure of its resistance to linear acceleration.

A further quantity, the *radius of gyration k* can be defined such that $k^2 = I/m$, where m is the total mass of the body, from which it follows that

$$I = mk^2 \tag{12.1}$$

The radius of gyration, like the moment of inertia, is defined with respect to a specified axis and carries the same double subscript as the corresponding moment of inertia. The radius of gyration is the particular measure of the distribution of the mass of the body, in relation to the specified axis, for angular motion.

From their definitions, the standard unit for moment of inertia is the kilogram (metre)^2, kg m^2, and that for the radius of gyration is the metre, m.

In the following sections the material of the body under consideration will be assumed to be homogeneous, and the mass per unit volume will be symbolised by ρ. The symbol t will be used to represent the uniform thickness of a thin plate.

12.2 Theorems

12.2.1 Perpendicular Axis Theorem for Thin Plates

A thin plate is one whose thickness t is so small as to be insignificant compared to other dimensions.

Consider the thin plate lying in the XOY plane as shown in figure 12.1. A small area of the plate δA is at a distance r from the axis $Z'OZ$ (the axis at O perpendicular to the plate). The mass of the small area of plate

$$\delta m = \rho t \, \delta A$$

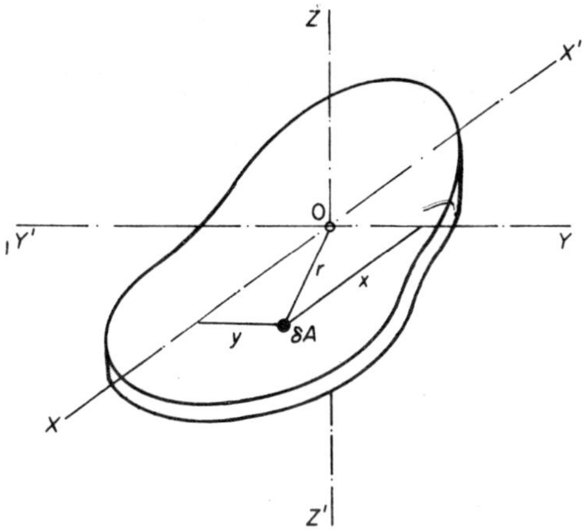

Figure 12.1

By definition the moment of inertia of the plate about $Z'Z$ is

$$I_{zz} = \Sigma(\delta m) r^2$$

Similarly by definition

$$I_{xx} = \Sigma(\delta m) y^2$$

and

$$I_{yy} = \Sigma(\delta m) x^2$$

Now $r^2 = x^2 + y^2$, therefore

$$I_{zz} = \Sigma\delta m(x^2 + y^2) = \Sigma(\delta m)x^2 + \Sigma(\delta m)y^2 = I_{yy} + I_{xx}$$

and

$$I_{zz} = I_{yy} + I_{xx} \text{ (thin plates)} \qquad (12.2)$$

The equation applies only to thin plates because any significant thickness amends the expressions for I_{xx} and I_{yy}. If the general particle is a distance z from the plane XOY its perpendicular distance from $X'X$ is $\sqrt{(y^2 + z^2)}$ and I_{xx} then becomes $\Sigma\delta m(y^2 + z^2)$; I_{yy} is similarly amended; I_{zz} remains the same and equation 12.2 cannot therefore apply.

Note that axes $X'X$ and $Y'Y$ are in the plane of the plate and $Z'Z$ is perpendicular to it. For example, I_{yy} is *not* equal to $I_{zz} + I_{xx}$.

Worked Example 12.1

Determine the moment of inertia of a circular disc radius R (a) about a diameter and (b) about its central axis. Confirm the relationship of equation 12.2.

Solution

(a) A y-axis is chosen to coincide with a diameter. If a strip is selected parallel to the axis, having very small width δx and distance x from the axis (see figure 12.2a) all elementary areas making up the strip are at the same distance from the $Y'Y$

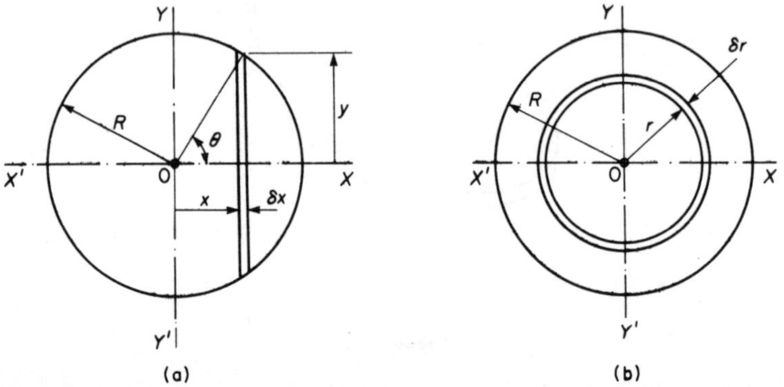

Figure 12.2

axis. Then for the strip, δI_{yy} = mass × $x^2 = 2\rho t y(\delta x)x^2$. For the whole plate the summation is expressed as an integration and

$$I_{yy} = 2\rho t \int_{-R}^{R} yx^2 \, dx$$

Now $y = R \sin\theta$ and $x = R\cos\theta$. It follows from the latter that $dx = -R \sin\theta \, d\theta$ and the lower and upper limits become, in terms of θ, π and 0 respectively.

Thus

$$I_{yy} = 2\rho t R^4 \int_{\pi}^{0} \sin\theta \cos^2\theta \, (-\sin\theta) d\theta$$

$$= 2\rho t R^4 \int_{0}^{\pi} \sin^2\theta \cos^2\theta \, d\theta$$

Now

$$\sin^2\theta\cos^2\theta = \tfrac{1}{4}\sin^2 2\theta = \frac{1 - \cos 4\theta}{8}$$

therefore

$$I_{yy} = \frac{\rho t R^4}{4}\left[\theta - \frac{\sin 4\theta}{4}\right]_0^{\pi}$$

$$= \frac{\rho t \pi R^4}{4}$$

The mass m of the disc is $\rho t\pi R^2$, therefore

$$I_{yy} = \frac{mR^2}{4} \tag{12.3}$$

and

$$k_{yy} = \sqrt{\left(\frac{I_{yy}}{\mathrm{m}}\right)} = \frac{R}{2}$$

(b) For the moment of inertia about a central z-axis select an elementary annular area radius r and small radial width δr as indicated in figure 12.2b.

$$\delta I_{zz} = \text{mass} \times r^2 = \rho t 2\pi r(\delta r) r^2$$

For the whole plate, by integration

$$I_{zz} = 2\pi\rho t \int_0^R r^3 \, \mathrm{d}r$$

$$= \frac{\pi\rho t R^4}{2}$$

$$= \frac{mR^2}{2} \tag{12.4}$$

and

$$k_{zz} = \sqrt{\left(\frac{I_{zz}}{m}\right)} = \frac{R}{\sqrt{2}}$$

Since $I_{xx} = I_{yy} = mR^2/4$ then

$$I_{zz} = I_{xx} + I_{yy}$$

and the relation of equation 12.2 is confirmed.

12.2.2 Parallel Axis Theorem for Bodies in General

Figure 12.3 shows a plan and side view of a solid body. Axes $X'X$, $Y'Y$ and $Z'Z$ are indicated, also an axis $Z_1'\,Z_1$ which is parallel to $Z'Z$ and distance d from it. Shown hatched is an elementary column of the body of very small cross-section; this has mass δm and is parallel to both $Z'Z$ and $Z_1{}'Z_1$.

By definition

$$I_{z_1 z_1} = \Sigma(\delta m) r_1{}^2 = \Sigma \delta m(x_1{}^2 + y_1{}^2)$$

and

$$I_{zz} = \Sigma(\delta m) r^2$$

Now $r^2 = (d + x_1)^2 + y_1{}^2 = d^2 + 2x_1 d + x_1{}^2 + y_1{}^2$ and

$$I_{zz} = \Sigma[(\delta m)d^2 + 2(\delta m)x_1 d + \delta m(x_1{}^2 + y_1{}^2)]$$

$$= d^2\, \Sigma\, \delta m + 2d\, \Sigma(\delta m)x_1 + I_{z_1 z_1}$$

Now $\Sigma\delta m = m$ the mass of the body; thus

$$I_{zz} = md^2 + I_{z_1z_1} + 2d\,\Sigma(\delta m)x_1$$

Now if $Z_1{}'Z_1$ passes through G, the centre of the mass of the body, then by definition $\Sigma(\delta m)x_1 = 0$ and it follows that

$$I_{zz} = (I_{z_1z_1})_G + md^2 \tag{12.5}$$

Note that $Z_1{}'\,Z_1$ at G must be parallel to $Z'Z$ and that the equation can only be used to relate moments of inertia if one axis passes through G.

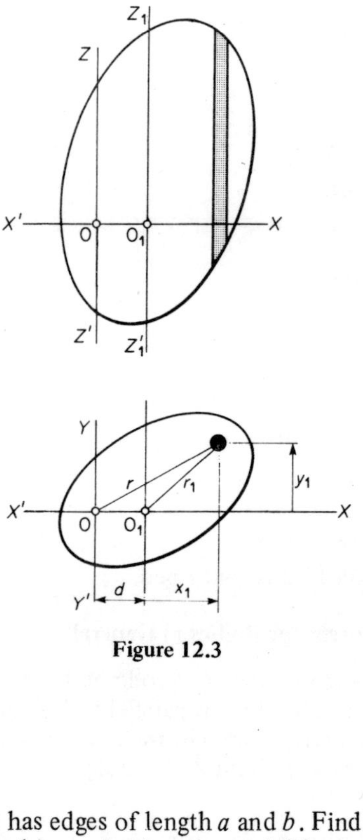

Figure 12.3

Worked Example 12.2

A rectangular thin plate has edges of length a and b. Find from first principles its moment of inertia about (a) an edge length a and (b) an axis parallel to this edge and passing through the centre of the plate. Confirm that equation 12.5 is satisfied.

Solution

See figure 12.4. For I_{xx} taking a strip parallel to $X'X$ distance y from it and of width δy

$$\delta I_{xx} = \rho ta(\delta y)y^2$$

and

$$I_{xx} = \int_0^b \rho tay^2 \, dy = \rho ta \frac{b^3}{3}$$

Now $m = \rho tab$; thus

$$I_{xx} = \frac{mb^2}{3} \tag{12.6}$$

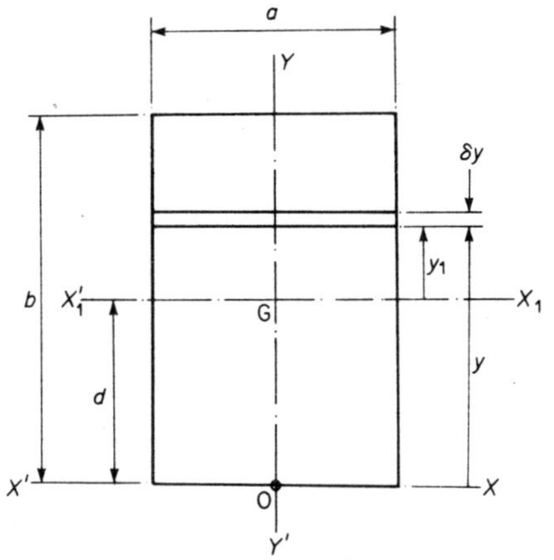

Figure 12.4

An axis parallel to $X'X$ and at the centre is $X_1{}'X_1$ and passes through G; the moment of inertia about this axis is termed $(I_{x_1x_1})_G$. Now

$$\delta(I_{x_1x_1})_G = \delta m(y_1)^2 = \rho ta(\delta y_1)y_1{}^2 \quad (\text{note } \delta y_1 = \delta y)$$

$$(I_{x_1x_1})_G = \int_{-b/2}^{+b/2} \rho ta(y_1)^2 \, dy_1 = \frac{\rho ta}{3}\left[\frac{b^3}{8} - \left(-\frac{b^3}{8}\right)\right]$$

$$= \frac{\rho tab^3}{12} = \frac{mb^2}{12} \tag{12.7}$$

$$(I_{x_1x_1})_G + md^2 = \frac{mb^2}{12} + m\left(\frac{b}{2}\right)^2$$

$$= \frac{mb^2}{3}$$

$$= I_{xx}$$

thus confirming that equation 12.5 is satisfied.

12.3 Standard Forms

12.3.1 Thin Plates

(1) *Circular Plate* (see equations 12.3 and 12.4)
About a diameter

$$I_{\text{dia}} = \frac{mR^2}{4}$$

about the polar axis

$$I_{\text{axis}} = \frac{mR^2}{2}$$

(2) *Rectangular Plate* (see equations 12.6 and 12.7)
About an axis $X'X$ coincident with an edge

$$I_{xx} = \frac{mb^2}{3}$$

where the edge perpendicular to $X'X$ has length b; about an axis $X_1{}'\,X_1$ parallel to $X'X$ and passing through G

$$(I_{x_1x_1})_{\text{G}} = \frac{mb^2}{12}$$

About an axis through G perpendicular to the plane: by use of the perpendicular axis theorem

$$(I_{zz})_{\text{G}} = (I_{x_1x_1})_{\text{G}} + (I_{yy})_{\text{G}} \quad \text{(see figure 12.4)}$$

$$= \frac{mb^2}{12} + \frac{ma^2}{12} = \frac{m}{12}(a^2 + b^2) \qquad (12.8)$$

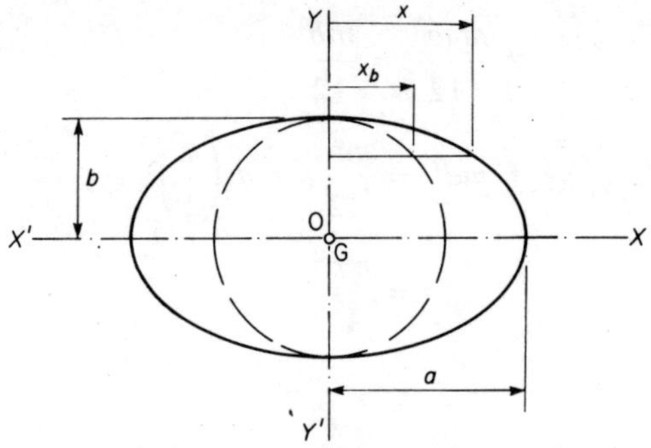

Figure 12.5

(3) *Elliptical Plate*

The elliptical plate shown in figure 12.5 has semi-major and semi-minor axes lengths a and b respectively. Axes $X'X$ and $Y'Y$ are chosen as shown to pass through G.

The elliptical shape is such that in the figure $x = (a/b)\, x_b$ where x_b refers to the circular shape radius b.

It follows that strips parallel to the x-axis are all increased in mass by the ratio a/b as compared with the corresponding strips of the circular shape. Since I_{xx} for a circular plate radius b, mass m_b, is $m_b b^2/4$, then for the elliptical plate

$$(I_{xx})_G = m_b \times \frac{a}{b} \times \frac{b^2}{4}$$

But the mass of the ellipse $m = m_b\,(a/b)$, therefore

$$(I_{xx})_G = m\frac{b^2}{4} \tag{12.9}$$

Similarly

$$(I_{yy})_G = \frac{ma^2}{4}$$

Thus

$$(I_{zz})_G = \frac{m}{4}(a^2 + b^2) \tag{12.10}$$

(4) *Triangular Plate*

In figure 12.6a choose an axis $Y'Y$ passing through the vertex and parallel to the

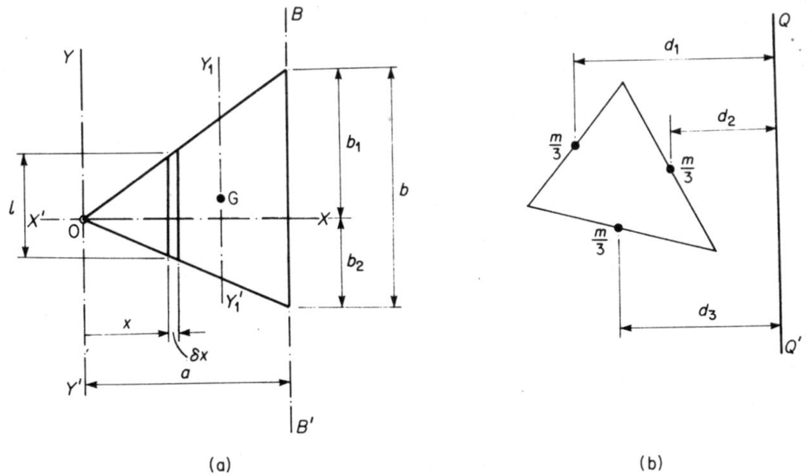

Figure 12.6

side length b. For I_{yy} select an elementary strip parallel to $Y'Y$, then

$$\delta I_{yy} = (\delta m)x^2$$

$$\delta m = \rho t l \delta x$$

where $l = (x/a)b$; therefore

$$I_{yy} = \int_0^a \rho t \frac{b}{a} x \times x^2 \, dx = \rho t b \frac{a^3}{4}$$

$m = \frac{1}{2} \rho t b d$, thus

$$I_{yy} = \frac{ma^2}{2} \tag{12.11}$$

By the parallel axis theorem

$$I_{yy} = (I_{y_1 y_1})_G + md^2$$

therefore

$$(I_{y_1 y_1})_G = \frac{ma^2}{2} - m\left(\frac{2a}{3}\right)^2 = \frac{ma^2}{18} \tag{12.12}$$

and

$$I_{B'B} = \frac{ma^2}{18} + m\left(\frac{a}{3}\right)^2 = \frac{ma^2}{6} \tag{12.13}$$

It can be shown that the moment of inertia of a thin triangular plate can be found *about any axis* by replacing the plate (mass m) by three particles each having mass $m/3$ placed at the mid points of the sides. Thus in figure 12.6b

$$I_{Q'Q} = \frac{m}{3}(d_1{}^2 + d_2{}^2 + d_3{}^2)$$

For example, applying this to figure 12.6a

$$I_{B'B} = \frac{m}{3}\left[\left(\frac{a}{2}\right)^2 + \left(\frac{a}{2}\right)^2 + 0\right] = \frac{ma^2}{6}$$

$$I_{yy} = \frac{m}{3}\left[\left(\frac{a}{2}\right)^2 + \left(\frac{a}{2}\right)^2 + a^2\right] = \frac{ma^2}{2}$$

$$(I_{y_1 y_1})_G = \frac{m}{3}\left[\left(\frac{a}{6}\right)^2 + \left(\frac{a}{6}\right)^2 + \left(\frac{a}{3}\right)^2\right] = \frac{ma^2}{18}$$

as previously determined. Also

$$I_{xx} = \frac{m}{3}\left[\left(\frac{b_1}{2}\right)^2 + \left(\frac{b_2}{2}\right)^2 + \left(\frac{b}{2} - b_2\right)^2\right]$$

and obviously depends on the shape of the triangle in relation to the axis $X'X$.

12.3.2 Three-dimensional Bodies of Symmetrical Form

The most convenient method is to split the solid body into very thin laminae, which can be regarded as thin plates whose moments of inertia are known.

(1) *Rectangular Prism*

The body is shown in figure 12.7 and the three perpendicular axes lying in the planes of symmetry are shown – all passing through G.

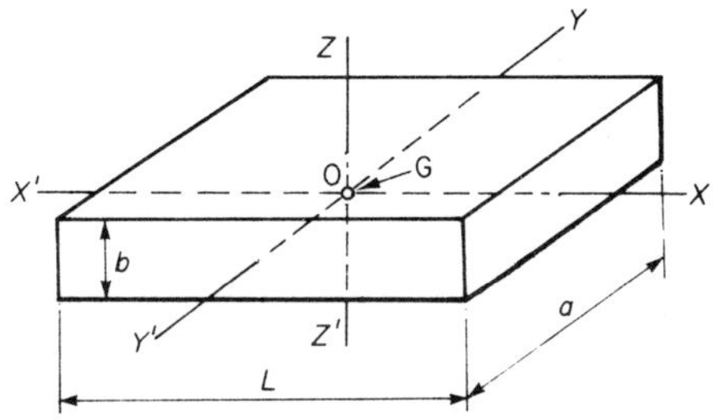

Figure 12.7

For a thin plate mass δm parallel to the XOY plane, from equation 12.8

$$(\delta I_{zz})_G = \frac{\delta m}{12}(a^2 + L^2)$$

Thus

$$(I_{zz})_G = \Sigma \frac{\delta m}{12}(a^2 + L^2)$$

$$= \frac{(a^2 + L^2)}{12} \Sigma\, \delta m = \frac{m}{12}(a^2 + L^2) \qquad (12.14)$$

Similarly

$$(I_{xx})_G = \frac{m}{12}(a^2 + b^2)$$

and

$$(I_{yy})_G = \frac{m}{12}(b^2 + L^2)$$

[Note: $(I_{zz})_G$ is *not* equal to $(I_{xx})_G + (I_{yy})_G$]

(2) *Circular Cylinder*

See figure 12.8. Three perpendicular axes lying in the planes of symmetry are

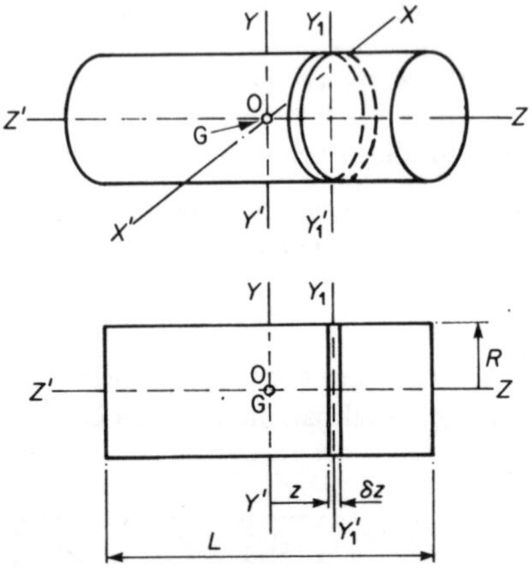

Figure 12.8

shown, all passing through G. For a thin plate thickness δz distance z from the XOY-plane

$$I_{y_1 y_1} = \delta m \frac{R^2}{4}$$

($Y_1' Y_1$ is a diameter) and $\delta m = \rho \pi R^2 \, \delta z$. Therefore by the parallel axis theorem

$$\delta I_{yy} = I_{y_1 y_1} + (\delta m) z^2$$

and since $Y_1' Y_1$ passes through the centre of mass of the thin plate

$$\delta I_{yy} = \delta m \frac{R^2}{4} + \rho \pi R^2 z^2 \delta z$$

and

$$I_{yy} = \frac{R^2}{4} \int_0^m \mathrm{d}m + \rho \pi R^2 \int_{-L/2}^{+L/2} z^2 \, \mathrm{d}z$$

$$= \frac{mR^2}{4} + \frac{\rho \pi R^2 L^3}{12}$$

But $m = \rho \pi R^2 L$; also $Y'Y$ passes through the mass centre G of the body and thus I_{yy} is correctly titled $(I_{yy})_G$; therefore

$$(I_{yy})_G = \frac{mR^2}{4} + \frac{mL^2}{12} \tag{12.15}$$

By symmetry

$$(I_{xx})_G = \frac{mR^2}{4} + \frac{mL^2}{12}$$

The moment of inertia of the thin plate about $Z'Z$ is

$$\delta I_{zz} = \frac{(\delta m)R^2}{2}$$

therefore

$$(I_{zz})_G = \frac{mR^2}{2} \tag{12.16}$$

The result of equation 12.15 can be generalised into the so-called *cylinder theorem,* applicable to cylinders of arbitrary, but uniform, cross section. A point O is chosen within the cylinder and an axis $Z'OZ$ is taken parallel to the length; axes $X'OX$ and $Y'OY$ define the plane perpendicular to the axis $Z'OZ$ at O. The cylinder is now replaced by (1) a thin plate in the XOY-plane having as its boundary the trace of the cylinder in the XOY plane and having mass equal to that of the cylinder, and (2) a thin rod in the $Z'OZ$-axis having ends coinciding with the ends of the cylinder and mass equal to that of the cylinder. The moments of inertia I_{xx} and I_{yy} of the cylinder are then given by

$$(I_{xx})_{\text{cylinder}} = (I_{xx})_{\text{plate}} + (I_{xx})_{\text{rod}} \tag{12.17a}$$

$$(I_{yy})_{\text{cylinder}} = (I_{yy})_{\text{plate}} + (I_{yy})_{\text{rod}} \tag{12.17b}$$

For example a cylinder having elliptical cross-section with semi-major and semi-minor axes a and b in the x- and y-directions respectively and length L will have

$$(I_{xx})_{\text{cylinder}} = \frac{mb^2}{4} + \frac{mL^2}{12}$$

in relation to an axis $X'X$ passing through the centre.

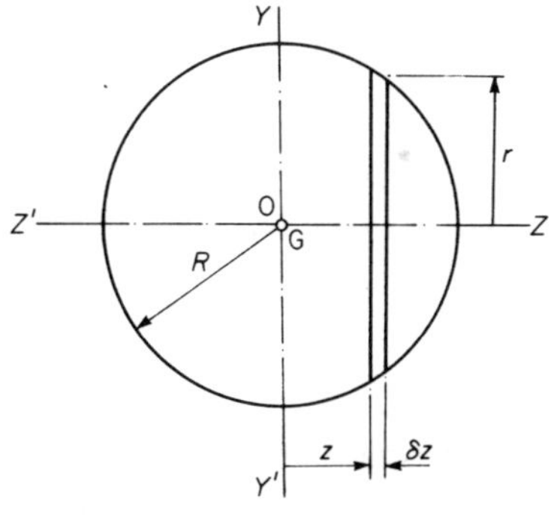

Figure 12.9

(3) *Sphere*

For the thin plate shown in figure 12.9

$$r = \sqrt{(R^2 - z^2)}$$

$$\delta I_{zz} = \frac{(\delta m)r^2}{2} = \frac{\rho\pi r^2(\delta z)r^2}{2} = \frac{\rho\pi}{2} r^4 \delta z$$

$$I_{zz} = \int_{-R}^{+R} \frac{\rho\pi}{2} (R^2 - z^2)^2 \, dz$$

$$= \rho \frac{\pi}{2} \left[R^4 z - \tfrac{2}{3} R^2 z^3 + \tfrac{1}{5} z^5 \right]_{-R}^{+R}$$

$$= \frac{8}{15} \rho\pi R^5$$

but $m = (4/3)\, \rho\pi R^3$, therefore

$$I_{zz} = \frac{2}{5} mR^2 \qquad (12.18)$$

12.3.3 Closing Note

Moments of inertia have been calculated only for axes which are parallel to or perpendicular to the principal linear dimensions of the bodies considered. Rigid bodies have moments of inertia about other axes, for example $Q'Q$ in figure 12.10,

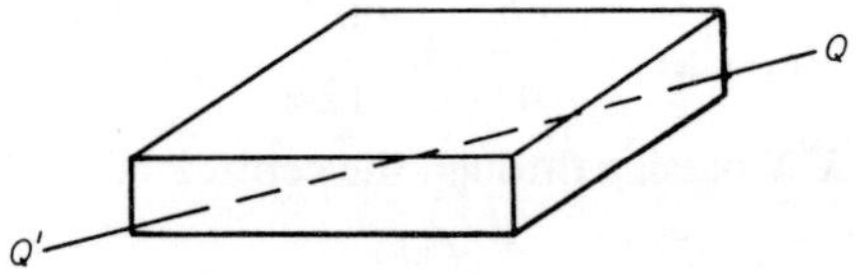

Figure 12.10

but their determination requires the use of a related quantity, the product of inertia. However, these are not required for the problems coming within the scope of this book.

12.4 General Method for Calculating *I* for Complex Bodies

(1) Make good sketches of the body; it is usually more satisfactory to use three views of the body in order to indicate all the dimensions required.

(2) Break up the body into standard shapes such as cylinders, rectangular prisms, etc., whose moments of inertia are known. Draw separate diagrams for each of these standard shapes selecting an axis through the centre of mass, parallel to the axis about which the moment of inertia of the whole body is required.
(3) Note that moments of inertia about a given axis are additive, that is, the sum of the moments of inertia of the parts is equal to the moment of inertia of the whole. Thus in figure 12.11a the moment of inertia of the body about $X'X$ = moments of inertia of block A about $X'X$ + moment of inertia of block B about $X'X$. This statement can also be extended to cover the removal of part of a body (see figure 12.11b). Thus

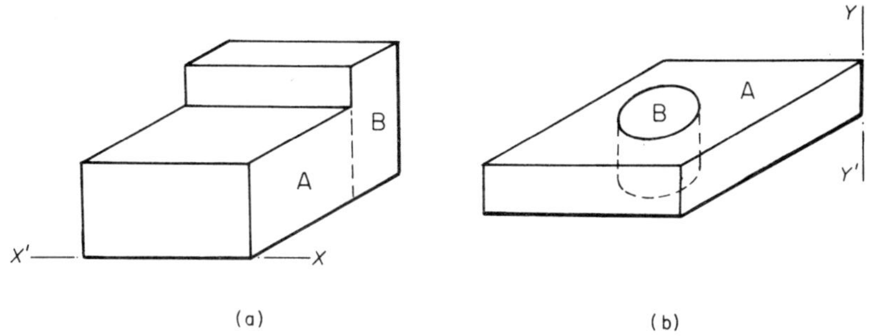

Figure 12.11

$$I_{yy} \text{ of the body without the hole } = I_{yy} \text{ of A } + I_{yy} \text{ of B}$$

therefore

$$I_{yy} \text{ of A } = I_{yy} \text{ of the body without the hole } - I_{yy} \text{ of B}$$

(4) Determine the moment of inertia of each part about the axis through its mass-centre and then use the parallel axis theorem to find its moment of inertia about the specified axis. Use the additive property to obtain the moment of inertia of the whole body.

Worked Example 12.3

Figure 12.12a shows a rigid body made up of a rectangular block A and circular cylinder B. Find the moments of inertia of the body about axes (a) P′P (b) R′R and (c) for the two parallel axes $X'X$ and $Z'Z$ through the mass-centre G. The density of the material is ρ kg/m^3.

Solution

Draw separate diagrams for parts A and B choosing x-, y- and z directions at G_A and G_B the respective mass-centres as indicated in figures 12.12b and c.

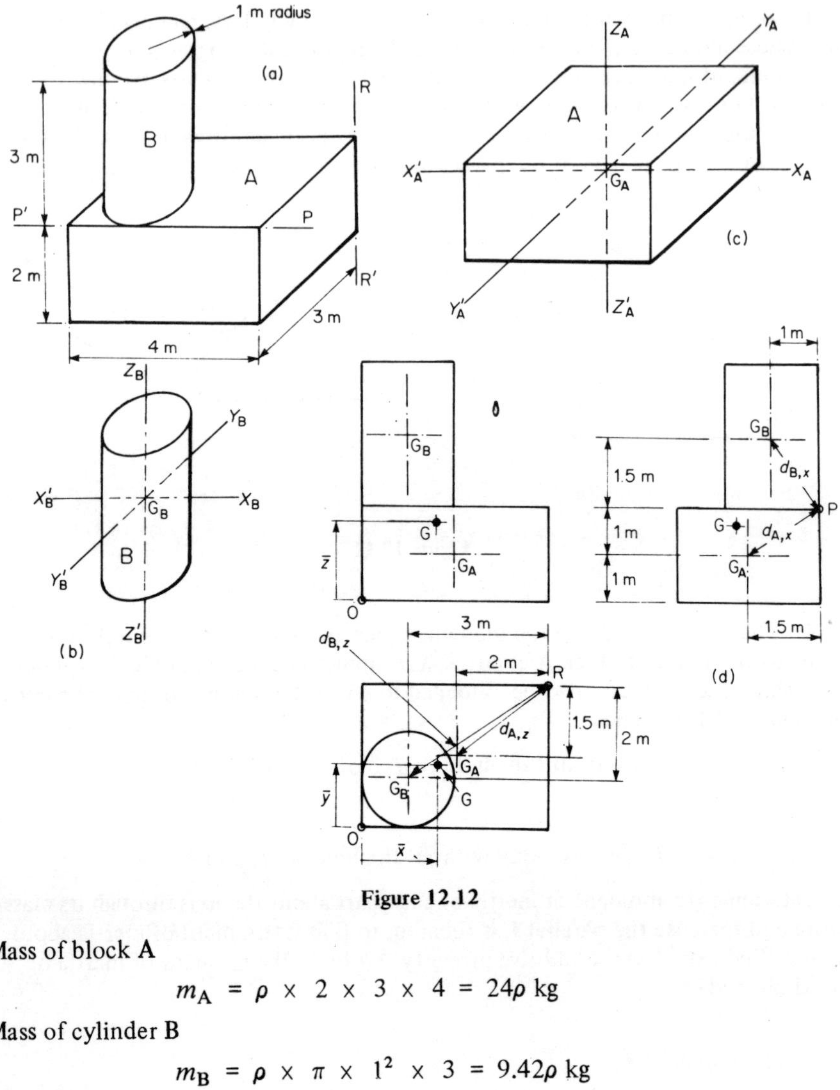

Figure 12.12

Mass of block **A**

$$m_A = \rho \times 2 \times 3 \times 4 = 24\rho \text{ kg}$$

Mass of cylinder **B**

$$m_B = \rho \times \pi \times 1^2 \times 3 = 9.42\rho \text{ kg}$$

Mass of body

$$m = 33.42\rho \text{ kg}$$

(a) For axis **P′P**: For the block **A** the axis parallel to **P′P** is $X_A{'}X_A$ and

$$(I_{xx})_A = \frac{m_A}{12}(3^2 + 2^2) \text{ (see equation 12.14)}$$

$$= \frac{13}{12} m_A \text{ kg m}^2$$

The distance between the parallel axes $X_A{}'X_A$ and P′P is (see figure 12.12d) $d_{A,x}$ where $d_{A,x}{}^2 = 1^2 + 1.5^2 = 3.25\ m^2$; thus

$$(I_{P'P})_A = (I_{xx})_A + m_A d_{A,x}{}^2 = m_A \left(\frac{13}{12} + 3.25\right) = 104\rho \text{ kg m}^2$$

For cylinder B

$$(I_{xx})_B = \frac{m_B}{4} 1^2 + \frac{m_B}{12} 3^2 \qquad \text{(see equation 12.15)}$$

$$= m_B \text{ kg m}^2$$

$$d_{B,x}{}^2 = 1^2 + 1.5^2 = 3.25 \text{ m}^2$$

$$(I_{P'P})_B = (I_{xx})_B + m_B d_{B,x}{}^2 = m_B (1 + 3.25) = 40.0\rho \text{ kg m}^2$$

For the whole body

$$I_{P'P} = (I_{P'P})_A + (I_{P'P})_B = 144.0\rho \text{ kg m}^2$$

(b) For axis R′R (parallel to the z-axes of A and B). Block A

$$(I_{zz})_A = \frac{m_A}{12}(4^2 + 3^2) = \frac{25}{12} m_A \text{ kg m}^2$$

$$d_{A,z}{}^2 = 2^2 + 1.5^2 = 6.25 \text{ m}^2$$

$$(I_{R'R})_A = (I_{zz})_A + m_A d_{A,z}{}^2 = m_A \left(\frac{25}{12} + 6.25\right) = 200\rho \text{ kg m}^2$$

Cylinder B

$$(I_{zz})_B = \frac{m_B}{2} 1^2 \text{ (see equation 12.16)} = \tfrac{1}{2} m_B \text{ kg m}^2$$

$$d_{B,z}{}^2 = 3^2 + 2^2 = 13$$

$$(I_{R'R})_B = (I_{zz})_B + m_B d_{B,z}{}^2 = m_B (\tfrac{1}{2} + 13) = 127.2\rho \text{ kg}$$

For the whole body

$$I_{R'R} = (I_{R'R})_A + (I_{R'R})_B = 327.2\rho \text{ kg m}^2$$

(c) In order to determine the moments of inertia about axes through the mass centre G this must first be located. Choosing an origin at the bottom left-hand corner as indicated in figure 12.12d then if the coordinates of G are $\overline{x}$, $\overline{y}$, and $\overline{z}$, and the coordinates of G_A and G_B carry corresponding subscripts, then (see section 7.3)

$$\bar{x} = \frac{m_A\bar{x}_A + m_B\bar{x}_B}{m_A + m_B} = \frac{24 \times 2 + 9.42 \times 1}{33.42} = 1.72 \text{ m}$$

$$\bar{y} = \frac{m_A\bar{y}_A + m_B\bar{y}_B}{m_A + m_B} = \frac{24 \times 1.5 + 9.42 \times 1}{33.42} = 1.36 \text{ m}$$

$$\bar{z} = \frac{m_A\bar{z}_A + m_B\bar{z}_B}{m_A + m_B} = \frac{24 \times 1 + 9.42 \times 3.5}{33.42} = 1.70 \text{ m}$$

The required moments of inertia can now be determined by application of the parallel axis theorem.

Between axis P′P and the x-axis through G there is a perpendicular distance d_x given by

$$d_x^2 = 0.3^2 + 1.36^2 = 1.94 \text{ m}^2$$

and

$$I_{P'P} = (I_{xx})_G + md_x^2$$

therefore

$$(I_{xx})_G = 144.0\rho - 33.42\rho \times 1.94 = 79.2\rho \text{ kg m}^2$$

Between axis R′R and the z-axis through G

$$d_z^2 = 2.28^2 + 1.64^2 = 7.89 \text{ m}^2$$

and

$$I_{R'R} = (I_{zz})_G + md_z^2$$

therefore

$$(I_{zz})_G = 327.2\rho - 33.42\rho \times 7.89 = 63.6\rho \text{ kg m}^2$$

The values $(I_{xx})_G$ and $(I_{zz})_G$ could, of course, be calculated without reference to $I_{P'P}$, etc., as for example

$$(I_{zz})_G = (I_{zz})_A + m_A(0.28^2 + 0.14^2) + (I_{zz})_B + m_B(0.36^2 + 0.72^2)$$
$$= 63.2\rho \text{ kg m}^2$$

(The difference is due to rounding-off errors.)

12.5 Summary

(1) Moment of inertia $I_{xx} = \Sigma(\delta m)r^2$, r being the perpendicular distance from the particle mass δm to the axis $X'X$. Also

$$I_{xx} = mk_{xx}^2 \qquad (12.1)$$

where k_{xx} is the radius of gyration of the body about the axis $X'X$, and m is its mass.

(2) Perpendicular axis theorem – applies only to thin plates

$$I_{zz} = I_{yy} + I_{xx} \text{ (thin plates)} \qquad (12.2)$$

where the axis $Z'Z$ is perpendicular to the plane of the plate.

(3) Parallel axis theorem – applies to thin plates and three-dimensional bodies

$$I_{zz} = (I_{z_1 z_1})_G + md^2 \qquad (12.5)$$

$(I_{z_1 z_1})_G$ is the moment of inertia of the body about an axis $Z_1{}'Z_1$, parallel to $Z'Z$ and passing through the centre of mass; d is the distance between $Z'Z$ and $Z_1{}'Z_1$.

(4) For complex bodies use the additive property.

(5) Use the cylinder theorem, equation 12.17, for bodies of constant cross-sectional area.

Problems

12.1 Show that the moment of inertia of the segment ABC of the thin plate in

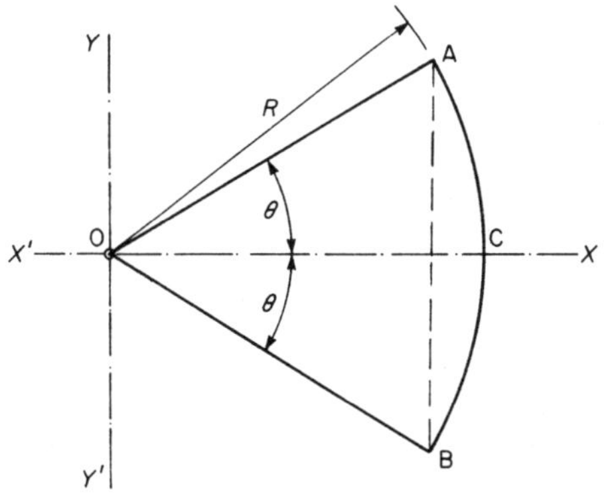

Figure 12.13

figure 12.13 about $Y'Y$ is

$$I_{yy} = \frac{\rho t R^4}{4}\left(\theta - \frac{\sin 4\theta}{4}\right)$$

(See worked example 12.1.) Hence, by using the standard form for the triangle OAB, show that for the whole plate

$$I_{yy} = \frac{\rho t R^4}{4}\left(\theta + \frac{\sin 2\theta}{2}\right)$$

12.2 Show that I_{zz} (where $Z'Z$ is an axis perpendicular to the plate passing through the apex) for the plate in figure 12.13 is

$$I_{zz} = \frac{\rho t R^4}{2} \theta$$

Hence by making use of the result of problem 12.1 show that

$$I_{xx} = \frac{\rho t R^4}{4} \left(\theta - \frac{\sin 2\theta}{2}\right)$$

12.3 Find the moment of inertia of a semicircular thin plate about an axis through its centre of mass parallel to the straight edge. Hence find the moment of inertia about an axis through G perpendicular to the plate.

12.4 Determine I_{yy} for the plate in figure 12.14. Take $t = 0.1$ m and $\rho = 1000$ kg/m^3.

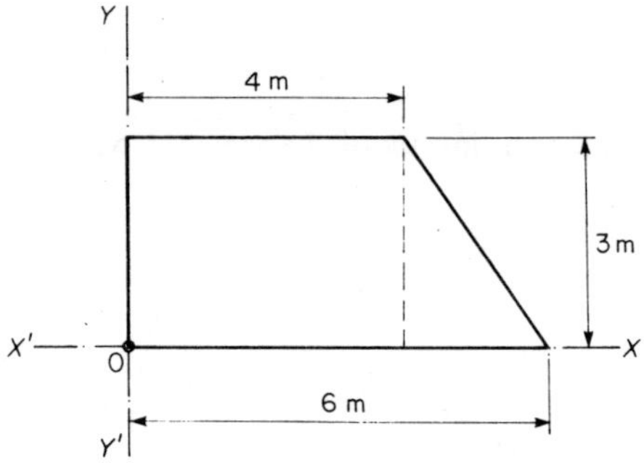

Figure 12.14

12.5 Calculate I_{xx} for the plate in figure 12.14.

12.6 Find I_{zz} for the thin plate in figure 12.15. Take $t = 0.1$ m and $\rho = 1000$ kg/m^3.

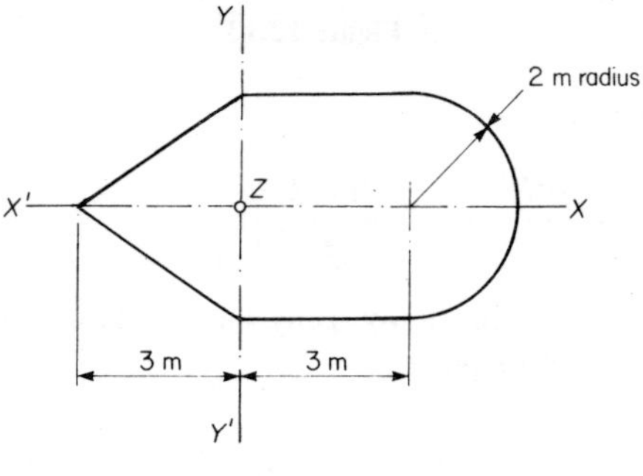

Figure 12.15

12.7 Find the moments of inertia of the solid block in figure 12.16 about $Z_1'Z_1$,

$Z_2'Z_2$ at the centre of the rear face, and $Z_3'Z_3$. Mass density is ρ kg/m^3.

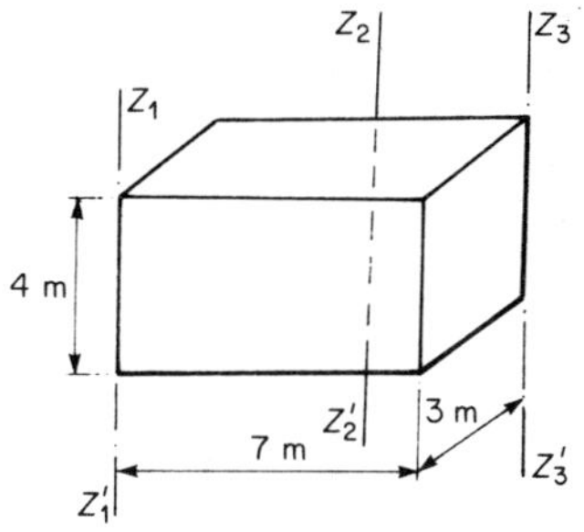

Figure 12.16

12.8 Determine I_{xx} for the body in figure 12.17. Take ρ = 1000 kg/m^3.

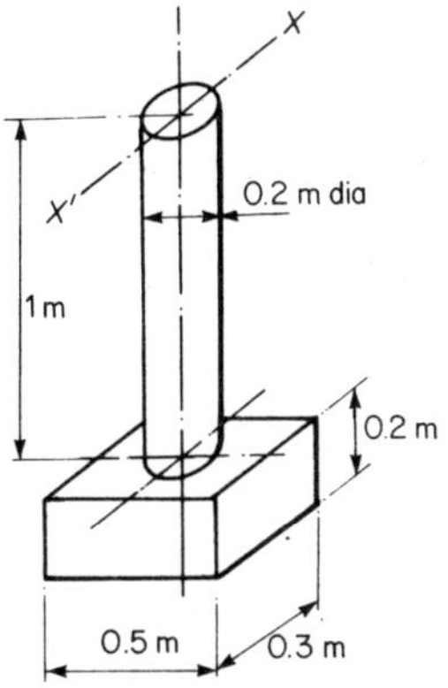

Figure 12.17

12.9 Find I_{yy} and I_{xx} for the body in figure 12.18. What is $I_{y_1y_1}$?

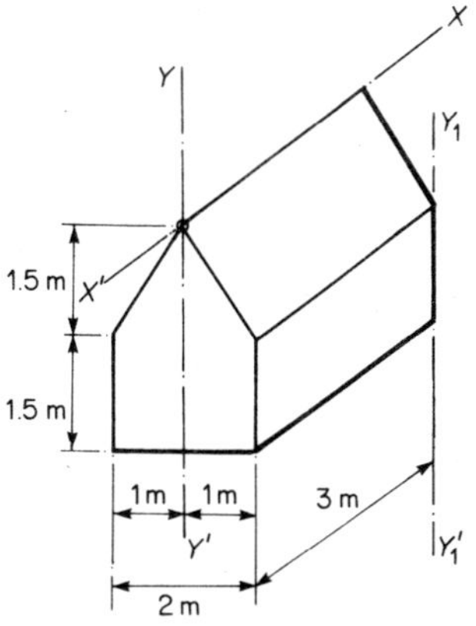

Figure 12.18

12.10 For the body in figure 12.18 determine $(I_{xx})_G$ and $(I_{zz})_G$.

12.11 A ring is formed by cutting a length from a cylindrical tube. It is 0.8 m outside diameter, 0.6 m inside diameter and is 0.3 m thick. Find its moment of inertia about (a) its central axis and (b) about a diameter on an end face. Take $\rho = 1000$ kg/m^3.

12.12 Figure 12.19 shows the cross-section of a flywheel. Calculate its moment

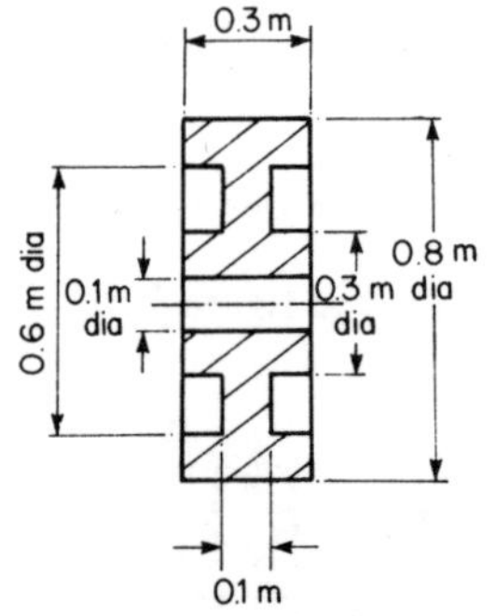

Figure 12.19

of inertia about its central axis if $\rho = 1000$ kg/m^2 (and see problem 12.11). What proportion of the moment of inertia is due to the rim (the outer ring) of the flywheel?

12.13 Verify the results for the standard moments of inertia given in the appendix.

13 Kinetics of a Rigid Body

In chapter 9 we derived the principle of the motion of the mass centre. For a particle system, and hence for a rigid body, we were able to relate the acceleration of the mass centre to the external forces through equations 9.9, namely

$$\Sigma F_x = m\bar{a}_x \tag{9.9a}$$

$$\Sigma F_y = m\bar{a}_y \tag{9.9b}$$

Given the external forces we were able to calculate, for a rigid body, the magnitude and direction of the acceleration of this one point, the mass centre. We were careful to note that these equations did *not* imply that the resultant of the external forces passed through the mass centre. Thus the equations could be applied to rigid bodies undergoing translation but were insufficient to describe any rotational motion of these bodies.

Since the kinematic behaviour of a rigid body, which can be translation, fixed-axis rotation or general plane motion, is completely determined by the resultant of the external forces, we now consider how this behaviour is related to these external forces. In particular we examine how to determine the external forces required to bring about a desired motion and the motion brought about by a given set of applied forces.

13.1 Equations of Motion

We first consider the set of effective forces for the body mass m shown in figure 13.1a which is moving with general plane motion parallel to the reference plane XOY. At some instant the position of the mass centre G is $(\bar{x}, \bar{y})$ and that of a typical particle P is (x, y) where $x = \bar{x} + x'$ and $y = \bar{y} + y'$. If the rectangular components of the acceleration of P are a_x and a_y then

$$a_x = \frac{d^2(\bar{x} + x')}{dt^2} = \bar{a}_x + \frac{d^2x'}{dt^2}$$

$$a_y = \frac{d^2(\bar{y} + y')}{dt^2} = \bar{a}_y + \frac{d^2y'}{dt^2}$$

where $\bar{a}_x$ and $\bar{a}_y$ are the components of the acceleration of the mass centre G.

If the mass of the particle P is δm then for the particle the components $\delta F_{x,\text{eff}}$ and $\delta F_{y,\text{eff}}$ of the effective force δF_{eff} are respectively $(\delta m)a_x$ and $(\delta m)a_y$ as shown in figure 13.1b. This effective force at P is equivalent to an equal effective force at G, having the same components, together with a couple as shown in figure 13.1c, the moment of the couple being

$$\delta M_{G,\text{eff}} = x'(\delta m)a_y - y'(\delta m)a_x$$

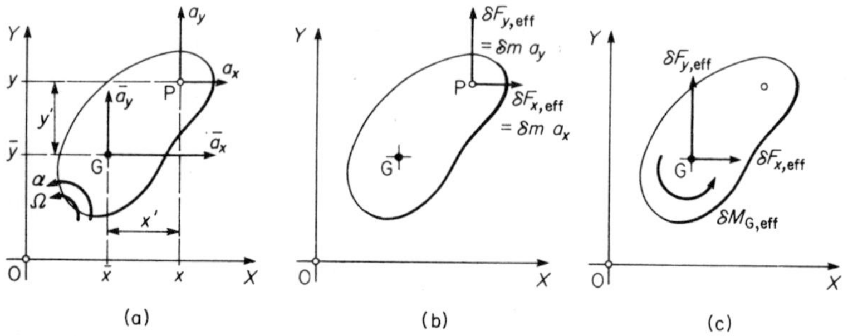

Figure 13.1

Summing over all the particles, the set of effective forces for the body is equivalent to a single resultant passing through G with components $F_{x,\text{eff}}$ and $F_{y,\text{eff}}$ and a couple having moment $M_{G,\text{eff}}$

$$\begin{aligned} F_{x,\text{eff}} &= \Sigma\delta F_{x,\text{eff}} = \Sigma(\delta m)a_x \\ &= \Sigma\delta m(\bar{a}_x + \mathrm{d}^2x'/\mathrm{d}t^2) \\ &= \bar{a}_x\,\Sigma\delta m + \Sigma(\delta m)\frac{\mathrm{d}^2x'}{\mathrm{d}t^2} \end{aligned}$$

similarly

$$\begin{aligned} F_{y,\text{eff}} &= \Sigma\delta F_{y,\text{eff}} \\ &= \bar{a}_y\,\Sigma\delta m + \Sigma(\delta m)\frac{\mathrm{d}^2y'}{\mathrm{d}t^2} \end{aligned}$$

By definition of the mass centre G

$$\Sigma(\delta m)x' = \Sigma(\delta m)y' = 0$$

and it follows that

$$\Sigma(\delta m)\frac{\mathrm{d}^2x'}{\mathrm{d}t^2} = \Sigma(\delta m)\frac{\mathrm{d}^2y'}{\mathrm{d}t^2} = 0$$

Therefore, since $\Sigma\delta m = m$ the effective force components are

$$\begin{aligned} F_{x,\text{eff}} &= m\bar{a}_x \\ F_{y,\text{eff}} &= m\bar{a}_y \end{aligned}$$

The couple

$$
\begin{aligned}
M_{G,\text{eff}} &= \Sigma\delta M_{G,\text{eff}} \\
&= \Sigma x'(\delta m)a_y - \Sigma y'(\delta m)a_x \\
&= \Sigma x'\delta m\left(\bar{a}_y + \frac{d^2y'}{dt^2}\right) - \Sigma y'\delta m\left(\bar{a}_x + \frac{d^2x'}{dt^2}\right) \\
&= \bar{a}_y\,\Sigma x'\delta m - a_x\,\Sigma y'\delta m \\
&\quad + \Sigma\delta m\left(x'\frac{d^2y'}{dt^2} - y'\frac{d^2x'}{dt^2}\right)
\end{aligned}
$$

From the definition of G the first two terms are zero. In the third summation the quantity in the brackets represents the net moment about G of the rectangular components of the acceleration of the typical particle at P relative to G (see figure 13.2a). This must equal the moment about G of the total acceleration of P relative to G, *or* the moment about G of any other set of components of ${}_G\boldsymbol{a}_P$. A

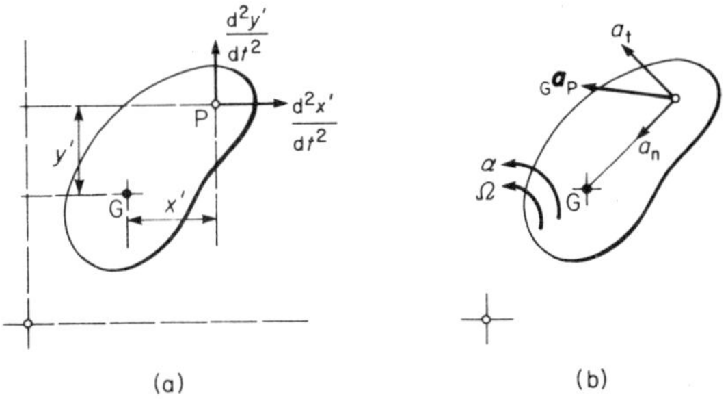

Figure 13.2

convenient set of such components is the tangential component $a_t = (GP)\alpha$ and the normal component $a_n = (GP)\Omega^2$, as in figure 13.2b, where α and Ω are the angular acceleration and angular velocity of the body respectively and $GP = r$. Thus the moment of d^2x'/dt^2 and d^2y'/dt^2 about G can be replaced by the moment of a_n and a_t. Hence

$$\Sigma\delta m\left(x'\frac{d^2y'}{dt^2} - y'\frac{d^2x'}{dt^2}\right) = \Sigma(\delta m)a_t r$$

since the moment of a_n about G is zero.

Thus

$$M_{G,eff} = \Sigma\delta m(r^2\alpha)$$
$$= \alpha\,\Sigma(\delta m)r^2$$

The quantity $\Sigma(\delta m)r^2$ is recognised (see preceding chapter) as the moment of inertia of the body about an axis passing through G perpendicular to the plane XOY, and denoted symbolically as $I_{G'G}$.

Hence

$$M_{G,eff} = I_{G'G}\alpha$$

The set of effective forces has therefore been reduced to a force - couple set, consisting of an effective force at the mass centre having components $m\bar{a}_x$ and $m\bar{a}_y$, together with a couple having moment $I_{G'G}\alpha$.

For a particle system we have the result expressed by equation 9.6 that the set of effective forces for a particle system is equivalent to the set of external forces. We can therefore write immediately

$$\Sigma F_x = m\bar{a}_x \tag{13.1}$$
$$\Sigma F_y = m\bar{a}_y \tag{13.2}$$
$$\Sigma M_G = I_{G'G}\alpha \tag{13.3}$$

where ΣF_x and ΣF_y are the x- and y-components of the resultant of the external forces and ΣM_G is the sum of the moments of the external forces about G. Signs, of course, must be taken into account. These three equations are the equations of motion for a rigid body in general plane motion.

The external force set must therefore reduce to a force passing through G having components ΣF_x and ΣF_y, plus a couple having moment ΣM_G. The result is summarised in figure 13.3. Figure 13.3a shows the body accelerations, figure 13.3b the external force set acting on the body, and figure 13.3c the equivalent force - couple set showing the relations between the force components, the couple and the body accelerations.

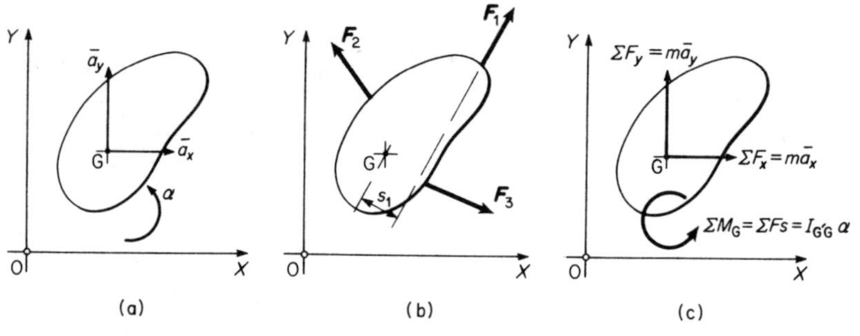

Figure 13.3

Worked Example 13.1

A body having mass 2 kg and for which $I_{G'G}$ is 1 kg m^2 has angular acceleration $\alpha = 10$ rad/s^2 anticlockwise and the components of the acceleration of G are $\bar{a}_x = 5$ m/s^2 and $\bar{a}_y = 10$ m/s^2 as shown in figure 13.4a. Determine the magnitude, direction and line of action of the resultant of the external forces.

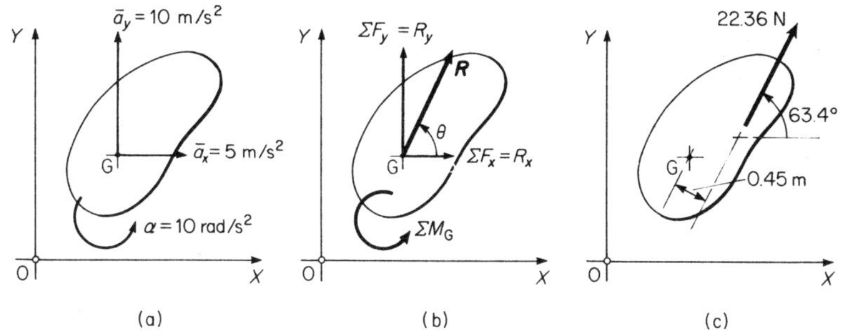

Figure 13.4

Solution

The required equivalent force - couple set is shown in figure 13.4b, $\boldsymbol{R}$ being equal to the resultant of the external forces. Using equations 13.1 to 13.3

$\Sigma F_x = m\bar{a}_x \qquad R_x = 2 \times 5 = 10$ N

$\Sigma F_y = m\bar{a}_y \qquad R_y = 2 \times 10 = 20$ N

and

$$R = 22.36 \text{ N } \angle\ 63.4°$$

$\Sigma M_G = I_{G'G}\alpha \qquad \Sigma M_G = 1 \times 10 = 10$ N m anticlockwise

This force - couple set is equivalent to a single force $\boldsymbol{R}$ which is such that its moment about G is anticlockwise and equal to $\Sigma M_G = 10$ N m. Its perpendicular distance from G is $\Sigma M_G/R = 10/22.36 = 0.45$ m. The required solution is shown in figure 13.4c.

Particular Conditions

(1) If the body is in equilibrium then $\Sigma F_x = 0$, $\Sigma F_y = 0$ and $\Sigma M_G = 0$, and if not at rest then its motion is such that the velocity of the mass centre and the angular velocity are constant.

(2) If the body is not in equilibrium and the motion is one of translation only, then $\alpha = 0$ and the external forces must reduce to a single resultant force passing through G.

(3) An important particular condition is that for which the body is not in equilibrium and the motion takes place about a fixed axis. If the axis passes through G then $\bar{a}_x = \bar{a}_y = 0$ and the external forces must reduce to a couple, moment ΣM_G.

13.1.1 General Fixed-axis Rotation

For general fixed-axis rotation the equations of motion, 13.1 to 13.3, can be written in a more convenient form. If the rigid body in figure 13.5a is rotating

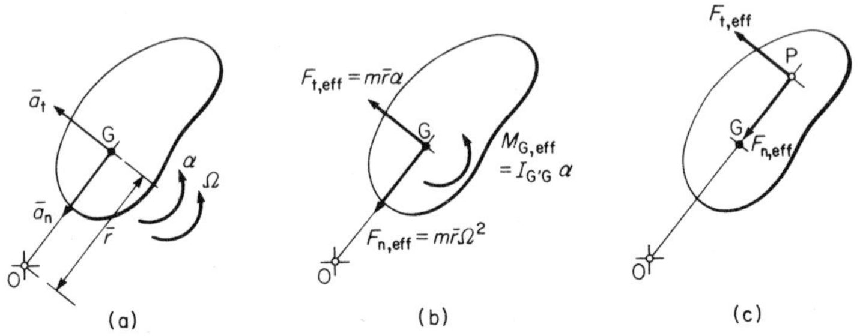

Figure 13.5

about an axis at O the acceleration of G, the mass centre, can be resolved into tangential and normal components $\bar{a}_t$ and $\bar{a}_n$ as shown. Since OG has a fixed length $\bar{r}$, then $\bar{a}_t = \bar{r}\,\alpha$ and $\bar{a}_n = \bar{r}\,\Omega^2$. The effective force couple set is now given by

$$F_{t,eff} = m\bar{r}\alpha$$

$$F_{n,eff} = m\bar{r}\,\Omega^2$$

$$M_{G,eff} = I_{G'G}\alpha$$

as shown in figure 13.5b.

The moment about O of the effective force - couple set is

$$\begin{aligned} M_{O,eff} &= F_{t,eff} \times \bar{r} + M_{G,eff} \\ &= m\bar{r}\alpha \times \bar{r} + I_{G'G}\alpha \\ &= \alpha(m\bar{r}^2 + I_{G'G}) \\ &= I_{O'O}\alpha \end{aligned}$$

by the parallel axis theorem.

Equating external and effective forces and couples

$$\Sigma F_t = m\bar{r}\alpha \tag{13.4a}$$

$$\Sigma F_n = m\bar{r}^2 \tag{13.4b}$$

$$\Sigma M_O = I_{O'O}\alpha \tag{13.4c}$$

These are alternative forms of the equations of motion which are applicable *only* to fixed-axis rotation.

13.1.2 Centre of Percussion

The effective force - couple set in figure 13.5b is equivalent to a single effective force, having the same components, acting through P on the line OG extended (see figure 13.5c). Equating moments about O for the effective forces in figures 13.5b and 13.5c

$$F_{t,eff} \times \bar{r} + M_{G,eff} = F_{t,eff}\,(OP)$$

Then

$$\begin{aligned} OP &= \bar{r} + M_{G,eff}/F_{t,eff} \\ &= \bar{r} + I_{G'G}\alpha/m\bar{r}\alpha \\ &= \bar{r} + k_{G'G}^2/\bar{r} \end{aligned} \tag{13.5}$$

where $k_{G'G}$ is the radius of the gyration of the body about an axis through G.

The resultant of the external forces producing fixed-axis rotation about O must therefore pass through P where GP = $k_{G'G}^2/\bar{r}$. The point P is called the *centre of percussion*; it will be referred to in later work.

13.1.3 Simple and Compound Pendulums

Consider a rigid body having mass m and radius of gyration $k_{G'G}$ about its mass centre G which can swing freely in a vertical plane about a frictionless pivot at O. It is deflected from its rest position and released so that it subsequently oscillates about its rest position. The body is then termed a *compound pendulum* and we wish to determine the periodic time of the oscillations if the amplitude of these is small.

A diagrammatic view of the body is shown in figure 13.6a giving the instantaneous position at some time t during the oscillation. The displacement θ anti-

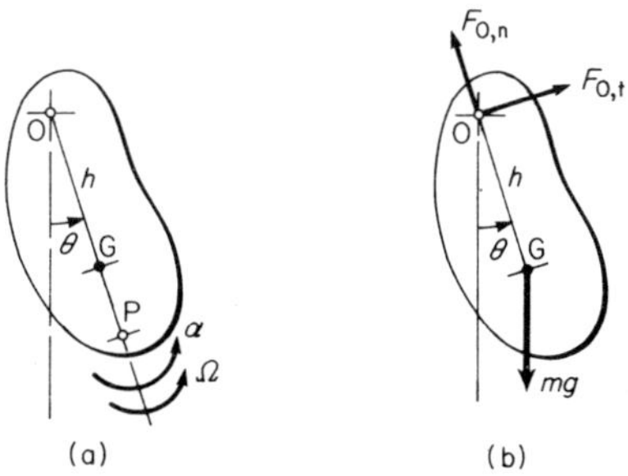

Figure 13.6

clockwise defines the positive senses of Ω and α. The free-body diagram of the body is shown in figure 13.6b, the forces $F_{O,t}$ and $F_{O,n}$ being the components of the reaction of the hinge at O on the body. From the free-body diagram it is clear that

$$\Sigma M_O = -\, mgh \sin\theta$$

and using equation 13.4c

$$\Sigma M_O = I_{O'O}\alpha$$

$$-mgh \sin\theta = m(k_{G'G}{}^2 + h^2)\alpha$$

hence

$$\alpha = -\frac{gh\sin\theta}{k_{G'G}{}^2 + h^2}$$

If the amplitude of the oscillation is small, then θ is small and $\sin\theta$ approximates to θ; the above equation may then be rewritten – writing $\ddot{\theta}$ for α as

$$\ddot{\theta} + \left(\frac{gh}{k_{G'G}{}^2 + h^2}\right)\theta = 0$$

This is the equation of simple harmonic motion, being of the form $\ddot{\theta} + \omega^2\theta = 0$ (see section 8.4). The periodic time is therefore

$$\tau = \frac{2\pi}{\omega} = 2\pi\sqrt{\left[\frac{(h^2 + k_{G'G}{}^2)}{gh}\right]} = 2\pi\sqrt{\left[\frac{(h + k_{G'G}{}^2/h)}{g}\right]}$$

(1) If h in the preceding equation is made a variable there is a particular value of h that makes τ a minimum; as an exercise show that this occurs when $h = k_{G'G}$.
(2) Suppose the body is suspended from an axis through P, the centre of percussion relative to O. Then h is replaced by $k_{G'G}{}^2/h$ and the periodic time

$$\tau = 2\pi\sqrt{\left(\frac{k_{G'G}{}^2/h + k_{G'G}{}^2h/k_{G'G}{}^2}{g}\right)} = 2\pi\sqrt{\left(\frac{k_{G'G}{}^2/h + h}{g}\right)}$$

which is the same as that for suspension at O. For this reason P is sometimes called the *centre of oscillation* with respect to the *centre of suspension* O.
(3) A *simple pendulum* consists ideally of a particle mass m suspended by a massless cord length l. For this system $k_{G'G} = 0$ and $\tau = 2\pi\sqrt{(l/g)}$. The simple pendulum having the same periodic time as the rigid body discussed already is the so called *equivalent simple pendulum* with length l_e. It can be seen that $l_e = h + k_{G'G}{}^2/h = OP$, the distance between O and the centre of percussion.

13.1.4 Inertia Force and Couple

The concept of reversed effective force or inertia force already discussed in the case of a particle can again be introduced to simplify the form of the equations of motion and permit an alternative method of problem solution.

If equations 13.1 to 13.3 are written in the form

$$\Sigma F_x - m\bar{a}_x = \Sigma F_x + (m\bar{a}_x)_{rev} = 0$$

$$\Sigma F_y - m\bar{a}_y = \Sigma F_y + (m\bar{a}_y)_{rev} = 0$$

$$\Sigma M_G - I_{G'G}\alpha = \Sigma M_G + (I_{G'G}\alpha)_{rev} = 0$$

then a rigid body in general plane motion can be considered to be in dynamic equilibrium (implying zero resultant force or couple) under the action of a force set comprising the external forces *and* the inertia force and couple. The inertia force and couple consist of

(1) $$(m\bar{a})_{\text{rev}} = -m\bar{a} \tag{13.6}$$

that is, a single force, magnitude $m\bar{a}$, passing through the mass centre G in the opposite sense to that of $\bar{\boldsymbol{a}}$;

(2) $$(I_{\text{G'G}}\alpha)_{\text{rev}} = -I_{\text{G'G}}\alpha \tag{13.7}$$

that is, a couple of magnitude $I_{\text{G'G}}\alpha$ in the opposite sense to that of α.

Note that the moment of inertia is always that about the mass-centre. We may, of course, use any sets of components of $(m\bar{a})_{\text{rev}}$ in equation 13.6, for example

$$\begin{Bmatrix} m\bar{a}_{x,\text{rev}} = -m\bar{a}_x \\ m\bar{a}_{y,\text{rev}} = -m\bar{a}_y \end{Bmatrix}$$

or

$$\begin{Bmatrix} m\bar{a}_{\text{t,rev}} = -m\bar{a}_{\text{t}} \\ m\bar{a}_{\text{n,rev}} = -m\bar{a}_{\text{n}} \end{Bmatrix}$$

or any other sets of components at G which may prove the most convenient in a particular problem. It also follows that if $\bar{a}$, the acceleration of the mass centre, is expressed in terms of some other vector components then to each such vector component there will be a corresponding component of the inertia force. For example if $\bar{\boldsymbol{a}} = \boldsymbol{a}_1 + \boldsymbol{a}_2 + \boldsymbol{a}_3$ then equation 13.6 can be replaced by the inertia force set

$$\begin{Bmatrix} (m\boldsymbol{a})_{1,\text{rev}} = -m\boldsymbol{a}_1 \\ (m\boldsymbol{a})_{2,\text{rev}} = -m\boldsymbol{a}_2 \\ (m\boldsymbol{a})_{3,\text{rev}} = -m\boldsymbol{a}_3 \end{Bmatrix}$$

the term $-m\boldsymbol{a}_1$ signifying a force, magnitude ma_1, passing through G in the opposite sense to that of $\boldsymbol{a}_1$.

To summarise the argument: if a body has the accelerations given in figure 13.7a the inertia force and couple are shown in figure 13.7b; in figure 13.7c the inertia

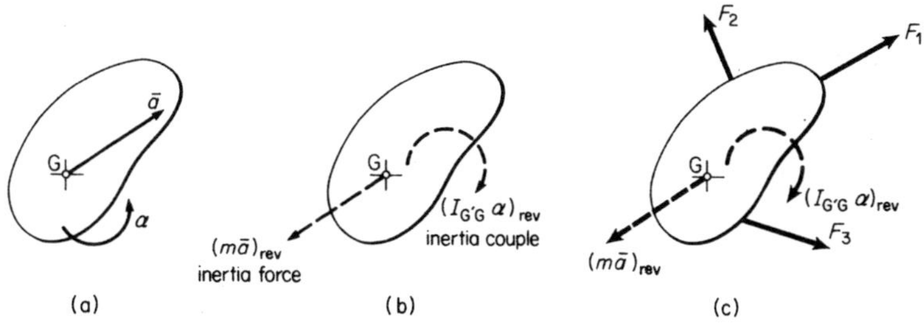

Figure 13.7

force and couple are shown together with the external forces in a free-body diagram. From the arguments already presented the body in this figure is in dynamic equilibrium, that is, the resultant force in any direction is zero and moments about *any* point sum to zero.

Worked Example 13.2

Solve worked example 13.1 using the inertia force method.

Solution

The solution is shown diagrammatically in figure 13.8. Figure 13.8a shows the body accelerations, figure 13.8b the inertia force - couple set and figure 13.8c the

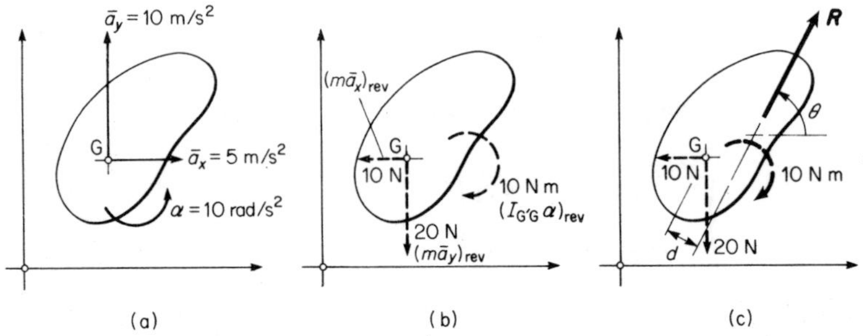

Figure 13.8

completed free-body diagram of the body including the inertia forces and the assumed external force R. Since the latter system is in dynamic equilibrium

$\Sigma F_x = 0 \qquad R_x - 2 \times 5 = 0$

$\Sigma F_y = 0 \qquad R_y - 2 \times 10 = 0$

$R_x = 10 \qquad R_y = 20$ giving $R = 22.36$ N $\angle$ 63.4°

$\Sigma M_G = 0 \qquad Rd - 10 = 0$

$Rd = 10$ giving $d = 0.45$ m

Worked Example 13.3

A hammer hinged at one end is allowed to fall and strikes a spring as indicated in figure 13.9a. At the particular instant when the hammer is vertical the force in the spring is 200 N. What is then the angular acceleration of the hammer and the horizontal reaction at the hinge? The hammer has mass 2 kg and radius of gyration about G of 0.5 m.

Solution

Figures 13.9b to 13.9e indicate the steps necessary to solve this problem – we need to end with a free-body diagram which includes all the external forces and couples

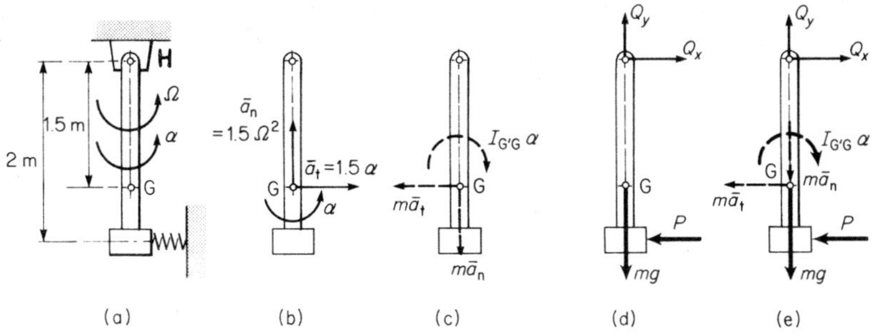

Figure 13.9

and the inertia force - couple set. In figure 13.9a the assumed directions of Ω and α are indicated; figure 13.9b shows the linear acceleration of G – the most convenient components in the case of a body rotating about a fixed axis are $\bar{a}_n$ and $\bar{a}_t$. Figure 13.9c shows the inertia force - couple set based upon the accelerations in figure 13.9b; figure 13.9d gives all the external forces in the free body – note the assumed reactions at the hinge, and figure 13.9e gives the completed free-body diagram with the combined external and inertia force - couple sets. The body in figure 13.9e is in dynamic equilibrium

$\Sigma F_x = 0 \qquad Q_x - m\bar{a}_t - P = 0$

and

$$Q_x = 2 \times 1.5\alpha + 200$$

$\Sigma M_H = 0 \qquad 1.5\, m\bar{a}_t + 2P + I_{G'G}\alpha = 0$

$$1.5 \times 2 \times 1.5\alpha + 2 \times 200 + 2(0.5)^2\alpha = 0$$

Thus

$$\alpha = -\ 80 \text{ rad/s}^2$$

and

$$Q_x = -\ 240 + 200 = -\ 40 \text{ N}$$

Worked Example 13.4

A car door is so hinged that it swings about a vertical axis. The door has mass m, its mass centre G is at a distance $\bar{r}$ from the hinge axis and its radius of gyration about a vertical axis through G is $k_{G'G}$. When the car is at rest the door is open at an angle ϕ to its closed position. If the car moves off with constant linear acceleration $\boldsymbol{a}$ determine the angular acceleration of the door when at an angle θ to its closed position, and hence the angular velocity with which it closes.

Solution

Figure 13.10a shows the given acceleration $\boldsymbol{a}$ of the hinge together with the assumed directions of θ, α and Ω in which the clockwise direction has been

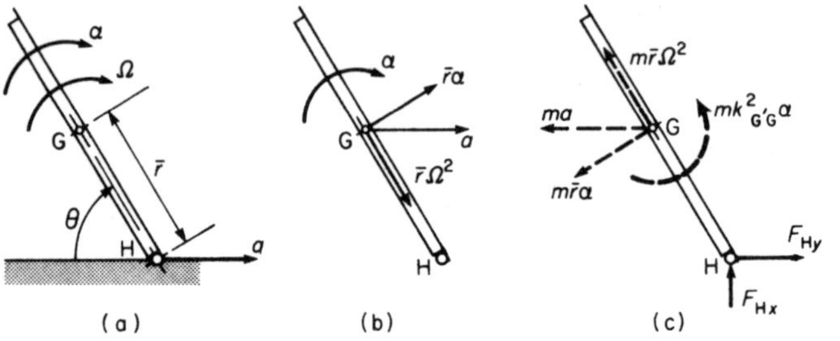

Figure 13.10

adopted as positive. The total linear acceleration of G is given by applying equation 11.3

$$_{E}a_{G} = {_{E}a_{H}} + {_{H}a_{G}}$$

in which $_{E}a_{H} = \boldsymbol{a}$ and $_{H}a_{G}$ has two components, $a_t = \bar{r}\alpha$ and $a_n = \bar{r}\Omega^2$. Figure 13.10b shows the angular acceleration and the three components of the linear acceleration of G.

Figure 13.10c shows the free-body diagram including the inertia force-couple set and the assumed external force, in component form, on the hinge.

The free body is in dynamic equilibrium and, choosing H as a moment centre

$$\Sigma M_H = 0 \qquad m\bar{r}\alpha \times \bar{r} + ma\bar{r}\sin\theta + mk_{G'G}{}^2\alpha = 0$$

and

$$\alpha = -\frac{a\bar{r}\sin\theta}{\bar{r}^2 + k_{G'G}{}^2}$$

Now

$$\alpha = \frac{d\Omega}{dt} = \Omega\frac{d\Omega}{d\theta}$$

and therefore

$$\Omega d\Omega = \alpha d\theta$$

Since the angular acceleration is always anticlockwise the angular velocity at closure Ω_c will also be anticlockwise when $\theta = 0$

$$\int_0^{-\Omega_c} \Omega\, d\Omega = -\int_\phi^0 \frac{a\bar{r}\sin\theta}{\bar{r}^2 + k_{G'G}{}^2}\, d\theta$$

$$\frac{\Omega_c^2}{2} = \frac{a\bar{r}}{\bar{r}^2 + k_{G'G}^2}[\cos\theta]_\phi^0$$

and

$$\Omega_c = \left[\frac{2a\bar{r}(1 - \cos\phi)}{\bar{r}^2 + k_{G'G}^2}\right]^{\frac{1}{2}}$$

13.2 Impulse and Momentum

In chapter 10 we developed the impulse - momentum equations for a particle. The discussion was extended to a particle system and it was found that the mass centre of the system behaved like a particle in which the mass of the system was concentrated. However this result did not imply any statement about the line of action of the linear impulse and linear momentum vectors. Thus the equations could be applied to rigid bodies undergoing translation but were insufficient to describe motion involving rotation.

In order to extend these equations to rigid bodies undergoing general plane motion we introduce two further quantities.

13.2.1 Angular Momentum and Angular Impulse of a Particle

If at some instant a particle mass m is moving in a plane XOY with velocity $\boldsymbol{v}$ under the action of a force $\boldsymbol{F}$ as in figure 13.11 then its linear momentum is a vector $m\boldsymbol{v}$ in the direction of $\boldsymbol{v}$.

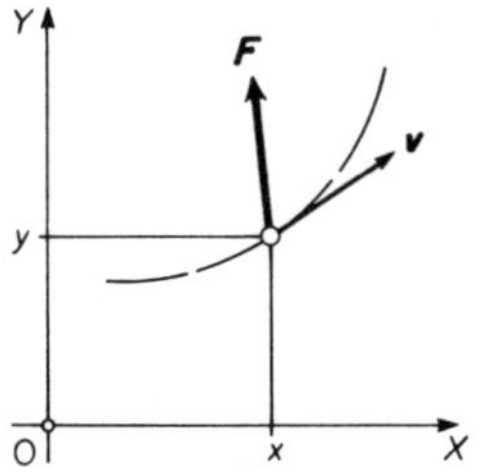

Figure 13.11

The *angular momentum* of the particle about the fixed point O is the moment of this momentum vector about O. This is given the symbol H_O, the suffix O referring to the point about which the angular momentum is defined. In terms of rectangular components

$$H_O = mv_y x - mv_x y$$

the anticlockwise sense being taken as positive.

The time derivative of the linear momentum is the force on the particle, that is

$(\mathrm{d}/\mathrm{d}t)\,(m\boldsymbol{v}) = \boldsymbol{F}$. If now we obtain the time derivative of angular momentum

$$\frac{\mathrm{d}H_\mathrm{O}}{\mathrm{d}t} = m\frac{\mathrm{d}v_y}{\mathrm{d}t}x + mv_y\frac{\mathrm{d}x}{\mathrm{d}t} - m\frac{\mathrm{d}v_x}{\mathrm{d}t}y - mv_x\frac{\mathrm{d}y}{\mathrm{d}t}$$

$$= F_y x - F_x y + m(v_y v_x - v_x v_y)$$

$$= F_y x - F_x y$$

$$= M_\mathrm{O}$$

the moment about O of the force on the particle. Rewriting this as

$$M_\mathrm{O}\,\mathrm{d}t = \mathrm{d}H_\mathrm{O}$$

and integrating between times t_1 and t_2

$$\int_{t_1}^{t_2} M_\mathrm{O}\mathrm{d}t = H_{\mathrm{O}2} - H_{\mathrm{O}1}$$

or

$$H_{\mathrm{O}1} + \int_{t_1}^{t_2} M_\mathrm{O}\mathrm{d}t = H_{\mathrm{O}2} \qquad (13.8\mathrm{a})$$

The *angular impulse* is defined as the time integral of the moment of the force F about O, that is

$$\int_{t_1}^{t_2} M_\mathrm{O}\mathrm{d}t$$

and is written in a shortened form as $(\text{Ang Imp}_\mathrm{O})_{1-2}$. Equation 13.8a can then be written

$$H_{\mathrm{O}1} + (\text{Ang Imp}_\mathrm{O})_{1-2} = H_{\mathrm{O}2} \qquad (13.8\mathrm{b})$$

13.2.2 Angular Impulse - Angular Momentum Equation for a Rigid Body

The rigid body, mass m, of figure 13.1 is shown again in figure 13.12a. If the rectangular components of the velocity of a typical particle P, mass δm, are v_x and v_y then the component linear momenta are as shown in the figure. These momenta are equivalent to equal momenta passing through G as shown in figure 13.12b plus a couple having a moment equal to the sum of the moments of the component momenta about G (and therefore an angular momentum) denoted by

$$\delta H_\mathrm{G} = x'\,(\delta m)v_y - y'\,(\delta m)v_x$$

Now

$${}_\mathrm{E}v_\mathrm{P} = {}_\mathrm{E}v_\mathrm{G} + {}_\mathrm{G}v_\mathrm{P}$$

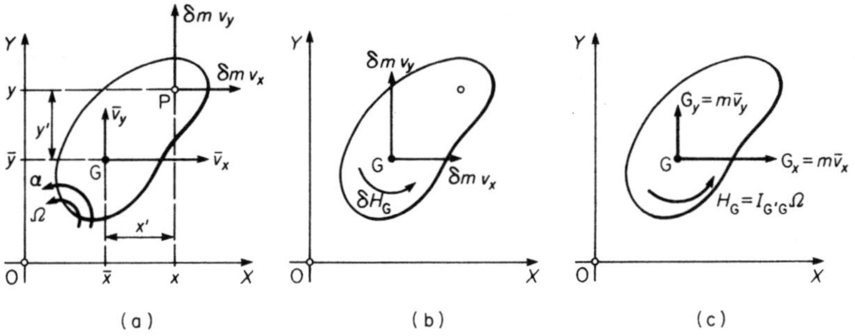

Figure 13.12

(where E is a fixed point on the Earth's surface) and for a rigid body ${}_{\mathrm{G}}v_{\mathrm{P}}$, with magnitude (GP)$\Omega$, has rectangular components $+\ \Omega x'$ in the y direction and $-\ \Omega y'$ in the x direction. It follows that

$$v_y = \bar{v}_y + \Omega x'$$

and

$$v_x = \bar{v}_x - \Omega y'$$

Summing over the body for the total linear momentum x-component at G

$$\begin{aligned}\Sigma(\delta m)v_x &= \Sigma\delta m\,(\bar{v}_x - \Omega y') \\ &= \bar{v}_x\ \Sigma\delta m - \Omega\Sigma(\delta m)y' \\ &= m\bar{v}_x\end{aligned}$$

since by the definition of the mass-centre the second term is zero; similarly

$$\Sigma(\delta m)v_y = m\bar{v}_y$$

Summing the angular momenta for the body

$$\begin{aligned}H_{\mathrm{G}} &= \Sigma\delta H_{\mathrm{G}} \\ &= \Sigma x'\delta m\,(\bar{v}_y + \Omega x') - \Sigma y'\delta m\,(\bar{v}_x - \Omega y') \\ &= \bar{v}_y\,\Sigma x'\delta m - \bar{v}_x\,\Sigma y'\delta m + \Omega\,\Sigma\delta m\,[(x')^2 + (y')^2]\end{aligned}$$

The first two terms on the right are zero (from the definition of the mass-centre) and $\Sigma\delta m\ [(x')^2 + (y')^2] = \Sigma\delta m\ (\mathrm{GP})^2$ is recognised as the moment of inertia of the body about G, $I_{\mathrm{G'G}}$; therefore

$$H_{\mathrm{G}} = I_{\mathrm{G'G}}\Omega \tag{13.9}$$

The momenta for the whole body (see figure 13.12c) are thus equivalent to linear momentum vectors

$$G_x = m\bar{v}_x \tag{13.10}$$

and

$$G_y = m\bar{v}_y \tag{13.11}$$

passing through the mass-centre, and a momentum couple having angular momentum about the mass-centre

$$H_G = I_{G'G}\Omega$$

The *linear momentum* **G** of the body is thus described by its rectangular components G_x and G_y and the *angular momentum* H_G of the body about G is given by $H_G = I_{G'G}\Omega$.

Now since from equations 13.1 to 13.3

$$\Sigma F_x = \mathrm{d}G_x/\mathrm{d}t \qquad \Sigma F_x \mathrm{d}t = \mathrm{d}G_x$$

$$\Sigma F_y = \mathrm{d}G_y/\mathrm{d}t \qquad \Sigma F_y \mathrm{d}t = \mathrm{d}G_y$$

and

$$\Sigma M_G = \mathrm{d}H_G/\mathrm{d}t \qquad \Sigma M_G \mathrm{d}t = \mathrm{d}H_G$$

where ΣF is the resultant external force on the body and ΣM_G is the resultant external moment on the body about G, after integrating we may write

$$G_{x1} + \Sigma\,(\mathrm{Imp}_x)_{1-2} = G_{x2} \tag{13.12}$$

$$G_{y1} + \Sigma\,(\mathrm{Imp}_y)_{1-2} = G_{y2} \tag{13.13}$$

$$H_{G1} + \Sigma\,(\mathrm{Ang\ Imp}_G)_{1-2} = H_{G2} \tag{13.14}$$

where

$$\Sigma\,(\mathrm{Imp}_x)_{1-2} = \Sigma \int_{t_1}^{t_2} F_x\,\mathrm{d}t \quad or \quad \int_{t_1}^{t_2} \Sigma F_x\,\mathrm{d}t$$

$$\Sigma\,(\mathrm{Imp}_y)_{1-2} = \Sigma \int_{t_1}^{t_2} F_y\,\mathrm{d}t \quad or \quad \int_{t_1}^{t_2} \Sigma F_y\,\mathrm{d}t$$

and

$$\Sigma\,(\mathrm{Ang\ Imp}_G)_{1-2} = \Sigma \int_{t_1}^{t_2} M_G\,\mathrm{d}t \quad or \quad \int_{t_1}^{t_2} \Sigma M_G\,\mathrm{d}t$$

It is usually more convenient to calculate the angular impulse about a fixed point such as O rather than G. The equivalent set of momenta is then (see figure 13.12c) G_y along OY, G_x along OX, and a momentum couple having angular momentum H_O being equal to the sum of H_G and the moments of G_x and G_y about O (due to changing their position). That is

$$H_O = m\bar{v}_y\bar{x} - m\bar{v}_x\bar{y} + I_{G'G}\Omega \tag{13.15}$$

Note that the first two terms on the right constitute the moment of the linear momentum vector $m\boldsymbol{v}$, passing through G, about O.

The corresponding angular impulse–angular momentum equation is

$$H_{O1} + \Sigma\,(\text{Ang Imp}_O)_{1-2} = H_{O2} \tag{13.16}$$

where

$$\Sigma\,(\text{Ang Imp}_O)_{1-2} = \Sigma \int_{t_1}^{t_2} M_O\, dt \quad \textit{or} \quad \int_{t_1}^{t_2} \Sigma M_O\, dt$$

Equations 13.12 to 13.14 and 13.16 are the *linear* and *angular impulse–momentum* equations for a rigid body and describe the changes in its velocity between two instants of time. It should be noted that H_O given by equation 13.15 is only safely applicable either to the mass centre (in which case it reduces to equation 13.9) or to a fixed point. The choice of some other moving point is subject to certain restrictions and should be avoided at this stage. With this in mind equation 13.16 (and 13.15) should be used in preference to equation 13.14. The equations are applicable to any type of motion but prove most useful during impulsive actions.

To illustrate equation 13.15 figure 13.13 shows a body mass m having

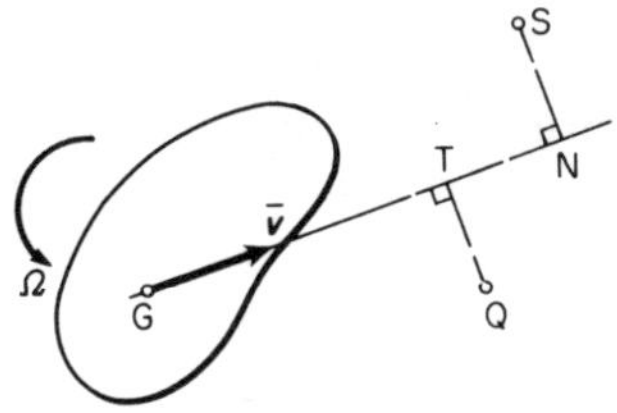

Figure 13.13

angular velocity Ω and linear velocity $\bar{\boldsymbol{v}}$ at G. Its angular momentum about fixed points S and Q is

(1) about S

$$H_S = I_{G'G}\Omega + m\bar{\boldsymbol{v}}\,(\text{SN}) \qquad \text{(anticlockwise)}$$

(2) about Q

$$H_Q = I_{G'G}\Omega - m\bar{\boldsymbol{v}}\,(\text{QT}) \qquad \text{(anticlockwise)}$$

The signs are both positive in (1) because both Ω and the moment of $m\bar{\boldsymbol{v}}$ (through G) about S are both anticlockwise. The sign for $m\bar{\boldsymbol{v}}$ (QT) is negative in (2) by the same argument. H_Q could of course be written

$$H_Q = m\bar{\boldsymbol{v}}\,(\text{QT}) - I_{G'G}\Omega \qquad \text{(clockwise)}$$

Particular Conditions

(1) If the motion is one of translation only then $\Omega = O$ and there is no angular

impulse or angular momentum about G.
(2) If the motion is fixed axis rotation about an axis through G then $\bar{v}_x = \bar{v}_y = 0$ and the system momenta reduce to a momentum couple having moment $I_{G'G}\Omega$.
(3) Fixed-axis rotation: If the motion is fixed axis rotation about an axis through Q distance $\bar{r}$ from G then, since in this case $\bar{v}_t = \bar{r}\Omega$, the application of equation 13.15 gives

$$H_Q = m\bar{v}_t\bar{r} + I_{G'G}\Omega$$

$$= m\bar{r}^2\Omega + I_{G'G}\Omega = I_{Q'Q}\Omega \qquad (13.17)$$

Worked Example 13.5

A hammer, mass 2 kg, is hinged at one end and is hanging freely with its mass-centre G at a distance 0.8 m below the hinge S. The hammer is struck on its head at a distance 1.2 m from the hinge by a horizontal force magnitude 2000 N which acts for 0.2 s. Find

(a) the angular impulse of the external forces about (i) the hinge and (ii) G
(b) the angular velocity of the hammer at the end of the impact.
The radius of gyration about G is 0.4 m.

Solution

The free-body diagram showing the external forces during the impact is given in figure 13.14. The positive direction of Ω has been taken as clockwise.

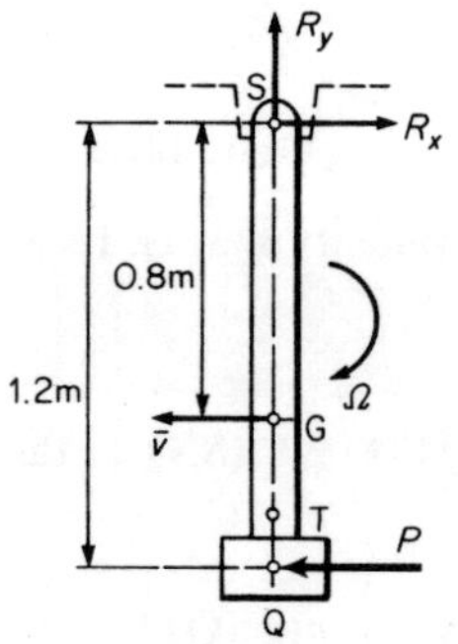

Figure 13.14

(a) Now

$$\Sigma\,(\text{Ang Imp}_O)_{1-2} = \Sigma \int_{t_1}^{t_2} M_O\,dt = \int_{t_1}^{t_2} \Sigma M_O\,dt$$

(i) About the hinge S the resultant external moment $\Sigma M_S = 1.2\,P$. Therefore the angular impulse about S is

$$\Sigma\,(\text{Ang Imp}_S)_{1-2} = 1.2 \int_0^{0.2} P\,dt$$

$$= 1.2 \times 2000 \times 0.2$$
$$= 480 \text{ N m s, clockwise}$$

(ii)

$$\Sigma\,(\text{Ang Imp}_G)_{1-2} = \Sigma \int_{t_1}^{t_2} M_G\,dt = \int_{t_1}^{t_2} \Sigma M_G\,dt$$

and

$$\int_{t_1}^{t_2} \Sigma M_G\,dt = \int_{t_1}^{t_2} (0.4P + 0.8\,R_x)\,dt$$

This cannot be calculated until $\int R_x\,dt$ is known.

(b) Since S is a fixed hinge then from equation 13.17 the angular momentum about S is

$$H_S = (I_{G'G} + m\bar{r}^2)\Omega = m(k_{G'G}{}^2 + \bar{r}^2)\Omega$$
$$= 2\,(0.4^2 + 0.8^2)\Omega = 1.6\Omega$$

$$H_{S1} = 0$$

since $\Omega_1 = 0$.

$$H_{S2} = 1.6\Omega_2$$

Thus, using the angular impulse - momentum equation 13.16

$$H_{S1} + \Sigma(\text{Ang Imp}_S)_{1-2} = H_{S2}$$
$$0 + 480 = 1.6\Omega_2$$

therefore

$$\Omega_2 = 300 \text{ rad/s clockwise}$$

To find $\int R_x\,dt$ apply the angular impulse - angular momentum equation 13.16 about a fixed point Q directly below S in order to eliminate P.

Now

$$\Sigma\,(\text{Ang Imp}_Q)_{1-2} = \int_0^{0.2} 1.2\,R_x\,dt$$

From equation 13.15 the angular momentum about Q is

$$H_Q = I_{G'G}\Omega - 0.4\, m\bar{v}$$

and

$$\bar{v} = 0.8\Omega$$

Thus

$$H_Q = I_{G'G}\Omega - 0.4\,\text{m}\,0.8\Omega = m(k_{G'G}{}^2 - 0.32)\Omega$$
$$= 2\,(0.4^2 - 0.32)\Omega = -0.32\Omega$$

$$H_{Q1} = 0, \quad H_{Q2} = -0.32\Omega_2 = -0.32 \times 300 = -96$$

Applying equation 13.16

$$H_{Q1} + \Sigma\,(\text{Ang Imp}_Q)_{1-2} = H_{Q2}$$

$$0 + 1.2 \int_0^{0.2} R_x\,\text{d}t = -96$$

Thus

$$\int_0^{0.2} R_x\,\text{d}t = -80\text{ N s}$$

(that is, actually R_x is to the left).

Thus in (a) (ii)

$$\int_{t_1}^{t_2} \Sigma M_G\,\text{d}t = \int_0^{0.2} (0.4 \times 2000)\,\text{d}t + 0.8\,(-80)$$

and hence

$$\Sigma\,(\text{Ang Imp}_G)_{1-2} = \int_{t_1}^{t_2} \Sigma M_G\,\text{d}t = 0.4 \times 2000 \times 0.2 - 64$$
$$= 96\text{ N m s} \quad \text{(clockwise)}$$

(Check that this agrees with the equation

$$H_{G1} + \Sigma\,(\text{Ang Imp}_G)_{1-2} = H_{G2}$$

using the already calculated value of Ω_2.)

13.2.3 Centre of Percussion

It is to be noted that in the preceding worked example involving fixed axis rotation R_x can in certain circumstances be zero although $\boldsymbol{P}$ is not. For this to occur it can be shown as follows that $\boldsymbol{P}$ must be applied at the centre of percussion of the hinged body.

In general terms (see figure 13.14) the angular momentum about a point T is

$$H_T = I_{G'G}\Omega - (TG) \times m\,(SG)\Omega$$
$$= m\,(k_{G'G}{}^2 - (TG) \times (SG))\Omega$$

and from the angular impulse - momentum equation 13.16

$$H_{T2} - H_{T1} = \Sigma\,(\text{Ang Imp}_T)_{1-2}$$

thus if P is applied at point T

$$m\,(k_{G'G}{}^2 - (TG) \times (SG))(\Omega_2 - \Omega_1) = \int_{t_1}^{t_2} (ST)\,R_x\,dt$$

and R_x will be zero if

$$k_{G'G}{}^2 = (TG) \times (SG)$$

that is when

$$TG = \frac{k_{G'G}{}^2}{SG}$$

which conforms to the definition (see section 13.1.2) of T as the centre of percussion.

Thus in the previous worked example R_x will be zero if P is made to act at point T such that

$$TG = \frac{0.4^2}{0.8} = 0.2 \text{ m}$$

Worked Example 13.6

A ladder length L has radius of gyration $k_{G'G}$ about G. It falls as indicated in figure 13.15a, turning without slipping about A; its angular velocity, just before

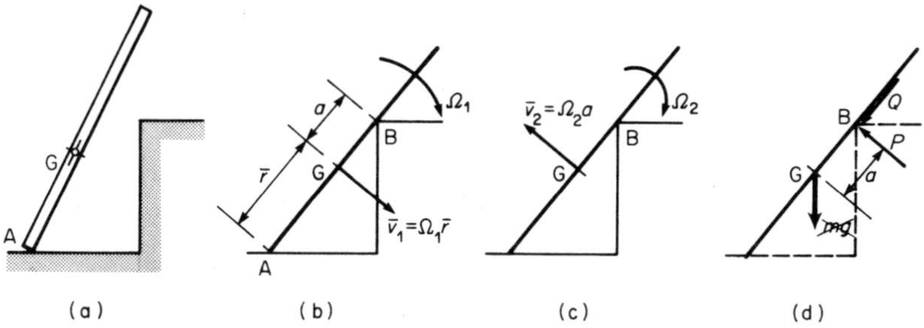

Figure 13.15

it hits the step, is Ω_1. If it hits the step and then continues to rotate about it without slipping
(a) find its angular velocity just after the impact is completed.
(b) What linear impulse is exerted by the step on the ladder?

Note: inherent in this type of problem is the assumption that the point of impact (the step in this case) has no elasticity, that is the point of impact must deflect in a 'plastic' manner, and it follows that the ladder does not rebound off the step.

It is assumed that the impact is of such short duration that the body, the ladder in this case, does not have any displacement during the impact.

A further but perhaps more reasonable assumption is that during the impact only the mutual impulsive forces are of importance. They will, in general, be much greater than the weight of the body which is then usually ignored. The ladder is further assumed to have lost contact with the ground as soon as it strikes the step.

Solution

Figure 13.15b shows the situation just before impact, with the ladder rotating about A; figure 13.15c shows the situation just after completion of the impact when it is rotating about B; figure 13.15d shows the free-body diagram of external forces during the impact (remember that we are neglecting mg compared to P and Q).

(a) Taking B as the moment-centre (in order to eliminate P and Q)

$$\Sigma\,(\text{Ang Imp}_\text{B})_{1-2} = \int_{t_1}^{t_2} \Sigma M_\text{B}\,\text{d}t = 0 \qquad (\text{since } \Sigma M_\text{B} = 0)$$

From equation 13.15 the angular momentum about B just before the impact is

$$H_{\text{B}1} = I_{\text{G}'\text{G}}\Omega_1 - m\bar{v}_1 a \qquad \text{(clockwise)}$$

and just afterwards is

$$H_{\text{B}2} = I_{\text{G}'\text{G}}\Omega_2 + m\bar{v}_2 a \qquad \text{(clockwise)}$$

Applying the angular impulse – momentum equation 13.16

$$H_{\text{B}1} + \Sigma\,(\text{Ang Imp}_\text{B})_{1-2} = H_{\text{B}2}$$

$$(I_{G'G}\Omega_1 - am\bar{r}\,\Omega_1) + 0 = (I_{G'G}\Omega_2 + ama\Omega_2)$$

$$\Omega_1\, m\,(k_{\text{G}'\text{G}}{}^2 - a\bar{r}) = \Omega_2\, m\,(k_{\text{G}'\text{G}}{}^2 + a^2)$$

therefore

$$\Omega_2 = \left(\frac{k_{\text{G}'\text{G}}{}^2 - a\bar{r}}{k_{\text{G}'\text{G}}{}^2 + a^2}\right)\Omega_1$$

(Note that for Ω_2 to be zero $k_{\text{G}'\text{G}}{}^2 = a\bar{r}$, and $a = k_{\text{G}'\text{G}}{}^2/\bar{r}$ is such that B is the centre of percussion of the ladder with respect to A.)

(b) Knowing Ω_2 the linear impulse $\int P\mathrm{d}t$ could be found using equation 13.16 with A as moment centre. A more direct method is to use the linear impulse-momentum equation 13.12

$$G_{x1} + \Sigma\,(\mathrm{Imp}_x)_{1-2} = G_{x2}$$

Taking the direction of $\bar{v}_1$ in figure 13.15b as the positive x direction

$$G_{x1} = m\bar{v}_1 = m\Omega_1\bar{r}$$

$$G_{x2} = -\,m\bar{v}_2 = -\,m\Omega_2 a$$

$$\Sigma\,(\mathrm{Imp}_x)_{1-2} = -\int_{t_1}^{t_2} P\mathrm{d}t$$

Thus

$$m\Omega_1\bar{r} - \int_{t_1}^{t_2} P\mathrm{d}t = -\,m\Omega_2 a$$

$$\int_{t_1}^{t_2} P\mathrm{d}t = m\,(\Omega_1\bar{r} + \Omega_2 a)$$

$$= m\Omega_1\left(\bar{r} + \frac{a\,(k_{G'G}^{\ 2} - a\bar{r})}{k_{G'G}^{\ 2} + a^2}\right)$$

$$= m\Omega_1\,\frac{k_{G'G}^{\ 2}\,(a + \bar{r})}{(k_{G'G}^{\ 2} + a^2)}$$

13.2.4 Impact of Rigid Bodies

Draw a free-body diagram for each body to show all external forces and couples that act during the impact (although it may be necessary at times to ignore the non-impulsive forces) and use the linear and angular impulse-momentum equations 13.12, 13.13 and 13.16 for each body. Note the equality of the impulsive forces between the bodies.

During the impact of rigid bodies we may still use a coefficient of restitution e. This is again defined by either equation 10.10

$$e = \frac{\text{Imp during restitution}}{\text{Imp during deformation}}$$

in which Imp is the linear impulse exerted on the body at the point of contact, or equation 10.11

$$e = -\,\frac{(\text{relative velocity after impact})}{(\text{relative velocity before impact})}$$

in which the relevant velocities are the linear velocities of the bodies at the point of contact.

Worked Example 13.7

Figure 13.16a shows the schematic diagram of a body A (of mass 0.2 kg) striking a body B (mass 20 kg; $k_{G'G} = 2$ m) which is hinged at S and is initially at rest.

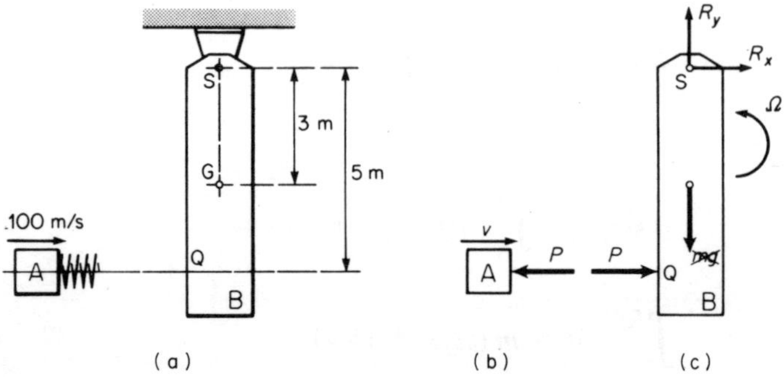

Figure 13.16

The impact is cushioned by a spring such that the coefficient of restitution e is 0.8. Find the velocities of A and B immediately after the completion of the impact. Assume that B is smooth at the point of impact.

Solution

Figures 13.16b and 13.16c show the free-body diagrams of A and B. $\boldsymbol{P}$ is the impulsive force in the spring during the impact and thus the force exerted by one on the other; gravitational forces are disregarded. Positive directions have been chosen as shown for Ω and v.

The first step is to find the common velocity, that is when the linear velocity of body A, v_A, is equal to the horizontal component v_Q of the linear velocity of the point of impact Q on body B. Let this common velocity be denoted by $v_{A2} = v_{Q2} = v_2$, and let it occur at time t_2. The impact begins at time t_1.

(There is no necessity to complicate our solution by a consideration of the vertical component of the linear velocity of the point Q because there can be no changes in the vertical components of the velocities of the mass-centres of either body during the impact.)

Note that for B

$$v_Q = 5\Omega$$

For A, applying the linear impulse – momentum equation 13.12 during deformation (that is until time t_2)

$$G_{x1} + \Sigma\,(\text{Imp}_x)_{1-2} = G_{x2}$$

$$0.2 \times 100 - \int_{t_1}^{t_2} P_\text{d}\,\text{d}t = 0.2\,v_2 \qquad (13.18)$$

(P_d being the value of P during deformation).

For B, applying the angular impulse - momentum equation 13.16 during deformation and using S as moment - centre (in order to eliminate R_x, R_y and mg)

$$H_{\text{S}1} + \Sigma\,(\text{Ang Imp}_\text{S})_{1-2} = H_{\text{S}2}$$

Now

$$H_\text{S} = (I_{\text{G}'\text{G}} + m\bar{r}^2)\Omega \qquad \text{(see equation 13.17)}$$

$$= 20\,(2^2 + 3^2)\Omega = 260\Omega$$

$$H_{\text{S}1} = 0$$

since $\Omega_1 = 0$

$$H_{\text{S}2} = 260\Omega_2$$

and

$$\Omega_2 = \frac{v_{\text{Q}2}}{5} = \frac{v_2}{5}$$

thus

$$H_{\text{S}2} = 52v_2$$

also

$$\Sigma\,(\text{Ang Imp}_\text{S})_{1-2} = \int_{t_1}^{t_2} + 5P_\text{d}\,\text{d}t$$

Substituting values

$$0 + 5\int_{t_1}^{t_2} P_\text{d}\,\text{d}t = 52v_2$$

or

$$\int_{t_1}^{t_2} P_\text{d}\,\text{d}t = 10.4v_2 \qquad (13.19)$$

Using this with equation 13.18 gives

$$20 - 10.4v_2 = 0.2v_2$$

and

$$v_2 = \frac{20}{10.6} = 1.89 \text{ m/s}$$

Thus

$$\Omega_2 = \frac{v_2}{5} = 0.377 \text{ rad/s}$$

Let $v_Q = v_{Q3}$ and $v_A = v_{A3}$ at the end of the impact (at time t_3).

For A, applying the linear impulse - momentum equation 13.12 during restitution

$$0.2v_2 - \int_{t_2}^{t_3} P_r \mathrm{d}t = 0.2v_{A3} \tag{13.20}$$

(P_r being the value of P during restitution; P_r is not necessarily the same function of time as P_d).

For B, applying the angular impulse - momentum equation 13.16 about S

$$260\Omega_2 + \int_{t_2}^{t_3} 5P_r \mathrm{d}t = 260\Omega_3$$

or

$$260\,\frac{v_2}{5} + 5\int_{t_2}^{t_3} P_r \mathrm{d}t = 260\,\frac{v_{Q3}}{5} \tag{13.21}$$

Combining this with equation 13.20 to eliminate

$$\int_{t_2}^{t_3} P_r \mathrm{d}t \text{ gives}$$

$$52v_2 + (v_2 - v_{A3}) = 52v_{Q3}$$

or

$$53v_2 = v_{A3} + 52v_{Q3} \tag{13.22}$$

Using equation 10.11 to define e

$$e = -\frac{(v_{A3} - v_{Q3})}{(v_{A1} - v_{Q1})} = -\frac{(v_{A3} - v_{Q3})}{100}$$

and thus

$$v_{Q3} - v_{A3} = 80 \tag{13.23}$$

Combining equations 13.22 and 13.23 gives

$$v_{A3} = -\ 76.6 \text{ m/s}$$

$$v_{Q3} = +\ 3.4 \text{ m/s}$$

and thus

$$\Omega_3 = +\ 0.68 \text{ rad/s}$$

If equation 10.10 is used to define e, then

$$e = \frac{\int_{t_2}^{t_3} P_r \mathrm{d}t}{\int_{t_1}^{t_2} P_d \mathrm{d}t}$$

and from equation 13.19

$$\int_{t_2}^{t_3} P_r \mathrm{d}t = 0.8 \times 10.4 \times v_2 = 8.32 v_2$$

Substitution in equation 13.21 gives

$$52 v_2 + 5 \times 8.32 v_2 = 52 v_{Q3}$$

and

$$v_{Q3} = \frac{(52 + 5 \times 8.32)}{52} \times 1.89 = 3.4 \text{ m/s}$$

as before.

Worked Example 13.8

Wheels A and B are rotating on horizontal parallel shafts. Wheel A is initially rotating at 500 rad/s clockwise and wheel B at 200 rad/s anticlockwise.

(a) They are pushed together and after a time t_2 the speed of A is 400 rad/s clockwise; what is the speed of B?
(b) After being pushed together for a total time t_3 slipping ceases between the two wheels; what are their final speeds?
(c) If the wheels are pushed together by a constant force of 500 N and the tangential force on each wheel is due to slipping friction what is the coefficient of friction (assumed constant) if t_2 is 2 s?
(d) What is then t_3?

Wheel A has $(I_{G'G})_A = 2$ kg m^2 and outside radius 1 m; B has $(I_{G'G})_B = 1.5$ kg m^2 and outside radius 0.5 m.

Solution

Figure 13.17a shows a schematic diagram of the system and figures 13.17b and 13.17c show the free-body diagrams of the external forces. The latter include a designation for positive directions of Ω and also the assumed directions of the tangential forces T (these have been deliberately taken as opposite to the physical direction – which can be argued from the relative peripheral velocities of the two wheels – to demonstrate that the mathematics will take care of this). The vertical components Y of the reactions at the bearings are also shown.

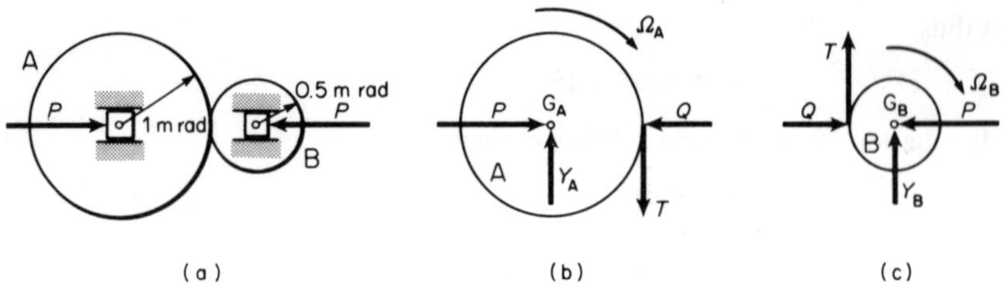

Figure 13.17

The method is to apply the angular impulse - momentum equation 13.16 to each body and thus eliminate $\int T\,dt$; it is convenient in each case to use the centres as origin since this also eliminates P, Y and Q.

(a) For A

$$\Sigma\,(\text{Ang Imp}_G)_{A,1-2} = \int_0^{t_2} \Sigma M_{G_A} \mathrm{d}t = \int_0^{t_2} T \times 1\mathrm{d}t = \int_0^{t_2} T\,\mathrm{d}t$$

Note that it should be left in this form since T may be a variable.

From equation 13.17 the angular momentum about the (fixed) centre of A is

$$(H_G)_A = (I_{G'G})_A \Omega_A = 2\Omega_A$$

Applying the angular impulse - momentum equation 13.16

$$(H_G)_{A1} + \Sigma\,(\text{Ang Imp}_G)_{A,1-2} = (H_G)_{A2}$$

$$2\,(+\ 500) + \int_0^{t_2} T\,\mathrm{d}t = 2\,(+\ 400)$$

and

$$\int_0^{t_2} T\,\mathrm{d}t = -\ 200 \text{ N m s} \tag{13.24}$$

(which indicates that physically T is in the opposite direction to that shown in figure 13.17).

For B

$$\Sigma\,(\text{Ang Imp}_G)_{B,1-2} = \int_0^{t_2} 0.5\,T\,\mathrm{d}t = 0.5 \int_0^{t_2} T\,\mathrm{d}t = -\ 100 \text{ N m s}$$

Applying the angular impulse - momentum equation 13.16

$$1.5\,(-\ 200) - 100 = 1.5\Omega_{B2}$$

Thus

$$\Omega_{B2} = -\ 266.7 \text{ rad/s}$$

(b) When slipping ceases the peripheral velocities will be equal – but take careful note that the angular velocities will be in opposite directions. Let these be Ω_{A3} and Ω_{B3}. Thus

$$1.0\Omega_{A3} = -0.5\Omega_{B3} \qquad (13.25)$$

Applying the angular impulse - momentum equation 13.16 to A for the whole of the slipping period

$$2(+500) + \int_0^{t_3} T\,dt = 2\Omega_{A3} \qquad (13.26)$$

and to B

$$1.5(-200) + \int_0^{t_3} 0.5T\,dt = 1.5\Omega_{B3} \qquad (13.27)$$

Eliminating $\int_0^{t_3} T\,dt$ between equations 13.26 and 13.27

$$1600 = 2\Omega_{A3} - 3\Omega_{B3}$$

which together with equation 13.25 gives

$$\Omega_{B3} = -400 \text{ rad/s}$$

$$\Omega_{A3} = +200 \text{ rad/s}$$

(c) From the free-body diagrams $P = Q$ since the centres of mass have no accelerations. Thus $Q = 500$ and is constant and it follows that $T = \mu Q$ (see chapter 5) and is also constant.

From equation 13.24 with T constant

$$\int_0^{t_2} T\,dt = \int_0^{2} T\,dt = 2T = -200$$

Thus

$$T = -100$$

and

$$100 = \mu\, 500$$

(the sign of T is unnecessary here since we are only concerned with the magnitude of the frictional force) hence

$$\mu = 0.2$$

(d) From equation 13.26 inserting values for T and Ω_{A3}

$$2(+500) + \int_0^{t_3} (-100)\,dt = 2(200)$$

giving

$$t_3 = 6 \text{ s}$$

13.3 Work and Energy

It was found in chapter 10 that the work – energy equations for a particle, in its various forms, could be applied to a rigid body in translation since the kinetic energy was simply $m\bar{v}^2/2$, and the work of external forces (or for conservative forces the corresponding potential energies) was easily calculated. We now extend the discussion to rigid bodies in general plane motion and also briefly discuss certain cases concerned with energy of deformation such as bodies subject to large impulsive forces or incorporating springs.

13.3.1 Kinetic Energy in General Plane Motion

Consider again the rigid body in figure 13.12a that is moving with general plane motion parallel to the reference plane XOY. It has been shown in section 13.2.2 that the rectangular components of the velocity of a typical particle P at (x, y) are

$$v_y = \bar{v}_y + \Omega x'$$

and

$$v_x = \bar{v}_x - \Omega y'$$

The kinetic energy of the particle mass δm is

$$\delta T = \tfrac{1}{2}(\delta m)v^2 = \tfrac{1}{2}(\delta m)(v_y{}^2 + v_x{}^2)$$

$$= \tfrac{1}{2}\delta m\,[(\bar{v}_y + \Omega x')^2 + (\bar{v}_x{}^2 - \Omega y')^2]$$

Summing over all particles the kinetic energy of the body is

$$T = \Sigma\tfrac{1}{2}\,\delta m\,(\bar{v}_y{}^2 + 2\bar{v}_y\Omega x' + \Omega^2 x'^2 + \bar{v}_x{}^2 - 2\bar{v}_x\Omega y' + \Omega^2 y'^2)$$

$$= \frac{\bar{v}_y{}^2 + \bar{v}_x{}^2}{2}\Sigma\delta m + \frac{\Omega^2}{2}\Sigma\delta m(x'^2 + y'^2) + \bar{v}_y\Omega\Sigma(\delta m)x' - \bar{v}_x\Omega\Sigma(\delta m)y'$$

From the definition of the centre of mass the last two terms are zero and

$$T = \tfrac{1}{2}m\bar{v}^2 + \tfrac{1}{2}I_{\mathrm{G'G}}\Omega^2 \qquad (13.28)$$

since

$$\bar{v}_y{}^2 + \bar{v}_x{}^2 = \bar{v}^2$$

where $\bar{v}$ is the velocity of the mass-centre, and

$$\Sigma\delta m(x'^2 + y'^2) = I_{\mathrm{G'G}}$$

is the moment of inertia of the body about an axis through G perpendicular to the plane XOY.

For the special case of fixed-axis rotation about an axis at Q, distance $\bar{r}$ from

$G, \bar{v} = \Omega\bar{r}$ and equation 13.28 becomes

$$\begin{aligned} T &= \tfrac{1}{2} m\Omega^2 \bar{r}^2 + \tfrac{1}{2} I_{G'G}\Omega^2 \\ &= \tfrac{1}{2}(m\bar{r}^2 + I_{G'G})\Omega^2 \\ &= \tfrac{1}{2} I_{Q'Q}\,\Omega^2 \end{aligned} \tag{13.29}$$

13.3.2 Work of a Force and of a Torque or Couple

We reiterate here, for use in this chapter, definitions and equations already stated or derived in earlier chapters.

The work of a force, as already defined in chapter 10 is

$$U_{1-2} = \int_{s_1}^{s_2} F_s \, ds$$

where

$$F_s = F \cos \alpha \qquad \text{(see figure 10.11)}$$

This work may also be written

$$U_{1-2} = \int_{x_1}^{x_2} F_x \, dx + \int_{y_1}^{y_2} F_y \, dy$$

which is a useful form for basing calculations on the rectangular components.

The work of a torque or couple as derived in chapter 6 is

$$U_{1-2} = \int_{\theta_1}^{\theta_2} M d\theta$$

where M is the magnitude of the torque or the moment of the couple. If this is constant

$$U_{1-2} = M(\theta_2 - \theta_1) \tag{13.30}$$

The unit of work is the joule (J) equal to 1 N m.

13.3.3 Work During General Plane Motion

In chapter 10 it was verified that the total work done on a particle was the sum of the individual works of each separate force. This result may now be extended to include the work of each separate couple or torque as well as the work of each separate force acting on a rigid body or on a system of rigid bodies (including a mechanism).

Work is being done, or is positive, if a force component in the direction of the displacement of the point of application of the force has the same sense as the displacement, or if a torque has the same sense as the angular displacement of the body. We speak of (i) work being done on a mechanism and (ii) work being done by mechanism; by (i) we mean that positive work is done on the mechanism and by (ii) that negative work is done on the mechanism. The latter case occurs when the force component or the torque has opposite sense to that of the corresponding displacement.

13.3.4 Power

Power, as already defined in chapter 10, is the rate at which work is done, or the rate of energy transfer. The definition applies to the forces or torques acting on a single rigid body or on a system of connected rigid bodies (such as a mechanism). The power of a torque, moment M, acting on a body is given by

$$\text{power} = M\frac{d\theta}{dt} = M\Omega \tag{13.31}$$

If work is being done on a mechanism at a particular point we say that the power (rate of energy transfer) is into the mechanism at that point; we say that the power is outwards at the point if work is being done by the mechanism. It follows that power is inwards if a force component has the same sense as the velocity of the point of the mechanism or a torque has the same sense as the angular velocity, and vice versa. Power into and power out of a mechanism are not necessarily equal at a given instant, the difference being accounted for by the rate of change of mechanical energy (kinetic and/or potential) of the system.

Worked Example 13.9

The rod AB in figure 13.18 is 1 m long and is acted upon by two forces, which

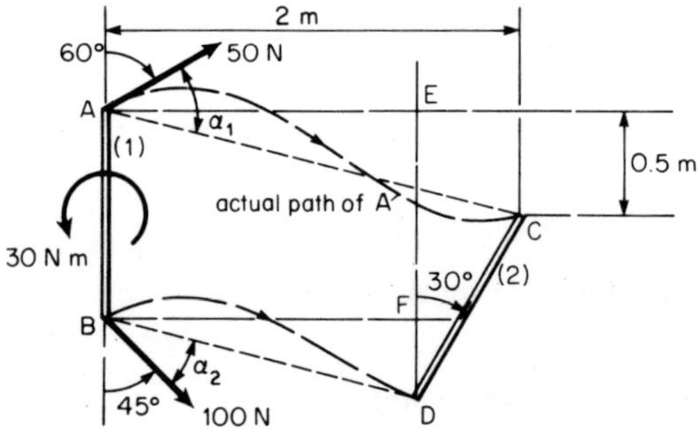

Figure 13.18

have constant magnitude and direction, and a torque. If the position of the rod changes from (1) to (2) determine

(a) the work done on the rod and
(b) the average power supplied to the rod if this work is done in 3 s.

Solution

(a) Since the forces have constant magnitude and direction any arbitrary path can be chosen for either force *as a device* for the calculation of the work of that force. This calculation is thus simplified if the straight line AC is chosen as the arbitrary path of the 50 N force and the straight line BD as the arbitrary path of the 100 N force. The work done by the forces and the torque is then

$$\Sigma U_{1-2} = 50 \times \cos\alpha_1 \times \text{AC} + 100 \times \cos\alpha_2 \times \text{BD} - 30 \times \left(\frac{30\pi}{180}\right)$$

Now

$$\text{AC} = \sqrt{(2^2 + 0.5^2)} = 2.06 \text{ m}$$

and

$$\alpha_1 = 30^\circ + \tan^{-1}\left(\frac{0.5}{2}\right) = 44.04^\circ$$

Also

$$\text{FD} = 0.5 + 1\cos 30^\circ - 1 = 0.366 \text{ m}$$

$$\text{BF} = 2 - 1\sin 30^\circ = 1.5 \text{ m}$$

$$\text{BD} = \sqrt{(1.5^2 + 0.366^2)} = 1.544 \text{ m}$$

and

$$\alpha_2 = 45^\circ - \tan^{-1}\left(\frac{0.366}{1.5}\right) = 31.3^\circ$$

$$\Sigma U_{1-2} = 50 \times \cos 44.04^\circ \times 2.06 + 100 \times \cos 31.3^\circ \times 1.544 - 30 \times 0.524$$
$$= 190.2 \text{ N m}$$

(b) Average power supplied equals the work done/s; that is:

$$\frac{190.2}{3} = 63.4 \text{ W}$$

An alternative solution would be to replace each force by an equal force at the centre of the rod plus an appropriate couple and to evaluate the work of the forces by using force components for an arbitrary path between the initial and final positions of the rod centre. This should then be summed with the work of the resultant couple.

13.3.5 Work - Energy Equation

In chapter 10 it was demonstrated that, for a rigid body moving under the action of conservative forces, namely external gravitational and elastic forces, the total mechanical energy, defined by $T + (V_g + V_s)$, was constant, since for a rigid body there was no work associated with internal forces. For the sake of clarity we shall now disregard all *external* springs in which case the total mechanical energy of a rigid body would be simply $(T + V_g)$. However if the body is one which contains a spring or can itself deform elastically we can ascribe internal potential energy or *strain energy* to the body (or particle system) this internal energy being associated with the internal conservative forces which do work during deformation. For example, if a system consists of two rigid bodies connected by a third deformable body, a massless linear spring, and this system were moving freely under gravity we could write for any two configurations of the system

$$[T + (V_g + V_e)]_1 = [T + (V_g + V_e)]_2$$

where T is the total kinetic energy of the two rigid bodies, V_g is the gravitational potential energy of the two rigid bodies and V_e is the internal strain energy of the spring given by $V_e = k\delta^2/2$, the latter (see section 10.6.2) being the negative of the work of the internal forces of the spring, that is equal to the work of the external forces on the spring, when it is deformed a length δ from its unstretched length.

If there were extraneous forces acting, namely applied external forces (including friction forces), the equation becomes

$$[T + (V_g + V_e)]_1 + \Sigma U_{1-2,\,\text{extr}} = [T + (V_g + V_e)]_2 \qquad (13.32)$$

where $\Sigma U_{1-2,\,\text{extr}}$ is the work done on the system by those extraneous forces.

Equation 13.32 is thus the equation to be applied if the deformation of the bodies is purely elastic, that is, upon recovery of shape the work of the internal forces is exactly equal (but opposite in sign) to that during deformation.

In most real situations the deformation is not purely elastic and furthermore, the quantity on the right hand side of equation 13.32 does not equal that on the left. We must then write

$$[T + (V_g + V_e)]_1 + \Sigma U_{1-2,\,\text{extr}} = [T + (V_g + V_e)]_2 + Q_{1-2} \qquad (13.33)$$

where V_e represents the strain energy (but may not be written as $k\delta^2/2$ without further justification) and Q_{1-2} represents energy that is said to be lost or dissipated as far as mechanical energy accounting is concerned. These energy losses usually occur in the form of heat, light and sound transferred to the surroundings.

Equation 13.33 is the general energy equation which can be used in all cases; it obviously covers the case of equation 13.32 by writing $Q_{1-2} = 0$ if losses can be disregarded.

As an example consider the case of direct central impact of two spheres A and B as described in section 10.4.1. The motion takes place in the horizontal plane and thus V_g remains constant; there are no extraneous forces.

Let the phases of the motion be defined as follows

(1) just before impact occurs, the sphere velocities are u_A and u_B,

(2) the spheres have maximum deformation and both have the same common velocity v,

(3) just after the end of the impact, sphere velocities are v_A and v_B.

(1) - (2) is the deformation phase; (2) - (3) is the restitution phase.

In the deformation phase (using the general energy equation 13.33)

$$(T + V_e)_1 = (T + V_e)_2 + Q_{1-2}$$

or

$$T_1 - T_2 = V_{e2} + Q_{1-2} \qquad (\text{since } V_{e1} = 0)$$

$$\begin{pmatrix}\text{reduction in kinetic energy}\\ \text{during deformation}\end{pmatrix} = \begin{pmatrix}\text{strain energy when}\\ \text{fully deformed}\end{pmatrix} + \begin{pmatrix}\text{energy dissipation}\\ \text{during deformation}\end{pmatrix}$$

Now

$$T_1 - T_2 = \tfrac{1}{2} m_A u_A{}^2 + \tfrac{1}{2} m_B u_B{}^2 - \tfrac{1}{2}(m_A + m_B)v^2$$

and since from momentum considerations

$$v = \frac{m_A u_A + m_B u_B}{m_A + m_B}$$

this can also be written

$$T_1 - T_2 = \tfrac{1}{2}\left(\frac{m_A m_B}{m_A + m_B}\right)(u_A - u_B)^2$$

In the restitution phase

$$(T + V_e)_2 = (T + V_e)_3 + Q_{2-3}$$

or

$$T_3 - T_2 = V_{e2} - Q_{2-3} \qquad (\text{since } V_{e3} = 0)$$

$$\begin{pmatrix}\text{gain in kinetic energy}\\ \text{during restitution}\end{pmatrix} = \begin{pmatrix}\text{strain energy when}\\ \text{fully deformed}\end{pmatrix} - \begin{pmatrix}\text{energy dissipation in}\\ \text{restitution}\end{pmatrix}$$

Now

$$T_3 - T_2 = \tfrac{1}{2} m_A v_A{}^2 + \tfrac{1}{2} m_B v_B{}^2 - \tfrac{1}{2}(m_A + m_B)v^2$$

$$= \tfrac{1}{2}\left(\frac{m_A m_B}{m_A + m_B}\right)(v_B - v_A)^2$$

For the whole impact, (applying the general energy equation 13.33 between conditions (1) and (3))

$$(T + V_e)_1 = (T + V_e)_3 + Q_{1-3}$$

$$T_1 = T_3 + Q_{1-3} \qquad (\text{since } V_{e1} = V_{e3} = 0)$$

$$T_1 - T_3 = (T_1 - T_2) - (T_3 - T_2) = Q_{1-2} + Q_{2-3} = Q_{1-3}$$

(net reduction in kinetic energy) = (energy dissipated)

From the definition of the coefficient of restitution

$$(v_B - v_A)^2 = e^2(u_B - u_A)^2$$

hence

$$Q_{1-3} = \tfrac{1}{2}\,\frac{m_A m_B}{(m_A + m_B)}\,(1 - e^2)(u_B - u_A)^2$$

= over-all reduction in kinetic energy

Note that e is a measure of the loss of mechanical energy, the loss being zero if $e = 1$. It is also of interest that

$$e^2 = \frac{(v_B - v_A)^2}{(u_B - u_A)^2} = \frac{\text{strain energy } \textit{less} \text{ losses incurred during restitution}}{\text{strain energy } \textit{plus} \text{ losses incurred during deformation}}$$

Worked Example 13.10

A uniform wheel, radius 0.8 m, has mass 1.5 kg and $k_{G'G} = 0.5$ m. It is placed on top of a slope, gradient $\sin^{-1}(1/10)$, and rolls downwards without slipping; find its linear and angular velocities after moving a distance of 20 m along the slope. Assume there are no losses of mechanical energy.

Solution

The general energy equation 13.33 is applied between the two positions of the wheel noting that strain energies are always zero and energy losses are zero; equation 13.28 is used to find the kinetic energy of the wheel.

From chapter 11 the instantaneous centre of rotation (if no slipping takes place) at any instant is the point of contact with the surface. Thus $\bar{v} = \Omega R$ where Ω is the angular velocity of the wheel and R its radius

$$\begin{aligned} T &= \tfrac{1}{2} m\bar{v}^2 + \tfrac{1}{2} I_{G'G}\Omega^2 \\ &= \tfrac{1}{2} mR^2\Omega^2 + \tfrac{1}{2} I_{G'G}\Omega^2 \\ &= \tfrac{1}{2} m(R^2 + k_{G'G}{}^2)\Omega^2 \\ &= \tfrac{1}{2} 1.5\,(0.8^2 + 0.5^2)\Omega^2 = 0.667\Omega^2 \end{aligned}$$

Taking the datum for gravitational potential energy at the initial position of the mass centre of the wheel

$$\begin{aligned} V_{g1} &= 0 \\ T_1 &= 0 \qquad (\Omega_1 = 0) \\ V_{g2} &= mgz = 1.5 \times 9.81 \times (-2) = -29.4 \\ T_2 &= 0.667\Omega_2{}^2 \end{aligned}$$

(The final position is $20 \times \frac{1}{10} = 2$ m below the initial position.)

$$\Sigma U_{1-2,\,\text{extr}} = 0$$

(there is no work for the external reaction at a rolling contact since the point of contact is always an instantaneous centre).

Applying the general energy equation 13.33: $T_1 + V_{g_1} = T_2 + V_{g2}$

$$0 + 0 = 0.667\Omega_2^2 - 29.4$$

Thus

$$\Omega_2 = 6.64 \text{ rad/s}$$

and

$$\bar{v}_2 = 0.8\Omega_2 = 5.31 \text{ m/s}$$

The general energy equation 13.33 can always be used to determine velocities in a potential field if information regarding energy losses and the work of the extraneous forces is known.

Worked Example 13.11

A hammer hinged at one end is allowed to drop from the horizontal position. A friction couple, moment M_F, resists motion at the hinge; the hammer has mass m, radius of gyration about G of $k_{G'G}$, and G is a distance $\bar{r}$ from the hinge. (a) What is the angular velocity of the hammer when it has turned through an angle θ?

(b) If $m = 20$ kg, $k_{G'G} = 0.4$ m, $\bar{r} = 0.7$ m and $M_F = 30$ N m find the angular velocity of the hammer when it reaches the vertical position. If at this instant the head of the hammer strikes a spring, having a constant 20000 N/m, attached to a rigid support find the maximum deflection of the spring

(i) if no losses occur and

(ii) if $e = 0.8$ between the rigid support of the spring and the hammer and the assumption is made that energy losses during deformation are twice those during the restitution phase. The strain energy can be described by $k\delta^2/2$.

Ignore the movement of the hammer during the impact.

Solution

(a) The method is to apply the general energy equation 13.33 between the two positions specified.

For fixed-axis rotation (see equation 13.29)

$$T = \tfrac{1}{2} I_{Q'Q}\Omega^2 = \tfrac{1}{2} m(k_{G'G}^2 + \bar{r}^2)\Omega^2$$

Figure 13.19 shows the original position (1) and a typical position (2). Taking the horizontal line through Q as a datum for z

$$V_{g1} = 0$$

$$T_1 = 0 \qquad (\Omega_1 = 0)$$

$$V_{g2} = mg\bar{z} = -mg\bar{r}\sin\theta$$

(note $\bar{z}$ is the position of G relative to the datum, the positive sense being upwards).

$$T_2 = \tfrac{1}{2} m(k_{G'G}^2 + \bar{r}^2)\Omega_2^2$$

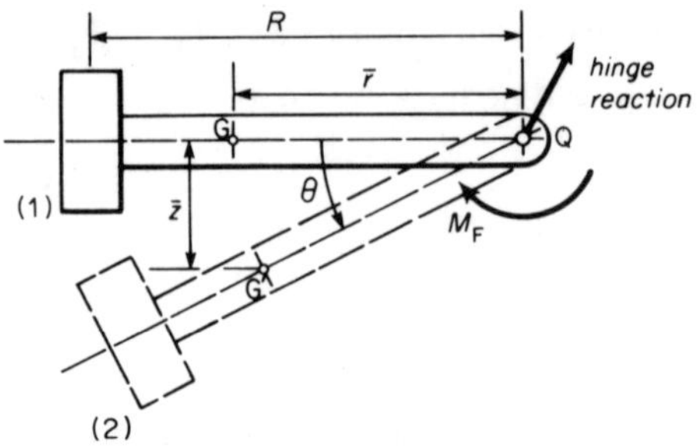

Figure 13.19

The extraneous forces in this case are the external friction couple and the hinge reaction at Q, the work of the latter being zero and thus

$$\Sigma U_{1-2,\,\text{extr}} = -M_F\theta$$

Applying the general energy equation 13.33

$$T_1 + V_{g1} + \Sigma U_{1-2,\,\text{extr}} = T_2 + V_{g2}$$

$$0 + 0 - M_F\theta = \tfrac{1}{2}m(k_{G'G}{}^2 + \bar{r}^2)\Omega_2{}^2 - mg\bar{r}\sin\theta$$

Hence

$$\Omega_2 = \sqrt{\left[\frac{2(mg\bar{r}\sin\theta - M_F\theta)}{m(k_{G'G}{}^2 + \bar{r}^2)}\right]}$$

(b) Applying the derived equation to the vertical position (denoted as stage 3)

$$\Omega_3 = \sqrt{\left[\frac{2(20g\,0.7\sin\pi/2 - 30\pi/2)}{20(0.4^2 + 0.7^2)}\right]} = 3.725 \text{ rad/s}$$

(i) Defining stage 4 when the spring is fully deflected (and therefore the hammer is stationary) and applying the general energy equation 13.33

$$T_3 + V_{e3} + V_{g3} + \Sigma U_{3-4,\,\text{extr}} = T_4 + V_{e4} + V_{g4} + Q_{3-4}$$

Now

$V_{e3} = 0$ (no deformation)

$V_{e4} = \tfrac{1}{2}k\delta^2 = \tfrac{1}{2}20000\,\delta^2$

$V_{g3} = V_{g4}$ (no movement of G)

$\Sigma U_{3-4,\,\text{extr}} = 0$ (no movement of the hammer)

$T_3 = \tfrac{1}{2}I_{Q'Q}\Omega_3{}^2$

$$T_4 = 0 \qquad \text{(no velocity)}$$

$$Q_{3-4} = 0 \qquad \text{(no losses)}$$

Substituting values

$$\tfrac{1}{2} I_{Q'Q}\Omega_3^2 = 10\,000\,\delta^2$$

and

$$\delta = 0.095 \text{ m}$$

(ii) Since energy is lost during the impact this loss has to be evaluated first before applying the general energy equation 13.33 to the deformation phase. Stage 5 is taken at the instant when the hammer loses contact with the spring in order to evaluate this energy loss. The use of impulse - momentum equations is usually necessary to determine the velocities at the end of the impact but because one of the bodies is fixed (the support) the definition of *e* is sufficient to provide a solution.

Using equation 10.11: *e* is defined as $-(v_H - v_S)/(u_H - u_S)$ where u is a velocity before the impact (stage 3), v a velocity after the impact (stage 5), and H and S refer to the head of the hammer and the support. The latter has no velocity so that

$$v_S = u_S = 0$$

and hence

$$v_H = -0.8\,u_H$$

also

$$u_H = \Omega_3 R \qquad \text{(see figure 13.19)}$$

$$v_H = \Omega_5 R$$

Thus

$$\Omega_5 = -0.8\Omega_3 = -2.98 \text{ rad/s}$$

Applying the general energy equation 13.33 between stages 3 and 5, noting that

$$V_{e3} = V_{e5} = 0$$

$$\Sigma U_{3-5,\,\text{extr}} = 0$$

and

$$V_{g3} = V_{g5}$$

it reduces to

$$\tfrac{1}{2} I_{Q'Q}\Omega_3^2 = \tfrac{1}{2} I_{Q'Q}\Omega_5^2 + Q_{3-5}$$

therefore

$$Q_{3-5} = 32.47 \text{ N m}$$

and since

$$Q_{3-5} = Q_{3-4} + Q_{4-5}$$

and we are given

$$Q_{3-4} = 2Q_{4-5}$$

then

$$Q_{3-4} = 21.65 \text{ N m}$$

Applying the general energy equation 13.33 between stages 3 and 4

$$\tfrac{1}{2}I_{Q'Q}\Omega_3^2 = V_{e4} + Q_{3-4}$$

therefore

$$V_{e4} = 68.54 \text{ N m}$$
$$= \tfrac{1}{2}\, 20000\, \delta^2$$

and

$$\delta = 0.083 \text{ m}$$

Special care must always be exercised to ensure taking into account the external work that might be done on or by the body by all the forces whose points of application move, as in the following worked example.

Worked Example 13.12

Th˯ tailgate of a lorry is hinged along its bottom horizontal edge as in figure 13.20a. When the lorry is moving forwards with velocity $v = 20$ m/s the tailgate

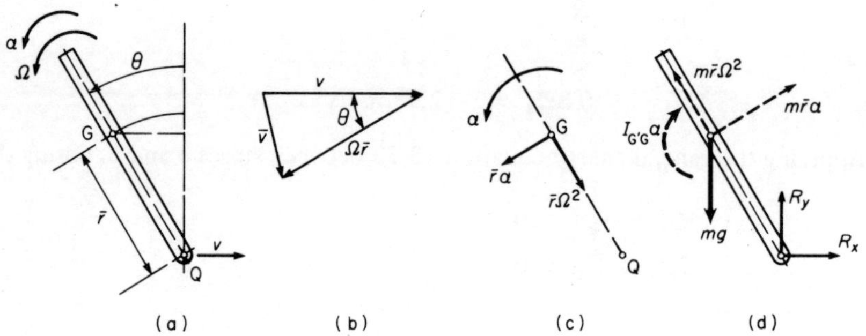

Figure 13.20

falls backwards. The tailgate has mass m = 50 kg, $k_{G'G}$ = 0.4 m and G is 0.7 m from the hinge.

(a) Obtain an expression for the kinetic energy of the tailgate when it has fallen through an angle θ, in terms of θ and its angular velocity Ω.

(b) Determine the angular velocity and acceleration of the tailgate at angle θ using equations of motion.
(c) Determine the work done on the tailgate through the hinge as θ varies from $\theta = 0$ to $\theta = 60°$.

Solution

(a) We require the absolute linear velocity of G when the angular velocity is Ω.

Now ${}_{E}v_{G} = {}_{E}v_{Q} + {}_{Q}v_{G}$ in which ${}_{E}v_{Q} = v$ and ${}_{Q}v_{G} = \Omega\bar{r}$ and so, from figure 13.20b, the absolute velocity of G

$$\bar{v} = [v^2 + (\Omega\bar{r})^2 - 2v\Omega\bar{r}\cos\theta]^{1/2}$$

Applying equation 13.28

$$\begin{aligned} T_\theta &= \tfrac{1}{2} I_{G'G}\Omega^2 + \tfrac{1}{2} m\bar{v}^2 \\ &= \tfrac{1}{2} mk_{G'G}{}^2\Omega^2 + \tfrac{1}{2} m(v^2 + \Omega^2\bar{r}^2 - 2v\bar{r}\Omega\cos\theta) \\ &= \tfrac{1}{2} m(k_{G'G}{}^2 + \bar{r}^2)\Omega^2 + \tfrac{1}{2} mv^2 - mv\bar{r}\Omega\cos\theta \qquad (13.34) \\ &= \tfrac{1}{2} 50(0.4^2 + 0.7^2)\Omega^2 + \tfrac{1}{2} 50 \times 20^2 - 50 \times 20 \times 0.7\Omega\cos\theta \\ &= 16.25\Omega^2 + 10000 - 700\Omega\cos\theta \qquad (13.35) \end{aligned}$$

(b) Now

$${}_{E}a_{G} = {}_{E}a_{Q} + {}_{Q}a_{G}$$

${}_{E}a_{Q} = 0$ since Q has a constant speed, and ${}_{Q}a_{G}$ has two components, magnitudes $\bar{r}\alpha$ and $\bar{r}\Omega^2$ as shown in figure 13.20c. From these are derived the inertia force - couple set that is shown together with all the external forces in figure 13.20d. By definition these are then in dynamic equilibrium

$$\Sigma M_Q = 0 \qquad -\bar{r}(m\bar{r}\alpha) - I_{G'G}\alpha + (mg)\bar{r}\sin\theta = 0$$

$$\alpha = \frac{mg\bar{r}\sin\theta}{m(k_{G'G}{}^2 + \bar{r}^2)} = \frac{g\bar{r}\sin\theta}{(k_{G'G}{}^2 + \bar{r}^2)}$$

To find Ω use

$$\alpha = \frac{d\Omega}{dt} = \frac{d\Omega}{d\theta}\left(\frac{d\theta}{dt}\right) = \Omega\frac{d\Omega}{d\theta} \qquad (13.36)$$

$$\alpha d\theta = \Omega d\Omega$$

$$\frac{g\bar{r}}{(k_{G'G}{}^2 + \bar{r}^2)}\int_{\theta_1=0}^{\theta_2=\theta}\sin\theta\, d\theta = \int_{\Omega_1=0}^{\Omega_2=\Omega}\Omega\, d\Omega$$

$$\frac{g\bar{r}}{(k_{G'G}{}^2 + \bar{r}^2)}[-\cos\theta]_0^\theta = \frac{\Omega^2}{2}$$

Thus

$$\Omega = \sqrt{\left[\frac{2g\bar{r}\,(1 - \cos\theta)}{k_{G'G}{}^2 + \bar{r}^2}\right]} \tag{13.37}$$

(c) Taking stages 1 and 2 when $\theta = 0$ and $\theta = 60°$ respectively and applying the general energy equation 13.33

$$T_1 + V_{g1} + \Sigma U_{1-2,\,\mathrm{extr}} = T_2 + V_{g2} + Q_{1-2}$$

$$Q_{1-2} = 0$$

(no information is given so losses are assumed to be zero)

From equation 13.35

$$T_1 = 10000 \qquad (\Omega_1 \text{ is zero})$$

$$T_2 = 16.25\Omega_2{}^2 - 700\Omega_2\,(0.5) + 10\,000$$

nd from equation 13.37

$$\Omega_2 = \sqrt{\left[\frac{2 \times 9.81 \times 0.7\,(1 - 0.5)}{0.4^2 + 0.7^2}\right]} = 3.25 \text{ rad/s}$$

Thus

$$T_2 = 172 + 10000 - 1140 \text{ N m}$$

Now, taking a datum through Q

$$V_{g1} = mg\,(0.7) = 50 \times 9.81 \times 0.7 = 344 \text{ N m}$$

$$V_{g2} = mg\,(0.7 - 0.7\cos\theta) = 172 \text{ N m}$$

Applying equation 13.33

$$10000 + \Sigma U_{1-2,\,\mathrm{extr}} + 344 = 172 + 10000 - 1140 + 172$$

$$\Sigma U_{1-2,\,\mathrm{extr}} = -1140 \tag{13.38}$$

(which, by inspection of equation 13.34, is seen to be equal to $-mv\Omega_2\bar{r}\cos\theta_2$). This means that the tailboard actually does work on the lorry.

The value of $\Sigma U_{1-2,\,\mathrm{extr}}$ can alternatively be calculated directly since it is the work done at the hinge, and (see figure 13.20d) is

$$\Sigma U_{1-2,\,\mathrm{extr}} = \int_{s_1}^{s_2} R_x \mathrm{d}s$$

the limits s_2 and s_1 being the positions of the lorry as θ changes from 0 to $\cos^{-1}$ 0.5. The work of R_y is zero.

R_x can be expressed symbolically from figure 13.20d and by writing the relations between α, Ω and θ, and s and v, $R_x\mathrm{d}s$ can be written in terms of Ω and θ. The latter can then be eliminated by use of equation 13.37 and the resulting equation leads to

$$\int_{s_1}^{s_2} R_x \mathrm{d}s = -\, mv\, \Omega_2 \bar{r} \cos\theta_2$$

which agrees with equation 13.38.

Note that in this case the velocities of the body cannot be derived directly from equation 13.33 because of the presence of the unknown forces at the moving hinge. In cases where there are unknown forces it is safer to find accelerations (as in part (b) of this example) and integrate.

13.4 Problems Involving Friction

We have seen in chapter 5 that the magnitude of the friction force between two surfaces in contact has a limiting value μN, where N is the normal reaction between the two surfaces. When this limiting value is attained relative motion is impending. It follows that until bodies slide relative to each other the friction force can have a magnitude less than μN, the magnitude depending upon the other external forces and the motion of the bodies.

In the solution of problems involving friction the following considerations arise in the application of the equations of this chapter.

13.4.1 Impulse – Momentum Equations

Those problems where it is appropriate to use these equations to determine velocity changes with time involve friction forces that are known in magnitude and direction – in particular – limiting friction forces. The friction forces must be indicated on the free body diagram with correct senses as determined by those of relative velocities.

13.4.2 Work – Energy Equations

These equations can be used to determine velocity changes with distance, but care must be taken in evaluating the work of friction forces in accordance with our definition of the work of a force. For example, if a block is sliding on a fixed plane the friction force acting on the moving surface of the block is in the opposite sense to that of the motion; the work can be calculated and is a negative quantity.

On the other hand, at the line of contact of a rolling wheel and a fixed surface there is no friction work since there is no tangential displacement at the line of contact. However, at a line contact between a rotating cylinder (the axis being stationary) and a fixed surface, or between two parallel contra-rotating cylinders the identification of a particle displacement across a *line* of contact becomes difficult, and we cannot calculate the work (if any) at this point on the basis of our definition. The existence of a line of contact is clearly theoretical, and in practice there is certainly dissipation of mechanical energy in this region; this can be calculated by reference to the work of cylinder driving torques and known changes in the angular velocities of the cylinders, as determined from the equations of motion. thus

$$T_1 + (U_{1-2})_{\text{torque}} = T_2 + Q_{1-2}$$

13.4.3 Equations of Motion

If there is insufficient information on which to base the energy equation – in particular, knowledge of friction forces – or it is required to determine details of the motion between two end points, then the equations of motion have to be used, as outlined in the following.

General method of solution

(1) Assume directions for the mass-centre accelerations and angular accelerations of the bodies and the frictional forces. Relate the accelerations at points of contact to the body accelerations.
(2) Draw free-body diagram for each body. If the bodies have accelerations include the inertia force - couple set.
(3) Write down general equations of motion for each body based on the free-body diagrams. Note in this regard that for each body only three equations can be deduced from the free-body diagram, for example, two force equations and one moment equation – in which case any other moment equation will be redundant.
(4) It may not be known initially if slipping is occurring or not at the points of contact. Each possible set of circumstances must be considered, and if necessary, examined in turn. For each possibility being examined write down two extra equations, as in (b) following, for each contact point.
(5) Solve the resulting set of equations for each possibility considered.
(6) There will usually be more than sufficient equations to solve for the unknowns (recall that to solve for n unknowns n independent equations are sufficient). Note which equations are used for the solution and check that any remaining equations are also satisfied. The correct solution is that which satisfies all equations.

When considering the possibilities take regard of the following.

(a) The sense of a frictional force if slipping is taking place depends upon the relative velocities of the surfaces in contact. For example if two bodies A and B have absolute velocities v_A and v_B to the right then if $v_A > v_B$, the frictional force exerted by A on B, is limiting and is to the right, and conversely that exerted by B on A is to the left. If $v_B > v_A$ the situation is reversed. When the bodies have the same velocity slipping is not occurring and friction must not be assumed to be limiting.

In this context if the absolute accelerations of A and B are a_A and a_B to the right respectively, then if the bodies start from rest it follows that if $a_A > a_B$ that $v_A > v_B$. If the bodies do not start from rest information will be required about initial velocities (see worked example 13.8). If no other information is given, assume that initial velocities are zero.

(b) For a cylindrical body having specified accelerations when moving in contact with a flat surface the tangential acceleration of the point of contact with the surface can be calculated. Thus in figure 13.21 $a_Q = \bar{a} - r\alpha$ (see chapters 8 and

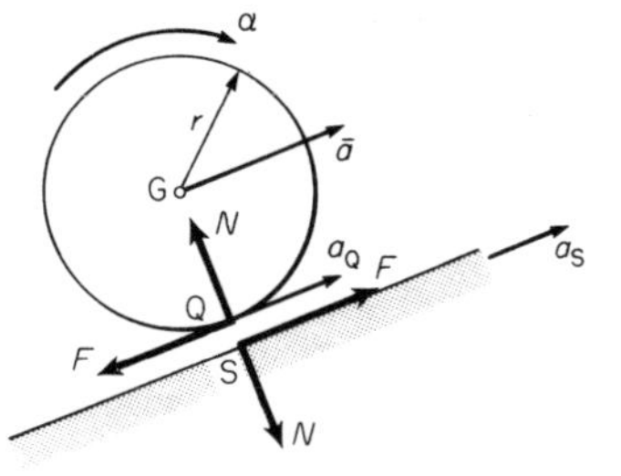

Figure 13.21

11). For this body the necessary conditions for the two available possibilities are
(1) for no slipping
 (i) $F < \mu N$
 (ii) $a_Q = a_S$
(2) for slipping
 (i) $F = \mu N$
 (ii) $a_Q \neq a_S$

(c) The relative values and directions of a_Q and a_S decide the direction of F and these must check. (This assumes of course – see (a) above – that the bodies start from rest, if not the velocities of the relevant points must be considered.)

Worked Example 13.13

The uniform block A in figure 13.22a is hinged as shown and is supported by the cylindrical body B. The coefficient of friction between the cylinder and the block is 1/2 and between the cylinder and the fixed surface is 1/6. If a horizontal force of $20g$ N is applied as shown find the accelerations of body B and state whether slipping is occurring or not at P and Q.

Solution

Figure 13.22b shows the assumed accelerations of body B. It follows from this that

$$a_P = a_G - 0.2\alpha \tag{13.39}$$

$$a_Q = a_G + 0.2\alpha \tag{13.40}$$

Figure 13.22c shows its free-body diagram with the inertia force - couple set based upon the accelerations in figure 13.22b. Directions have been assumed for the frictional forces F_Q and F_P.

General equations of motion for body A

From its free-body diagram, figure 13.22d, we can obtain three equations but

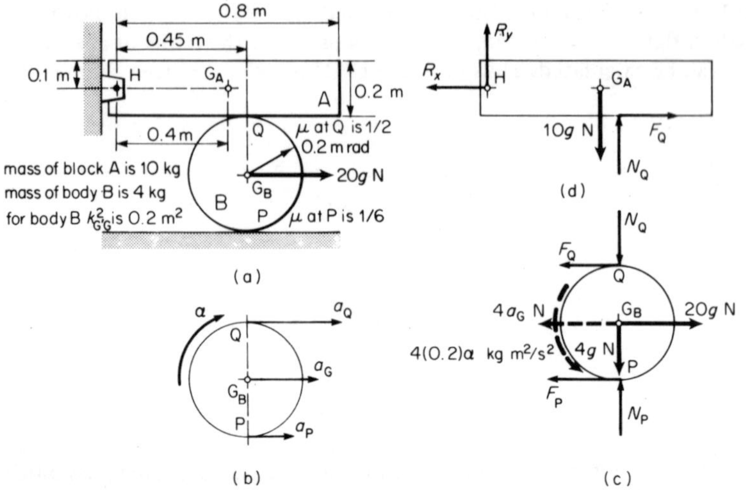

Figure 13.22

the only useful one in this case is by moments about the hinge

$$\Sigma M_H = 0 \qquad 0.1 \times F_Q + 0.45 \times N_Q - 0.4 \times 10g = 0 \qquad (13.41)$$

General equations of motion for body B (see figure 13.22c)

$$\Sigma F_y = 0 \qquad N_P - N_Q - 4g = 0 \qquad (13.42)$$

$$\Sigma F_x = 0 \qquad 20g - F_Q - F_P - 4a_G = 0 \qquad (13.43)$$

$$\Sigma M_P = 0 \qquad 0.2 \times 4a_G + 0.4 \times F_Q + 0.8\alpha - 0.2 \times 20g = 0 \qquad (13.44)$$

There are three possibilities to consider.

(1) Slipping occurs at Q but not at P, in which case

$$F_Q = N_Q/2 \qquad (13.45)$$

$$F_P < N_P/6 \qquad (13.46)$$

$$a_P = 0 \qquad (13.47)$$

For F_Q to be in the direction shown

$$a_Q > 0 \qquad (13.48)$$

(2) Slipping occurs at P but not at Q, in which case

$$F_P = N_P/6 \qquad (13.49)$$

$$F_Q < N_Q/2 \qquad (13.50)$$

$$a_Q = 0 \qquad (13.51)$$

For F_P to be in the direction shown

$$a_P > 0 \tag{13.52}$$

(3) Slipping occurs at both P and Q, in which case

$$F_P = N_P/6 \tag{13.53}$$

$$F_Q = N_Q/2 \tag{13.54}$$

$$a_P > 0 \tag{13.55}$$

$$a_Q > 0 \tag{13.56}$$

Condition (1): slipping occurs at Q but not at P

Equations 13.39 to 13.48 must be satisfied.

From equation 13.41 and using equation 13.45

$$4g = 0.05N_Q + 0.45N_Q$$

Thus

$$N_Q = 8g$$

$$F_Q = 4g$$

Using equations 13.39 and 13.47

$$a_G = 0.2\alpha$$

and with equation 13.44

$$0.8a_G + 0.4 \times 4g + 0.8\left(\frac{a_G}{0.2}\right) - 4g = 0$$

Thus

$$a_G = g/2$$

and

$$\alpha = 2.5g$$

Equation 13.40

$$a_Q = g/2 + g/2 = g$$

Equation 13.42

$$N_P = 8g + 4g = 12g\text{ N}$$

Equation 13.43

$$20g = 4g + F_P + 4 \times g/2$$

Therefore

$$F_P = 14g\text{ N}$$

Equations 13.39 – 13.44, 13.45 and 13.47 have been used.

We now check the unused equations.

Equation 13.48: $a_Q > 0$, is satisfied.

Equation 13.46: $F_P < N_P/6$, this cannot be satisfied and so this solution is impossible.

Condition (2): slipping occurs at P but not at Q

Equations 13.39 to 13.44 and 13.49 to 13.52 must be satisfied.

From equation 13.40 and 13.51

$$\alpha = -\frac{a_G}{0.2} \tag{13.57}$$

which with equation 13.44 gives

$$0.8a_G + 0.4F_Q - 4a_G - 4g = 0$$

or

$$F_Q = 10g + 8a_G \tag{13.58}$$

Using equations 13.49 and 13.42

$$F_P = \tfrac{1}{6}(N_Q + 4g)$$

which with equation 13.43 gives

$$20g = F_Q + \tfrac{1}{6}(N_Q + 4g) + 4a_G \tag{13.59}$$

From equation 13.41

$$N_Q = \frac{4g - 0.1F_Q}{0.45}$$

which with equation 13.59 gives

$$20g = F_Q + \frac{1}{6}\left(\frac{4g - 0.1F_Q}{0.45} + 4g\right) + 4a_G \tag{13.60}$$

Combining equations 13.60 and 13.58 gives

$$a_G = 0.7025g \text{ m/s}^2$$

$$F_Q = 15.62g \text{ N}$$

From equation 13.57

$$\alpha = -\ 3.51 \text{ rad/s}^2$$

From equation 13.41

$$N_Q = 5.42g$$

Equations 13.40 - 13.44, 13.49, 13.51 and 13.57 to 13.60 have been used.
Checking the unused equations
Equation 13.39: $a_P > 0$, is satisfied.
Equation 13.50: $F_Q < 0.5N_Q$, this cannot be satisfied and this solution is also impossible.
The only acceptable solution must be condition (3) which we proceed to check.

Condition (3): slipping occurs at both P and Q

Equations 13.39 to 13.44 and 13.53 to 13.56 must be satisfied.
Using equations 13.54 and 13.41

$$N_Q = 8g$$

$$F_Q = 4g$$

Using equations 13.43, 13.53 and 13.42

$$20g = 4g + \tfrac{1}{6}(8g + 4g) + 4a_G$$

Hence

$$a_G = 3.5g \text{ m/s}^2$$

Using equation 13.44

$$0.8 \times 3.5g + 0.4 \times 4g + 0.8\alpha - 4g = 0$$

Hence

$$\alpha = -\frac{g}{2} \text{ rad/s}^2$$

Equation 13.41 - 13.44 and 13.53 and 13.54 have been used. Checking the unused equations
Equation 13.39

$$a_P = 3.5g - 0.2(-g/2) = 3.6g \text{ m/s}^2$$

Equation 13.40

$$a_Q = 3.5g + 0.2(-g/2) = 3.4g \text{ m/s}^2$$

Equation 13.55: $a_P > 0$ is satisfied.
Equation 13.56: $a_Q > 0$ is satisfied.
All equations are satisfied and the solution is

$$a_G = +3.5g \text{ m/s}^2$$

$$\alpha = -g/2 \text{ rad/s}^2$$

and slipping occurs at both P and Q.

13.5 Composite Problems

Many problems can involve the use of equations of motion, impulse - momentum equations and work - energy equations. These problems must be subdivided into subsidiary problems solved by the most appropriate of the equations referred to. The subsidiary problems must then be connected by their end conditions in order to solve the over-all problem.

In general, if the subsidiary problem is concerned with displacement and over-all changes in velocities, the work - energy equations should be used; but be careful if there is the possibility of energy loss (usually during an impact) or external work of unknown extraneous forces. If the subsidiary problem is only concerned with changes in velocity the impulse - momentum equations should usually be used. For impacts use the impulse - momentum equations; if it is inferred that no mechanical energy is lost the work - energy equation can also be applied. Equations of motion describe the accelerations at a particular instant; velocities can then be found (not always simply) by integration. Impulse - momentum and work - energy equations, in contrast, relate conditions at two different instants.

Worked Example 13.14

A uniform beam, mass 100 kg, 10 m long and 1 m deep, is hinged on its top face at the centre of its length so that it lies initially stationary with its length horizontal. A body mass 1 kg, to the underside of which is fixed a massless spring of stiffness 2000 N/m is suspended at a position 4 m to the left of the hinge with the free end of the spring 3 m above the top of the beam. If this body is released find (a) the maximum deflection of the spring (b) the angle through which the beam turns before coming instantaneously to rest (c) the height through which the body rebounds. Assumptions are that firstly, the duration of the impact is so short that the beam does not move during the impact, for which $e = 0.5$; secondly any energy loss occurs in the spring as it extends after compressing elastically; thirdly the spring deflection is negligible compared to the height dropped by the mass.

Solution

The problem can be separated into at least four different phases.

(i) When the body drops its velocity will increase and its velocity just before it strikes the beam will determine what happens during the impact. Since the body drops in a potential field energy considerations should be used.
(ii) There is an impact. Use the impulse - momentum equations to find body velocities after the impact has ended.
(iii) The energy is a function of spring deflection. Use the general energy equation to determine the deflection.
(iv) Both the beam and the body move in a potential field after the impact. Energy considerations should again be used.
The datum for gravitational potential energy V_g is taken through the hinge throughout the problem.

The following situations should be noted; they are shown diagrammatically in figure 13.23.

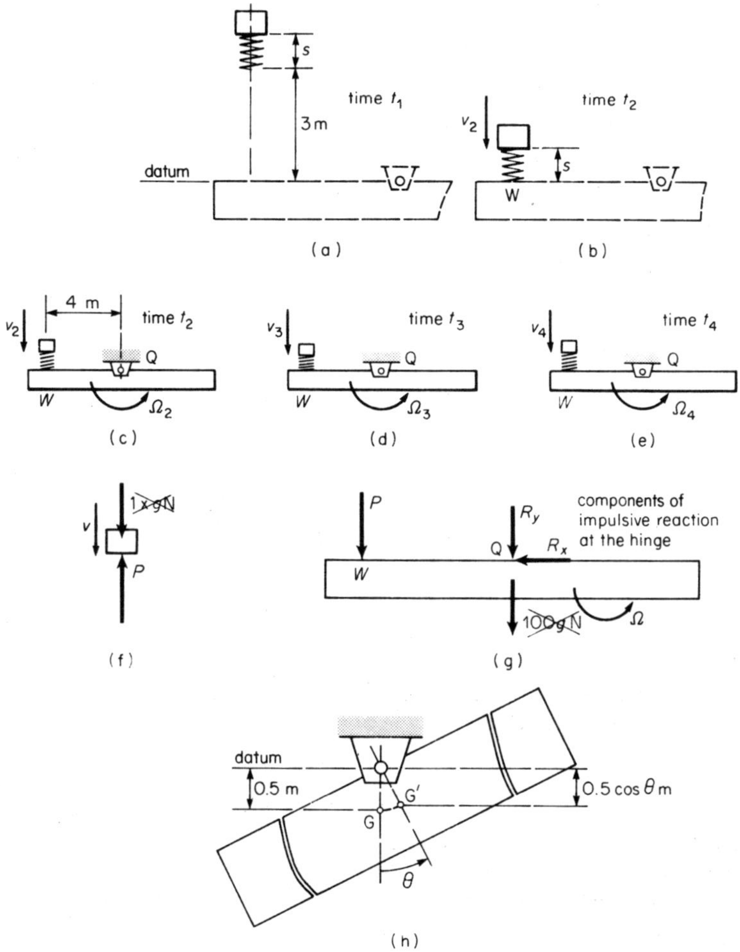

Figure 13.23

(1) The body is released from 3 m above the beam at time t_1.
(2) The body attains a velocity v_2 just before it strikes the beam (time t_2).
(3) At a particular instant t_3 the body having velocity v_3 has the same velocity as the adjacent point W on the beam; this will occur when the spring is fully deflected. (At any other instant they must be moving with respect to each other.)
(4) The body just loses contact with the beam at the end of the impact (time t_4).
(5) The beam comes instantaneously to rest (t_5).
(6) The body comes instantaneously to rest (t_6).

(i) Applying the general energy equation 13.33 to the body (and spring) between positions (1) and (2) (see figure 13.23a and 13.23b) and noting that

$$V_{e1} = V_{e2} = 0$$

$$T_1 + V_{g1} + \Sigma U_{1-2\text{, extr}} = T_2 + V_{g2} + Q_{1-2}$$

Now

$$T_1 = 0 \qquad \text{(no velocity)}$$

$$T_2 = \tfrac{1}{2} m\bar{v}^2$$

(since it has no rotation), thus

$$T_2 = \tfrac{1}{2} \times 1 \times v_2{}^2$$

$$V_{g1} = mg\bar{z} = 1 \times 9.81\,(3 + s)$$

$$V_{g2} = 1 \times 9.81 \times s$$

$$\Sigma U_{1-2\text{, extr}} = 0 \qquad Q_{1-2} = 0$$

Substituting values

$$0 + 9.81\,(3 + s) + 0 = \frac{v_2{}^2}{2} + 9.81s + 0$$

and

$$v_2 = \sqrt{(2 \times 9.81 \times 3)} = 7.67 \text{ m/s}$$

Figures 13.23c, 13.23d and 13.23e show the conditions at the times t_2, t_3 and t_4 with assumed directions for positive velocities. Note that at time t_3 the velocity of the point W equals the velocity of the body v_3, and

$$v_3 = 4\Omega_3 \tag{13.61}$$

Figures 13.23f and 13.23g show the free-body diagrams during the impact; P is the impulsive force in the spring assumed to be much larger than the gravitational forces which are ignored. P is denoted by P_d during deformation.

For the body, applying the linear impulse – momentum equation 13.13 between conditions 2 and 3

$$G_{y2} + \Sigma\,(\text{Imp}_y)_{2-3} = G_{y3}$$

$$1 \times v_2 + \int_{t_2}^{t_3} -P_d \mathrm{d}t = 1 \times v_3$$

(P_d has a negative sign because it is in the opposite direction to the positive direction of velocity.) Thus

$$1 \times 7.67 - \int_{t_2}^{t_3} P_d \mathrm{d}t = v_3 \tag{13.62}$$

We now have to consider the beam, and for this purpose we require $I_{G'G}$ (see chapter 12)

$$I_{G'G} = \frac{M}{12}(a^2 + b^2) = \frac{100}{12}(10^2 + 1^2)$$

and

$$I_{Q'Q} = I_{G'G} + 100\,(0.5)^2 = 866 \text{ kg m}^2$$

In order to eliminate R_x and R_y we use Q as origin for momentum purposes. For fixed-axis rotation the angular momentum about Q is given by equation 13.17

$$H_Q = I_{Q'Q}\Omega = 866\Omega$$

Applying the angular impulse – momentum equation 13.16, with Q as origin, between conditions 2 and 3

$$H_{Q2} + \Sigma\,(\text{Ang Imp}_Q)_{2-3} = H_{Q3}$$

$$866\,(0) + \int_{t_2}^{t_3} 4P_d \mathrm{d}t = 866\Omega_3 \qquad (13.63)$$

Combining equations 13.61, 13.62 and 13.63 gives

$$4\,(7.67 - v_3) = 866\,\frac{v_3}{4}$$

hence

$$v_3 = 0.139 \text{ m/s}$$

and

$$\Omega_3 = 0.0348 \text{ rad/s}$$

(ii) In order to find the final velocity we have to apply the equations again for the conditions 3 to 4 and also use the equations for e.

For the body, applying the linear momentum equation 13.13

$$1 \times v_3 + \int_{t_3}^{t_4} -P_r \mathrm{d}t = 1 \times v_4 \qquad (13.64)$$

P_r being the impulsive force in the spring during restitution.

For the beam, applying the angular momentum equation 13.16 again with Q as origin

$$866\,\Omega_3 + \int_{t_3}^{t_4} 4P_r \mathrm{d}t = 866\,\Omega_4 \qquad (13.65)$$

Combining equations 13.64 and 13.65 to eliminate $\int P_r \mathrm{d}t$, using equation 13.61 and writing $\Omega_3 = v_3/4$ and $\Omega_4 = v_{W4}/4$, where v_W is the velocity of point W, we obtain

$$54.1v_3 + (v_3 - v_4) = 54.1v_{W4} \tag{13.66}$$

Also (equation 10.11)

$$e\,(= 0.5) = -\frac{(v_4 - v_{W4})}{(v_2 - v_{W2})} \tag{13.67}$$

Now

$$v_{W2} = 0 \qquad \text{(the beam is stationary)}$$

$$v_2 = 7.67 \qquad v_3 = 0.139$$

and equations 13.66 and 13.67 combine to give

$$v_{W4} = 0.208 \text{ m/s}$$

hence

$$\Omega_4 = 0.052 \text{ rad/s}$$

and

$$v_4 = -3.63 \text{ m/s}$$

(meaning the body is actually moving upwards).

(iii) Using the general energy equation 13.33 *for the system* of beam and body between conditions (1) and (3) [or (2) and (3) if required, since no energy loss occurs between (1) and (3)].

$$T_1 + V_{g1} + V_{e1} + \Sigma U_{1-3,\,\text{extr}} = T_3 + V_{g3} + V_{e3} + Q_{1-3} \tag{13.68}$$

$$\Sigma U_{1-3,\,\text{extr}} = Q_{1-3} = V_{e1} = 0$$

where T_1 is the kinetic energy of the system at $t_1 = 0$; V_{g1} is the potential energy of body plus the potential energy of the beam, that is

$$V_{g1} = 1g\,(3 + s) + 100g\,(-0.5)$$

T_3 is the kinetic energy of body plus the kinetic energy of the beam at time t_3, that is

$$T_3 = \tfrac{1}{2} \times 1 \times v_3^2 + \tfrac{1}{2} I_{Q'Q}\Omega_3^2 \qquad \text{(see equation 13.29)}$$

$$= \frac{0.139^2}{2} + \frac{866}{2}(0.0348)^2 = 0.534 \text{ N m}$$

V_{g3} is the potential energy of the body plus the potential energy of the beam, that is

$$V_{g3} = 1g(s) + 100g\,(-0.5)$$

V_{e3} is the strain energy in the spring after elastic compression and

$$V_{e3} = \tfrac{1}{2} k\delta^2$$

δ being the maximum deflection. Note that we are told in the question to ignore the deflection of the spring compared with movement of the body [to be rigorous the first term of the equation for V_{g3} should be $1 \times g(s - \delta)$].

Substituting values in equation (13.68) gives

$$0 + g(3 + s) - 50g + 0 + 0 = 0.534 + g(s) - 50g + \tfrac{1}{2}k\delta^2 + 0$$

giving

$$\delta^2 = \frac{3g - 0.534}{k}$$

Now $k = 2000$ and thus

$$\delta = 0.12 \text{ m}$$

(iv) Applying the general energy equation 13.33 to the *beam alone* between conditions (4) and (5). [Condition (4) is shown in figure 13.23e and condition (5) in figure 13.23h.]

$$T_4 + V_{g4} = T_5 + V_{g5}$$

since

$$\Sigma U_{4-5,\,\text{extr}} = 0 \qquad Q_{4-5} = 0$$

$$T_4 = \tfrac{1}{2}I_{Q'Q}\Omega_4{}^2 = \tfrac{1}{2}866\,(0.052)^2 = 1.17 \text{ N m}$$

$$T_5 = 0 \qquad \text{(it is at rest)}$$

$$V_{g4} = 100g\,(-0.5)$$

$$V_{g5} = 100g\,(-0.5\cos\theta)$$

Thus

$$1.17 - 50g = 0 - 50g\cos\theta$$

and the answer to (b) is

$$\theta = 3.96°$$

Applying the general energy equation 13.33 to the *body alone* between conditions (4) and (6). In the latter condition the body is assumed to be instantaneously at rest having rebounded a distance h from its position at (4)

$$T_4 + V_{g4} = T_6 + V_{g6}$$

$$T_4 = \tfrac{1}{2} \times 1(v_4)^2 = \tfrac{1}{2}(3.63)^2 = 6.59 \text{ N m}$$

$$T_6 = 0$$

$$V_{g4} = 1g(s)$$

$$V_{g6} = 1g(s + h)$$

Thus

$$6.59 + g(s) = 0 + g(s + h)$$

giving the answer to (c)

$$h = 0.672 \text{ m}$$

Do the body and the beam become instantaneously to rest at the same time? (Check the energy of the system.)

13.6 Summary

(1) If a rigid body has angular acceleration α and its mass centre G has component accelerations $\bar{a}_x$ and $\bar{a}_y$ the external force set causing these accelerations is equivalent to a force ΣF having components $\Sigma F_x = m\bar{a}_x$ and $\Sigma F_y = m\bar{a}_y$ passing through G, and a couple having moment $\Sigma M_G = I_{G'G}\alpha$ (equations 13.1 – 13.3).
(2) The centre of percussion P of a body, mass centre G, rotating about a fixed point O is on the line OG extended so that $GP = k_{G'G}{}^2/\bar{r}$ where $\bar{r}$ = OG. It can be described in at least two ways; see sections 13.1.2 and 13.2.3.
(3) A compound pendulum is any rigid body which oscillates in a vertical plane under the action of gravity in an angular fashion about a fixed horizontal axis. The frequency of oscillation for small angular movements is derived as a solution of the equation (for simple harmonic motion) $\ddot{\theta} + \omega^2\theta = 0$. A simple pendulum is a particle on the end of a massless cord.
(4) The inertia force - couple set consists of a force $(-m\bar{a})$ or $(m\bar{a})_{rev}$ passing through G (or any set of the components of this force) and a couple $(-I_{G'G}\alpha)$ or $(I_{G'G}\alpha)_{rev}$ (equations 13.6 and 13.7 in section 13.1.4).
(5) The inertia force - couple set and the external force set are in dynamic equilibrium, that is, $\Sigma F_x = 0$, $\Sigma F_y = 0$ and $\Sigma M_A = 0$, where A is any point (section 13.1.4).
(6) The linear impulse - momentum equation $\boldsymbol{G}_1 + \Sigma\,(\mathbf{Imp})_{1-2} = \boldsymbol{G}_2$, or its component forms – equations 13.12 and 13.13 – applies to a rigid body where $\boldsymbol{G}$ is the linear momentum of the body which equals the mass times the velocity of the mass centre.
(7) $\boldsymbol{H}_O$, the angular momentum of a body about a fixed point O is the sum of the momentum couple $I_{G'G}\Omega$ and the moment of the linear momentum vector $m\bar{\boldsymbol{v}}$ (which passes through G) about O, taken in the same sense (equation 13.15).
(8) The angular impulse of a force or a couple about O, (Ang Imp_O), equals $\int M_O\,dt$, the time integral of the moment of the force or couple about O.
(9) The angular impulse - angular momentum equation for a rigid body is

$$H_{O1} + \Sigma\,(\text{Ang Imp}_O)_{1-2} = H_{O2}$$

(equation 13.16).
(10) Equations 13.12, 13.13, and 13.16 describe the changes in velocity of a rigid body between two instants of time.
(11) For fixed-axis rotation

$$H_{axis} = I_{axis}\,\Omega \tag{13.17}$$

(12) Impulsive forces are generally much larger than gravitational forces (weight) and the latter can usually be ignored during the impact.
(13) The choice of moment centre for calculation of angular momentum is restricted to any fixed point or the mass-centre G.
(14) For the impact of rigid bodies draw a free-body diagram of the external forces for both bodies and use equations 13.12, 13.13 and 13.16 for both. The coefficient of restitution is based on linear velocities at the point of impact, or the linear impulses at the point of impact.
(15) The kinetic energy of a rigid body

$$T = \frac{m\bar{v}^2}{2} + \frac{I_{G'G}\Omega^2}{2} \tag{13.28}$$

(16) The general energy equation is

$$(T + V_g + V_e)_1 + \Sigma U_{1-2,\,\text{extr}} = (T + V_g + V_e)_2 + Q_{1-2} \tag{13.33}$$

where V_g is the gravitational potential energy $mg\bar{z}$, $\Sigma U_{1-2,\,\text{extr}}$ is the work done on the body or system by the extraneous forces, Q_{1-2} is the mechanical energy lost or dissipated, V_e is the strain energy which can always be described by $k\delta^2/2$ if the deformation is elastic.

Equation 13.33 can be applied to a single body or a system of bodies.
(17) Problems involving friction must be examined for all possibilities of slipping or not slipping at each contact surface. There are two extra conditions to be taken into account regarding the frictional force F and the relative accelerations (or more exactly the relative velocities) of the sliding surfaces (section 13.4.3).
(18) Composite problems must be divided into subsidiary problems to be dealt with by appropriate equations. These are: work - energy equations, impulse - momentum equations and equations of motion. Clarify the end conditions of subsidiary problems before proceeding (section 13.5).

Problems

13.1 A body, mass 2 kg, radius of gyration about the mass centre 1.2 m, is acted upon by a single force of 200 N directed to the right whose line of action is 0.6 m above the mass-centre. What is its effect upon the body?

13.2 The body in figure 13.24 is standing on a smooth horizontal floor when the force $\boldsymbol{P}$ is applied. Find the minimum value of $\boldsymbol{P}$ required to cause the body to begin overturning. What then is $\bar{a}$?
Hint: assume a symbolic $\bar{a}$ and use a free-body diagram including the inertia force. If it is just about to overturn the reaction at one edge is zero.

13.3 A body has dimensions as in figure 13.24, but the floor is now rough with $\mu = 0.3$. If $\boldsymbol{P}$ is applied while the body is sliding to the left what is its value if the body is on the verge of tipping?

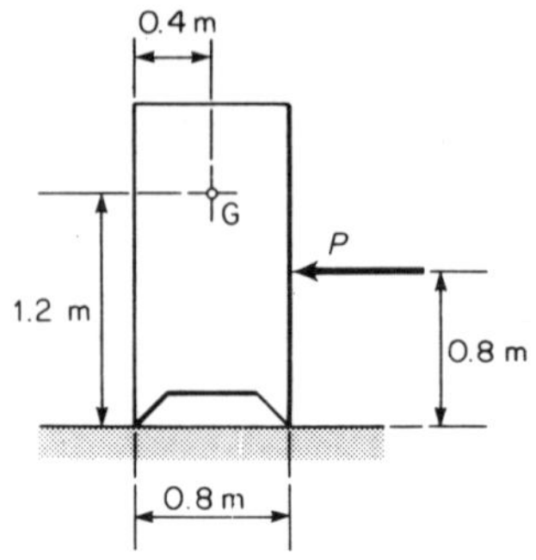

Figure 13.24

13.4 A ship, length 1000 m, mass 2×10^5 kg and radius of gyration about its mass centre 200 m, is being manoeuvred by two tugs with tow lines fixed to its bow and stern. At a particular instant the bow B is lying due north of the stern S; B has an absolute acceleration 5 m/s^2 $\angle -60°$ and S an absolute acceleration 2 m/s^2 $\angle -30°$. Find

(a) the angular acceleration of the ship
(b) its angular velocity and
(c) the component accelerations of its mass centre G which is on the line SB, 400 m from S.

If at the same instant the tow rope at S makes an angle $\angle 0°$ find the tensions in both ropes and the direction of the rope at the bow.
Hint: see chapter 11 regarding determination of the velocities and accelerations; from the ship's accelerations insert the inertia force - couple set into the free-body diagram and solve for equilibrium.

13.5 A car, mass m, has a wheel base (distance between the front and rear axles) of length $(b + c)$, its mass centre G being a distance c in front of the rear axle and height h above ground level. The car has rear wheel drive and is on the upward slope of a hill which can be approximated in shape to the arc of a circle of radius R ($\gg h$) in the vertical plane. If when the car has forward velocity v and forward acceleration a the front wheels leave the ground, show that the angle subtended at the centre of the arc between G and the vertical is given by

$$\theta = \cos^{-1}\left[\frac{1}{g\sqrt{(c^2 + h^2)}}\left(\frac{v^2 c}{R} + ha\right)\right] - \tan^{-1}\frac{h}{c}$$

Hint: draw a free-body diagram to include the inertia forces.

13.6 A uniform thin rod of length $3L$ is supported horizontally on two small rough pegs each at a distance $L/2$ from its mid-point.

(a) If one peg is suddenly removed show that the reaction of the other peg is suddenly increased by half its original value.

(b) If in the subsequent motion the rod begins to slip on the second peg when its inclination to the horizontal is $\tan^{-1} 0.1$ find μ.
Hint: draw the free-body diagram (after assuming symbolic velocity and accelerations) when having turned through an angle θ; solve for component reactions at the peg.

13.7 A particle of mass m is fixed at the centre of a massless rod of length L. The rod has rounded ends and is held in a vertical plane with one end against a smooth vertical wall and the other end on a smooth horizontal floor. If the rod is released find

(a) an expression for the angular acceleration of the rod when it is moving and making an angle $\angle\, \theta°$ with the horizontal. If the initial angle is $\angle\, \beta°$ use the expression to find further expressions for
(b) the angular velocity of the rod and
(c) the component accelerations of the particle in the x and y directions when the rod is inclined at angle $\angle\, \theta°$.
(d) At what angular velocity does the rod strike the ground?
Hint: draw a free-body diagram including inertia forces based on mass-centre component accelerations; find expressions for the total acceleration of the rod ends (chapter 11) and note the conditions on their components perpendicular to the contact face.

13.8 Figure 13.25 shows a body A hinged to a uniform block B through one

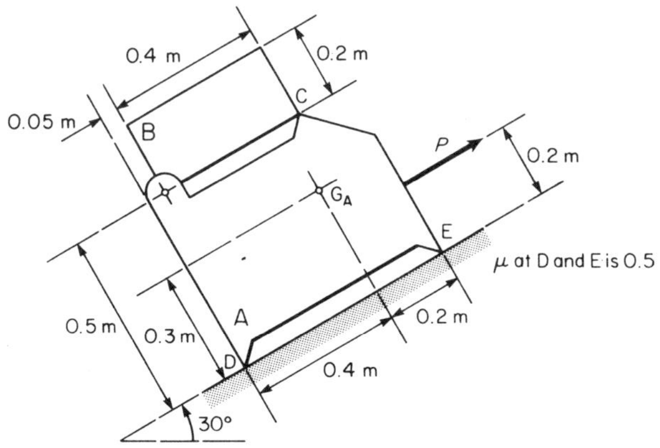

Figure 13.25

corner of the block. The whole assembly is being accelerated up the rough incline by the applied force ***P***.

Find the acceleration of the assembly which is just sufficient to reduce to zero the reactions between the bodies at C. Hence find the value of P and the normal reactions between the plane and body A at points D and E.

The coefficient of friction between the plane and A is 0.5, the mass of A is 50 kg and its centre of mass is at G_A. The uniform block B has a mass of 20 kg. *Hint*: draw free-body diagrams, including inertia effects, for each body.

13.9 Refer to problem 11.18. If the motion of the link is caused by application of a force F at B parallel to the guide find the magnitude and sense of F at the instant described. The link AB may be regarded as a uniform rod of mass 10 kg and the mechanism lies in the horizontal plane. Assume the guides are frictionless.

13.10 Refer to problem 11.7. If BC has a mass of 10 kg and its mass centre is at G, about which its radius of gyration is 0.5 m, find
(a) the external couple required on AB to maintain motion and
(b) the reaction of the hinge D on the mechanism.

Assume that links AB and CD have no mass and the mechanism moves in the horizontal plane.
Hint: draw a free-body diagram for BC inserting inertia effects. Note (see chapter 4) the known direction of the reaction at C. Solve for the reactions at B and C; then consider the equilibrium of links AB and CD.

13.11 A uniform door 2 m square, mass 10 kg, lies in the horizontal plane and is hinged along its left hand edge. If it is allowed to drop clockwise from the horizontal position find expressions for angular acceleration and angular velocity when it has turned through an angle θ. What is the total reaction at the hinge when $\theta = 45°$?
Hint: assume senses of α and Ω; draw a free-body diagram including inertia effects (use $\bar{a}_t$ and $\bar{a}_n$). Solve the resulting equations for α and integrate to find Ω.

13.12 Two rods AB and BC are connected by a hinge at B. End A is hinged to a fixed support and end C carries a small roller which can slide in a horizontal slot at the same level as end A. AB and BC have length L_1, L_2 and masses m_1, m_2 respectively. Their mass centres are distant r_1, r_2 from A and C respectively and their radii of gyration about the same points A and C are k_1 and k_2.

The rods are held in a horizontal position and then released. Show that the initial angular acceleration of BC is given by

$$\alpha_{BC} = \frac{g[m_1 r_1 + (m_2 r_2 L_1)/L_2]}{(L_2/L_1)m_1 k_1^2 + (L_1/L_2)m_2 k_2^2}$$

Hint: draw a free-body diagram for each rod to include inertia effect based upon assumed accelerations; relate the accelerations of the two rods at the hinge B.

13.13 A rear wheel drive vehicle has a wheel base length 5 m and its mass centre G is 3 m behind the front axle and 2 m above ground level. The mass of the body excluding wheels and axles is 1000 kg and each of the four wheels and its axle has mass 20 kg, outside radius 0.5 m and a radius of gyration about its centre $0.25\sqrt{2}$ m. μ between the road and wheels is 0.3 and the vehicle is on a horizontal road.

(a) Find the maximum acceleration that it can have before skidding just occurs at the driving wheels.
(b) What is then the total external driving couple required on the two rear wheels and the horizontal reaction of one wheel (and axle) on the vehicle frame at
(i) the rear and
(ii) the front?
Hint: draw free-body diagrams for the body and each wheel, inserting the inertia effects. Note an external couple on the rear wheels and its reaction on the body; if skidding is just about to take place $F = \mu N$ and the horizontal component of the acceleration of the point of the wheel in contact with the road is zero.

13.14 A wheel, with moment of inertia I about its centre, is fixed on a shaft of radius r and round the shaft is wound a length L of thin chain of mass m per unit length, one end being fixed to the shaft and the other carrying a body having mass equal to that of the whole chain. Initially the shaft is at rest with the attached body level with the axis. If the body is released show that the length of chain unwound after t seconds is

$$2L \sinh^2 (bt)$$

where

$$b = \tfrac{1}{2} \sqrt{\left(\frac{mgr^2}{I + 2mLr^2}\right)}$$

Hint: assume a length x has dropped at time t and at this instant draw free-body diagrams for (1) the mass and the chain length x and (2) the wheel and remainder of the chain. Include inertia effects; relate forces in the chain where it leaves the wheel and also relate the acceleration of the chain and the angular acceleration of the shaft. Hence deduce the equation of motion and the required solution. Alternatively consider the whole system.

13.15 Two shafts each 0.1 m diameter are carried in bearings so that they run parallel to each other 2 m apart in the same horizontal plane. The left hand shaft is rotated rapidly clockwise and the right hand anticlockwise. A uniform thin plank, longer than 2 m, is rested across the shafts with its length perpendicular to the shaft axes, whereupon limiting friction forces act on the plank proportional to the normal reactions at the shafts.

Show that the plank oscillates continuously with simple harmonic motion, with an amplitude equal to the initial displacement of the centre of the plank from a point mid-way between the shafts. If the relevant friction coefficient remains constant and equal to 1/4, determine the frequency of oscillation. If the initial displacement is 0.5 m find the minimum speed at which the shafts must be run to ensure that the motion of the plank is simple harmonic.
Hint: draw a free-body diagram at the instant t when the plank's mass centre G is at a distance x from mid-way between the shafts; express its acceleration in terms of x, insert inertia forces and determine the equation of motion; compare with simple harmonic motion. If simple harmonic motion is to be maintained there is a

relationship between the peripheral velocity of the shafts and the velocity of the plank.

13.16 A uniform rod of mass 2 kg, length 5 m, and of negligible thickness is placed in the vertical plane against a rough wall and on a rough floor ($\mu = 0.3$ for both surfaces). If the rod is placed at an angle of $\angle 30°$ to the floor find
(a) its instantaneous angular acceleration and
(b) the normal reaction at each end when it just begins to move.
Hint: see problem 13.7.

13.17 If in problem 11.16 the link AB is a uniform rod 8 m long of mass 2000 kg and DC has a mass of 500 kg find the force required along DC to maintain the motion when $t = 1$ s. What power is required to be supplied to DC at this particular instant to maintain motion?

Assume that friction is absent and that the mechanism lies in the horizontal plane.
Hint: draw separate free-body diagrams for AB and DC inserting inertia effects where necessary.

13.18 If in problem 11.14 roller C has mass of 2 kg and a force of 10 N is applied to it to the right, find the couple required on AB to maintain the motion when θ equals 60°. Assume links AB and BC are massless and friction is absent.

Calculate the values of power into and out of the mechanism. Why are they different? Check that the difference equals the rate of change of kinetic energy of the system (show that $dT/dt = mav$).
Hint: draw a free-body diagram for C inserting the inertia effects and noting the known direction of the force in BC; hence from the equilibrium of AB find the couple on AB; refer to chapter 4 if necessary.

13.19 If in problem 11.14 the mechanism lies in the horizontal plane, block C has mass 2 kg (no force now being applied) and BC is a uniform rod of mass 1 kg find the couple required on AB to maintain motion when $\theta = 60°$. Assume AB is massless and friction is absent. What power is then developed at AB? Is this going out of or into the mechanism?
Hint: draw a combined free-body diagram (include all inertia effects) of BC and C together. Take moments about a convenient point to eliminate one unknown reaction. Check the power by calculating the work done per second by all forces and couples on AB.

13.20 A uniform wooden beam, of mass 20 kg, 4 m long and 0.2 m by 0.5 m cross section is lying stationary on a horizontal frictionless surface with its 0.5 m side face downwards. A bullet of mass 0.1 kg is fired in the horizontal plane and perpendicular to the length of the beam so that it enters and lodges in the beam 0.5 m from one end. If the bullet's velocity is 2000 m/s what are the angular velocity of the beam and the linear velocity of its centre of mass just after the impact?
Hint: use the impulse – momentum equations after drawing a separate free-body diagram for both the bullet and the beam; note the equality of the impulsive forces.

13.21 Two uniform gears are rotating clockwise on parallel shafts. Gear A has 50 teeth, a moment of inertia about its mass centre G of 8.0 kg m^2 and a speed of + 200 rad/s. Gear B has 20 teeth, a moment of inertia about its mass centre of 2.0 kg m^2 and a speed of + 100 rad/s. If the gears are pushed into mesh what are their final speeds? Note that radii are proportional to number of teeth.
Hint: choose a sign convention and draw separate free-body diagrams; apply the angular impulse - momentum equation 13.16 to both bodies noting that when meshed peripheral velocities will be equal but opposite.

13.22 Three uniform discs A, B and C are freely pivoted and coupled together as shown in figure 13.26. A small electric motor which has a normal running speed

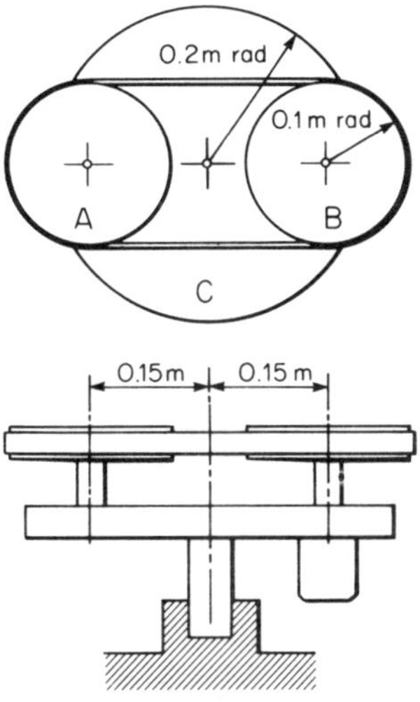

Figure 13.26

of 200 rev/m is fixed to disc C with its shaft coupled to disc B and B is connected to A by a belt. The masses of A, B and C are 5 kg, 5 kg and 10 kg respectively. If the motor is started when the system is at rest find the angular velocity of each disc when the motor has attained its normal running speed. Neglect the masses of the motor and shafts.
Hint: draw a free-body diagram for the whole system inserting any external forces and couples that act during the acceleration period; use impulse - momentum equations to relate ω_A and ω_C; note other relationship of ω_A and ω_C given by the motor speed.

13.23 A uniform solid cube of mass m and edge of length $2a$ rests on one face on a smooth horizontal table. It is given a horizontal impulse I at the mid-point of one edge of its top face and perpendicular to that edge. Show that the impulsive reaction at the table is $3I/5$, and find the initial angular velocity of the cube.

Show also that the cube will overturn in the subsequent motion if

$$I^2 > \frac{10M^2ga(\sqrt{2}-1)}{3}$$

Hint: at impact the body will slide along the surface with one edge in contact with the surface.

13.24 A uniform rectangular shaped body 3 m wide and 5 m high, of mass 1000 kg, is sliding to the right in the direction of its width along a horizontal frictionless surface with velocity V. It hits a horizontal bar A which is 1 m above the surface, maintains contact with this during the collision and continues to rotate about it without slipping after the impact is completed.

Find the value of V (m/s) in order that G, the centre of mass, shall just pass the vertical above A. For this value of V find the angular acceleration of the body when the impact is completed and the body is just starting to turn about A.
Hint: use equation 13.16 making a judicious choice of origin; use the general energy equation to relate velocities and draw a free-body diagram including inertia effects to find angular acceleration.

13.25 A tilt hammer is hinged at one end and carries a head at 0.5 m from the hinge. Its mass centre is 0.4 m from the hinge, its mass is 150 kg and when allowed to oscillate freely with small amplitude about its vertical position its frequency of oscillation is 0.7 Hz. The hammer is raised 45° from the horizontal position and allowed to fall to strike a pile of mass 1000 kg when in its horizontal position. Find
(a) the angular velocity just before the impact
(b) the linear velocity of the pile immediately after the impact
(c) the average impulsive force between the bodies if the impact lasts 0.01 s and
(d) the resistance of the earth surrounding the pile if the latter moves 0.05 m after the impact.

Ignore the resistance of the earth during the impact and assume $e = 0$ for the impact.
Hint: the equations relating to a compound pendulum are required. For (a) use the work - energy equation; for (b) and (c) use the impulse - momentum equations, for (d) use the work - energy equation. (Be careful about V_g for the hammer.)

13.26 Two equal uniform rods AB and BC, each of mass M and length L are freely jointed at B and are in line, moving perpendicular to their length with velocity u in a horizontal plane. The mass centre of AB is suddenly brought to rest by a plastic impact. Find the angular velocity of each rod immediately after the impact is completed and prove that 4/7 of the original kinetic energy is lost in the impact.
Hint: draw separate free-body diagrams for AB and BC inserting impulsive forces acting on each; use the impulse - momentum equations 13.12, 13.13 and 13.16; note the relationship of the linear velocity of the mass centre of BC to the angular velocities.

13.27 Figure 13.27 shows a uniform beam A hinged at one end. This is allowed to fall from the horizontal position and when vertical, strikes the cylindrical body B, which is initially rotating at 20 rad/s anticlockwise. If the beam rebounds until

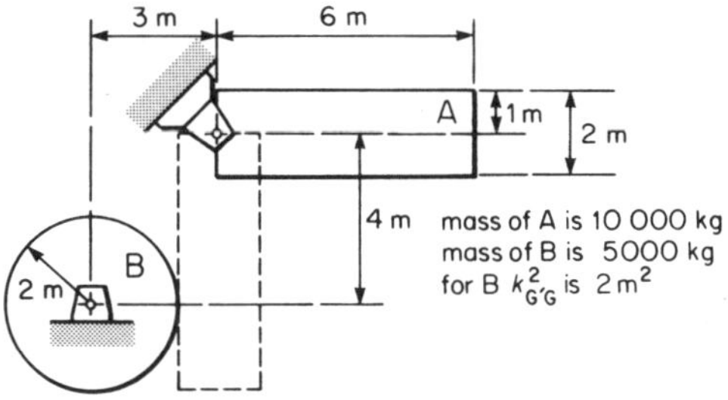

Figure 13.27

it makes an angle of 60° with the vertical and μ between the bodies during the impact is 0.5 find the speed of B at the completion of the first impact.
Hint: adopt a sign convention; use the work - energy equation to find the angular velocity of A just before and just after the impact; use the angular momentum – angular impulse during the impact – relating the impulsive forces on the two bodies. Use free-body diagrams.

13.28 (a) Check parts (b) and (d) of problem 13.7 by using work - energy methods.
(b) Check the equation for Ω in problem 13.11 using work - energy methods.

13.29 An electric motor drives the flywheel (moment of inertia $I_{G'G}$ of 2 kg m^2) of a machine with a constant torque M. The machine punches holes in 0.1 m thick plate during which the resisting force is 2000 N; during the withdrawal of the punch the resisting force is 500 N. The punching and withdrawal each take place during one half revolution of the flywheel.

The mass of the punch is negligible, the kinetic energy of the system being due solely to the flywheel. The speed of the latter at the beginning and end of a complete cycle is to be the same and equal to 20 rad/s. Find

(a) the work done by the machine in a complete cycle and hence
(b) the constant torque M delivered by the motor.

What is the flywheel speed at the end of the punching operation?
Hint: use work - energy considerations assuming no energy losses occur.

13.30 What energy is lost during the impact

(a) in problem 13.21 and
(b) in problem 13.20?

13.31 Two trucks run on parallel tracks inclined at $\sin^{-1} 0.1$ to the horizontal. The trucks are connected by a rope, of mass per unit length 3 kg/m, which passes round a 3 m diameter pulley at the top of the incline. The axis of the pulley is normal to the plane containing the trucks and the rope is supported to lie parallel with the track. The moment of inertia of the pulley about its axis, including the rope wrapped around it, is 80 kg m^2 and the length of rope exclusive of the amount wrapped round the pulley is 300 m. Each truck has mass 1000 kg when empty and each, in turn, is loaded at the top of the incline with 4000 kg of ballast and then descends under gravity pulling the empty truck up the other track.

Assuming the resistance to motion of each truck is constant at 0.25 N/kg of the total mass of the truck and its load, find the velocity of the trucks after they have travelled 200 m along the track and the constant torque to be applied to the pulley to bring them to rest after a further 100 m.
Hint: can be solved by using equations of motion but since all resisting forces are known it is simpler to use work – energy equations; apply equation 13.33 (do not forget the rope).

13.32 Refer to problem 4.22. Each frame has a radius of gyration $a/\sqrt{3}$ about its mass centre and the force $\boldsymbol{P}$ (at the value calculated in problem 4.22) is applied when $\theta = 30°$ and the platform is stationary. If $\boldsymbol{P}$ is kept constant until $\theta = 60°$ show that, neglecting friction, the velocity of the platform is then

$$\frac{(2 - \sqrt{3})(M + m)ag}{M/4 + 2m/3}$$

Hint: since friction is absent use work – energy equations; relate the velocities of the mass-centres of the platform and frames to the angular velocities of the frames.

13.33 A uniform cylinder of radius r and mass m is placed on a rough surface (coefficient of friction μ) which is inclined at an angle β to the horizontal. Show that the cylinder will roll without slipping if $\tan \beta < 3\mu$.
Hint: draw a free-body diagram inserting inertia effects based upon assumed accelerations, check all the relevant equations including the conditions for slipping or not slipping.

13.34 A uniform cylinder radius r and mass m rests on a rough horizontal surface for which the coefficient of friction is μ. If the surface is now moved horizontally with acceleration $\boldsymbol{a}$ perpendicular to the cylinder axis determine the maximum value of $\boldsymbol{a}$ if the cylinder is not to slip. What is then the linear acceleration of the cylinder?
Hint: as for problem 13.33.

13.35 The mass centre of body A in figure 13.28 is at G; A has mass 80 kg. The cylindrical body B has mass 20 kg, outside radius 1.5 m and radius of gyration 1 m about its centre. For the condition shown find the value of x at which slipping just ceases at P.
Hint: draw free-body diagrams for both bodies including inertia effects based upon assumed accelerations; note the special conditions if slipping just ceases.

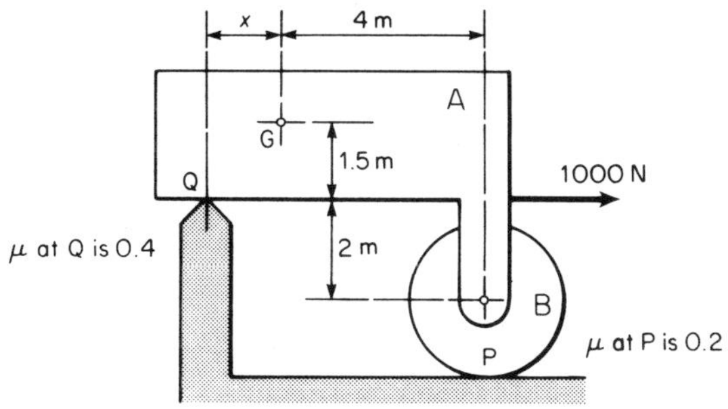

Figure 13.28

13.36 The cylindrical body B in figure 13.29 is in the centre of the top face of A and both are at rest when a force of 100g N is applied to A as shown. Decide whether B slips on A or not and give the accelerations of A and B.
Hint: draw free-body diagrams (including inertia effects) for both bodies; see worked example 13.13.

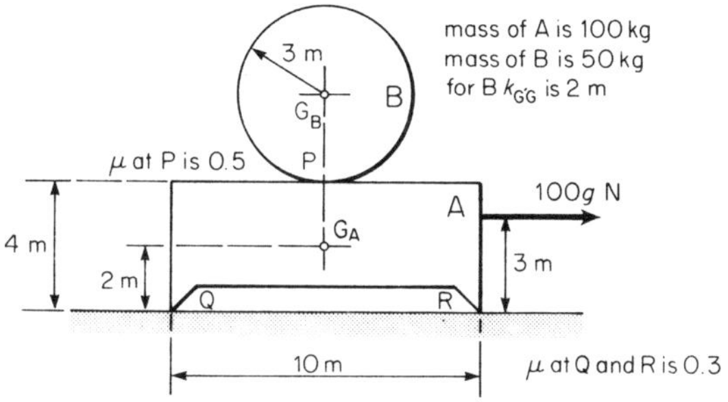

Figure 13.29

13.37 If in problem 13.33 $\beta < \tan^{-1} 3\mu$ show that after rolling a distance s the kinetic energy T is exactly equal to the reduction in potential energy, thus confirming that there is no work for the friction force at a rolling contact.
Hint: find the accelerations of the body, hence the velocities and kinetic energy.

13.38 If in problem 13.33 $\beta > \tan^{-1} 3\mu$, show that the body accelerations are

$$\bar{a} = g(\sin\beta - \mu\cos\beta)$$

$$\alpha = (2\mu g\cos\beta)/r$$

and hence that the kinetic energy T after moving a linear distance s is given by

$$T = mgs\left(\sin\beta - \mu\cos\beta + \frac{2\mu^2\cos^2\beta}{\sin\beta - \mu\cos\beta}\right)$$

What is the energy dissipated in this case?

13.39 A hammer of mass M_1 is hinged at one end at a distance L from its head; its mass centre G is at a distance r from the hinge and its moment of inertia about G is $I_{G'G}$. The hammer is dropped from the horizontal position and when vertical its head strikes a spring fastened to a body B of mass M_2, this body being initially stationary on a rough horizontal table such that the coefficient of friction is μ.

Taking $e = 1$ for the impact and assuming that the movement of both the hammer and the body is negligible during the impact find expressions for

(a) the angular velocity of the hammer just before the impact
(b) the linear velocity of the body when the spring is fully deformed
(c) the velocity of the body at the end of the impact and
(d) the distance the body moves after the impact.

Ignore the frictional effect of the table during the impact.
Hint: split into three problems
(1) dropping the hammer – use the work - energy equation
(2) the impact – use impulse - momentum equations etc
(3) motion after impact – find the acceleration and hence the distance, or alternatively use work - energy equations.

13.40 A thin uniform beam ABCD, 6 m long, rests horizontally on supports at B and C such that AB = CD. End A is raised by turning the beam about C and is then released.

(a) Assuming that there is no slipping at the supports, show that the end D will rise above the original horizontal position of the beam if BC is less than 3.46 m.
(b) If BC = 3 m and A is raised a vertical distance of 0.6 m by turning the beam about C, determine the vertical distance D will rise when the beam is released.
Hint: for (a) use the impulse - momentum equations after drawing a free-body diagram; you will have to assume that any impulsive reaction at C is very small compared with that at B. For (b) use the work - energy equations to find the angular velocity before impact, a momentum equation for angular velocity after, and again a work - energy equation to find the height to which D rises.

Appendix

Standard Integrals

$$\int \frac{dx}{a^2 - x^2} = \frac{1}{2a} \ln \left(\frac{a + x}{a - x} \right) \quad \text{(for } x < a\text{)}$$

$$\int \frac{dx}{a^2 + x^2} = \frac{1}{a} \tan^{-1} \left(\frac{x}{a} \right)$$

$$\int \frac{dx}{x^2 - a^2} = \frac{1}{2a} \ln \left(\frac{x - a}{x + a} \right) \quad \text{(for } x > a\text{)}$$

$$\int \frac{x dx}{a^2 - x^2} = -\tfrac{1}{2} \ln (a^2 - x^2) \quad \text{(for } x < a\text{)}$$

$$\int \frac{x dx}{a^2 + x^2} = \tfrac{1}{2} \ln (x^2 + a^2)$$

$$\int \sqrt{(a^2 + x^2)} \mathrm{d}x = \tfrac{1}{2} \left\{ x\sqrt{(a^2 + x^2)} + a^2 \ln [x + \sqrt{(a^2 + x^2)}] \right\}$$

$$\int \sqrt{(a^2 - x^2)} \mathrm{d}x = \tfrac{1}{2} \left[x\sqrt{(a^2 - x^2)} + a^2 \sin^{-1} \left(\frac{x}{a} \right) \right]$$

$$\int \frac{1}{\sqrt{(a^2 + x^2)}} \mathrm{d}x = \sinh^{-1} \left(\frac{x}{a} \right)$$

$$\int \frac{1}{\sqrt{(a^2 - x^2)}} \mathrm{d}x = \sin^{-1} \left(\frac{x}{a} \right)$$

Length of a Curve and Radius of Curvature

Length of a small segment of a curve

$$\mathrm{d}s = \left[1 + \left(\frac{\mathrm{d}y}{\mathrm{d}x} \right)^2 \right]^{\frac{1}{2}} \mathrm{d}x$$

or

$$\mathrm{d}s = \left[\left(\frac{\mathrm{d}r}{\mathrm{d}\theta} \right)^2 + r^2 \right]^{\frac{1}{2}} \mathrm{d}\theta$$

Radius of curvature

$$\rho = \frac{[1 + (\mathrm{d}y/\mathrm{d}x)^2]^{3/2}}{\mathrm{d}^2 y/\mathrm{d}x^2}$$

or

$$\rho = \frac{[r^2 + (\mathrm{d}r/\mathrm{d}\theta)^2]^{3/2}}{r^2 + 2(\mathrm{d}r/\mathrm{d}\theta)^2 + r(\mathrm{d}^2 r/\mathrm{d}\theta^2)^2}$$

Solutions of Particular Differential Equations

For

$$\frac{\mathrm{d}^2 y}{\mathrm{d}t^2} + Ky = 0$$

the solution is

$$y = A \sin(\sqrt{K}t + \phi)$$

For

$$\frac{\mathrm{d}^2 y}{\mathrm{d}t^2} - Ky = 0$$

the solution is

$$y = A\mathrm{e}^{\sqrt{K}t} + B\mathrm{e}^{-\sqrt{K}t}$$

Properties of Homogeneous Thin Plates

m is the mass of the body shown.

Plate	Area	Mass-centre	Moment of Inertia
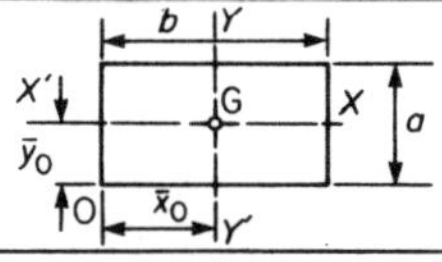 $y_0 =$ constant, rectangular	ab	$\bar{x}_0 = \frac{b}{2}$, $\bar{y}_0 = \frac{a}{2}$	$I_{yy} = \frac{mb^2}{12}$, $I_{xx} = \frac{ma^2}{12}$
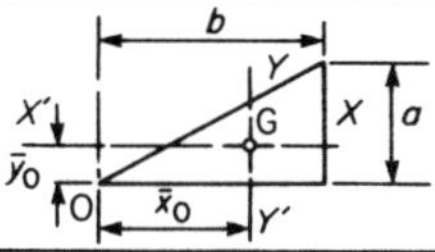 $y_0 =$ const x x_0, triangular	$\frac{ab}{2}$	$\bar{x}_0 = \frac{2b}{3}$, $\bar{y}_0 = \frac{a}{3}$	$I_{yy} = \frac{mb^2}{18}$, $I_{xx} = \frac{ma^2}{18}$
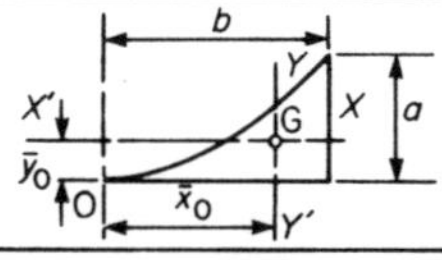 $y_0 =$ const x x_0^2, parabolic	$\frac{ab}{3}$	$\bar{x}_0 = \frac{3b}{4}$, $\bar{y}_0 = \frac{3a}{10}$	$I_{yy} = \frac{3mb^2}{80}$, $I_{xx} = \frac{37ma^2}{700}$
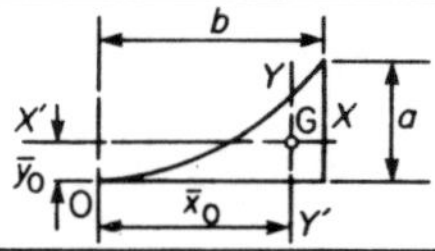 $y_0 =$ const x x_0^n	$\frac{ab}{(n+1)}$	$\bar{x}_0 = \frac{(n+1)b}{(n+2)}$, $\bar{y}_0 = \frac{(n+1)a}{2(2n+1)}$	$I_{yy} = \frac{mb^2(n+1)}{(n+2)^2(n+3)}$, $I_{xx} = \frac{ma^2(n+1)(7n^2+4n+1)}{12(2n+1)^2(3n+1)}$
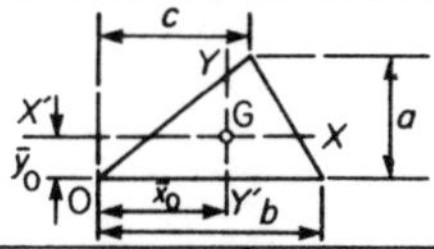 triangular	$\frac{ab}{2}$	$\bar{x}_0 = \frac{b+c}{3}$, $\bar{y}_0 = \frac{a}{3}$	$I_{yy} = \frac{m(b^2+c^2-bc)}{18}$, $I_{xx} = \frac{ma^2}{18}$
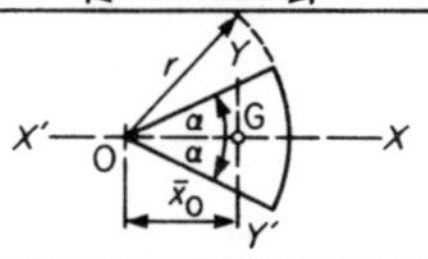 circular sector	$r^2\alpha$	$\bar{x}_0 = \frac{2r\sin\alpha}{3\alpha}$	$I_{yy} = \frac{mr^2}{4}\left[1+\frac{\sin 2\alpha}{2\alpha} - \frac{16}{9}\left(\frac{\sin^2\alpha}{\alpha^2}\right)\right]$, $I_{xx} = \frac{mr^2}{4}\left(1-\frac{\sin 2\alpha}{2\alpha}\right)$
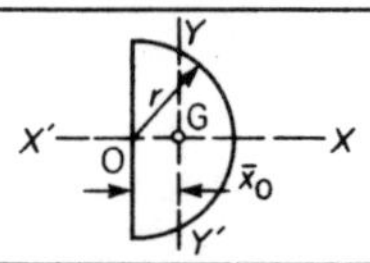 semicircular	$\frac{\pi r^2}{2}$	$\bar{x}_0 = \frac{4r}{3\pi}$	$I_{yy} = \frac{mr^2}{4}\left(1-\frac{64}{9\pi^2}\right)$, $I_{xx} = \frac{mr^2}{4}$
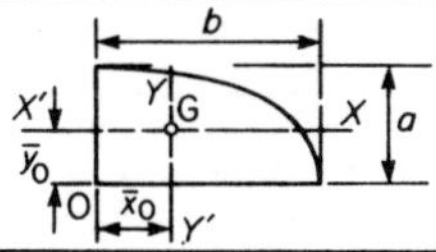 elliptical quadrant	$\frac{\pi ab}{4}$	$\bar{x}_0 = \frac{4b}{3\pi}$, $\bar{y}_0 = \frac{4a}{3\pi}$	$I_{yy} = \frac{mb^2}{4}\left(1-\frac{64}{9\pi^2}\right)$, $I_{xx} = \frac{ma^2}{4}\left(1-\frac{64}{9\pi^2}\right)$
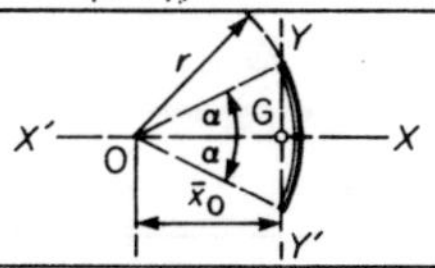 thin circular annulus	–	$\bar{x}_0 = \frac{r\sin\alpha}{\alpha}$	$I_{yy} = mr^2\left(\frac{1}{2}+\frac{\sin 2\alpha}{4\alpha} - \frac{\sin^2\alpha}{\alpha^2}\right)$, $I_{xx} = mr^2\left(\frac{1}{2} - \frac{\sin 2\alpha}{4\alpha}\right)$

Properties of Homogeneous Bodies

m is the mass of the body shown.

Body			Moment of Inertia
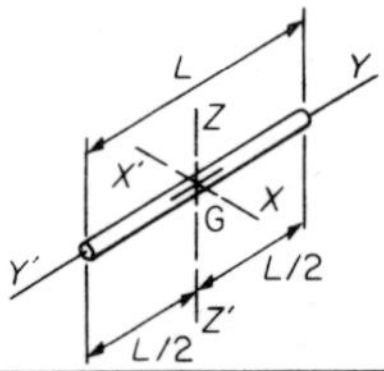	thin rod		$I_{yy} = 0$ $I_{xx} = I_{zz}$ $= \frac{1}{12} mL^2$
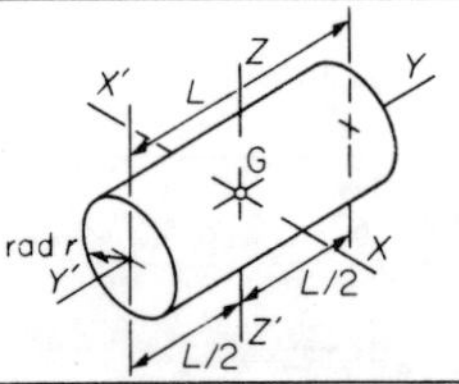	circular cylinder	volume $= \pi r^2 L$	$I_{yy} = \frac{1}{2} mr^2$ $I_{xx} = I_{zz}$ $= \frac{1}{4} mr^2 + \frac{1}{12} mL^2$
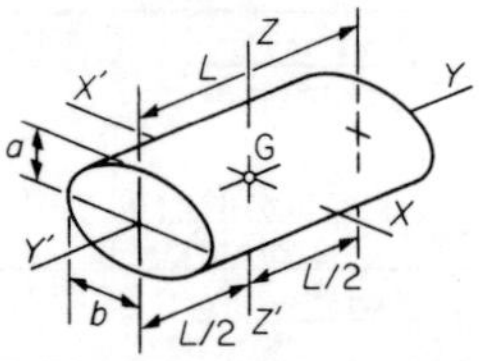	elliptical cylinder	volume $= \pi abL$ area of curved surface $= \pi (a+b) L$	$I_{yy} = \frac{1}{4} m(a^2+b^2)$ $I_{xx} = \frac{1}{4} ma^2 + \frac{1}{12} mL^2$ $I_{zz} = \frac{1}{4} mb^2 + \frac{1}{12} mL^2$
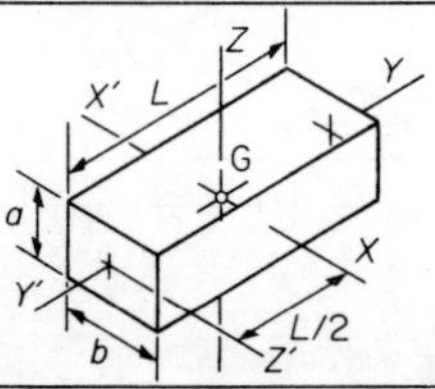	rectangular parallelepiped		$I_{yy} = \frac{1}{12} m(a^2+b^2)$ $I_{xx} = \frac{1}{12} m(a^2+L^2)$ $I_{zz} = \frac{1}{12} m(b^2+L^2)$
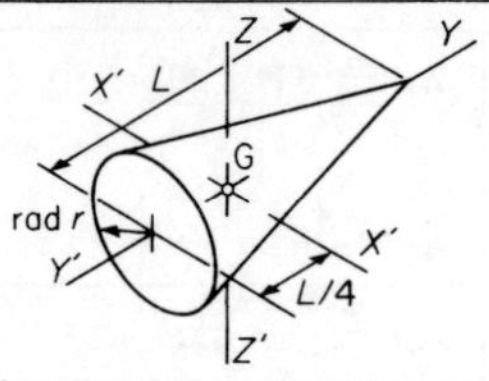	right circular cone	volume $= \frac{1}{3} \pi r^2 L$ area of curved surface $= \pi r \sqrt{(r^2+L^2)}$	$I_{yy} = \frac{3}{10} mr^2$ $I_{xx} = I_{zz}$ $= \frac{3}{20} mr^2 + \frac{3}{80} mL^2$
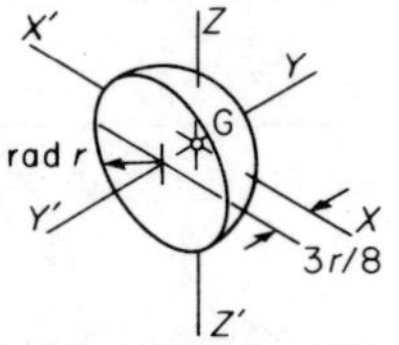	hemisphere	volume $= \frac{2}{3} \pi r^3$ area of curved surface $= 2\pi r^2$	$I_{yy} = \frac{2}{5} mr^2$ $I_{xx} = I_{zz}$ $= \frac{83}{320} mr^2$

Answers to Problems

Chapter 2

2.1 (a) (b) (e) (f) scalar; (c) (d) (g) (h) vector.
2.2 (a) $7.74 \angle 71.2°$; (b) $26.5 \angle -10.9°$; (c) $213 \angle 195.9°$.
2.3 (a) $2.83 \angle 28°$; (b) $17.3 \angle 120°$; (c) $595 \angle 35°$.
2.4 $3.1 \angle 142°$
2.5 (a) $a_x = 9.64$, $a_y = 11.5$; (b) $b_x = -17.3$, $b_y = 10$; (c) $c_x = -8.66$, $c_y = 5$.
2.6 $8.78 \angle -52.9°$.
2.7 $|a| = 8.37$; (a) 21°; (b) 45.8°.
2.8 $12.8 \angle -18.2°$.
2.9 $\boldsymbol{a} = 14.5 \angle 39.9°$, $\boldsymbol{b} = 6.2 \angle 186.2°$.

Chapter 3

3.1 20 N at 47°, 30 N at 29°.
3.2 $|\boldsymbol{P}| = |\boldsymbol{N}| = 1.13$ N.
3.3 37.4°, 6.69 N, 23.8 N.
3.4 (a) $37.8 \text{ N} \angle 126.6°$; (b) $37.8 \text{ N} \angle -53.4°$.
3.5 175400 N.
3.6 $|\boldsymbol{P}| = 127$ N, $|\boldsymbol{Q}| = 58.5$ N.
3.7 $\boldsymbol{P} = 133.6 \text{ N} \angle 153.9°$, $\boldsymbol{Q} = 66.8 \text{ N} \angle 63.9°$.
3.8 $T = 16.24$ N.
3.9 Note $\overrightarrow{OA} = \overrightarrow{OG} + \overrightarrow{GD} + \overrightarrow{DA}$ and $GC = 2\overrightarrow{GD}$, where D is the centre of AB.

Chapter 4

4.1 (a) $16.4 \text{ N} \angle 165.4°$, 29.3 N m anticlockwise; (b) $16.4 \text{ N} \angle 165.4°$, $1.79 \text{ m} \angle 75.4°$ from A.
4.2 $\boldsymbol{P} = 10$ N, $14.14 \text{ N} \angle 135°$.
4.3 (a) Yes; (b) $\boldsymbol{R}_1 = 425$ N, $\boldsymbol{R}_2 = 175$ N, $\boldsymbol{R}_3 = 1000$ N.
4.4 (a) Yes; (b) $\boldsymbol{R}_1 = 204$ N, $\boldsymbol{R}_2 = 1022$ N, $\boldsymbol{R}_3 = 162$ N.
4.5 (a) No; (b) $\boldsymbol{R}_A = 375$ N, $\boldsymbol{R}_B = 41.6$ N, $\boldsymbol{R}_C = 83.3$ N; (c) statically indeterminate.
4.6 A force $23.2 \text{ N} \angle 190.3°$ plus a couple 2.17 N m anticlockwise.
4.7 $\boldsymbol{F} = 115$ N; $\boldsymbol{R}_H = 140 \text{ N} \angle 64.3°$.
4.8 $\boldsymbol{P} = 68$ N; $\boldsymbol{R}_H = 207 \text{ N} \angle 109.1°$.
4.9 (a) 22.3° (40 kg), 49.5° (20 kg); (b) 11.3°.
4.11 (a) No; (b) 8.53°.
4.12 $\boldsymbol{P} = 343$ N, $\boldsymbol{R}_{Cx} = 98.1$ N→, $\boldsymbol{R}_{Cy} = 109$ N↑, $\boldsymbol{R}_D = 60.4$ N↑, $\boldsymbol{R}_H = 263$ N↑, $\boldsymbol{R}_E = 332$ N↑.
4.13 $\boldsymbol{P} = 1037$ N, $\boldsymbol{R}_D = 770$ N→, $\boldsymbol{R}_{Ex} = 870$ N←, $\boldsymbol{R}_{Ey} = 273$ N↓.
4.14 $\boldsymbol{R}_E = 0.53$ kN↑, $\boldsymbol{R}_{Ax} = 2.12$ kN→, $\boldsymbol{R}_{Ay} = 0.18$ kN↑; ED, 1.06 kN(T);

EC, 1.19 kN (S); DC, 1.41 kN (T); DA, 0.35 kN (S); CB, 0.71 kN (T); CA, 1.97 kN (S); BA, 0.71 kN (T).

4.15 BC, 2.73 kN (T); CD, 2.45 kN (S); BD, 1.73 kN (T); DE, 1.73 kN (S); AB, 4.46 kN (T); BE, 2.45 kN (S); AE, 1.73 kN (T). The force in AB would be reduced to 1.73 kN.

4.16 CE, 1.42 kN (T); ED, 2.58 kN (S); CD, 1.33 kN (S); AC, 1.38 kN (T); DF, 3.02 kN (S); AD, 0.72 kN (S).

4.17 Torque = 2.19 N m clockwise; R_A = 13.8 N $\angle$ 179.4°, R_D = 8.37 N $\angle$ 126.9°.

4.18 R_{Ay} = 0.354 kN↑, R_{Ax} = 2.6 kN→; R_{Cy} = 0.646 kN↓, R_{Cx} = 3.63 kN←; R_{By} = 0.354 kN↓, R_{Bx} = 1.9 kN→.

4.19 Torque = 1.53 kN m clockwise; R_D = 1.78 kN $\angle$ − 90°; R_E = 14.5 kN $\angle$ 200.6°; R_F = 7.7 kN $\angle$ 62.6°

4.20 Torque = 5 kN m clockwise; R_O 28 kN $\angle$ − 39.7°, R_D = 24.7 kN $\angle$ 243°, R_E = 10 kN $\angle$ 180° (graphical solution).

4.21 Torque = 522 N m anticlockwise.

4.22 $P = \sqrt{3}\,(M+m)g$

Chapter 5

5.4 $\mu > 0.5$

5.5 (a) No; (b) N_W, 147 to 131, F_W, 0 to 26.3, N_F, 490 to 464, F_F, 147 to 131; (c) No, the friction forces cannot be limiting simultaneously. (d) No. (e) 32.6°.

5.7 0.52 m.

5.10 (a) 334 N; (b) 864 N; (c) (i) 56.6°, (ii) 33.4°.

5.11 (a) 51.4°; (b) 68.2 N.

5.12 (a) 16.7°; (b) Yes; (c) (i) 232 N, (ii) 65.7 N.

5.13 3.34 m.

5.14 Belt first slips at B when T_1/T_2 = 1.702; 132 N m; Yes.

5.15 37.5°.

5.16 515 N.

5.17 (a) 27 N m; (b) 52.7 N m in direction of motion.

5.18 (a) 11.7 N m; (b) 3.45 N m; 0.08 m.

5.19 (a) 327 N m; (b) Yes; (c) 0.205 m.

5.20 125 N m.

5.21 r_1 = 0.126 m, r_2 = 0.378 m, M_f = 437 N m; P_{max} = 100 kN/m², M_f = 403 N m.

5.22 (a) 503 N; (b) 208 N.

5.23 $\mu_a > R_o\,\mu_o/r_a$.

5.24 (a) At the axle; (b) θ = 9.4°, P = 650 N.

Chapter 6

6.1 $P = \sqrt{3}\,(M+m)g$

Chapter 7

7.1 $\bar{x} = 1.67\text{m}, \bar{y} = 0.833$ m.
7.2 2.77 m.
7.3 (a) 0.266 m; (b) 0.197 m.
7.4 27.3°.
7.5 $\bar{x} = 1.54$ m, $\bar{y} = 0.937$ m.
7.6 $\bar{x} = 2$ m, $\bar{y} = \bar{z} = 0.85$ m.
7.7 $L = R\sqrt{(2/3)}$.
7.8 0.163 m.
7.9 (a) 0.35ρ g; (b) 1.96 m.
7.10 (a) 6280 kg; (b) 0.975 m.
7.11 $A = 0.141$ m^2, $V = 0.00471$ m^3.
7.14 Area $= KR^3/3$; $\bar{x} = 3R/4$; volume of revolution $= \pi KR^4/2$; 14 rad/s.

Chapter 8

8.1 (a) 23.1 s; (b) 46.7 s.
8.2 $v = 30$ m/s, $a = 5/6$ m/s^2, 540 m.
8.3 (a) 15 m/s; (b) 400/3 s, 200 s, 800/3 s.
8.4 (a) No; (b) No, it has a limiting value of 0.05 m/s^2; (c) 240.5 s; (d) 1360 m; (e) 1/45 m/s^2, 1/25 m/s^2.
8.5 (a) 10 m/s; (b) 1.35 s, 8.47 m; (c) It falls until it reaches a steady value of 10 m/s.
8.6 (a) 29.3 s, 170.7 s; (b) 100 s; (c) – 1.414 m/s^2, + 1.414 m/s^2; (d) – 50 m/s; (e) + 690 m, – 4023 m; (f) 4713 m; (g) 63.4 s, 236.6 s; (h) 9426 m.
8.7 $v = 20.6$ m/s $\angle$ 61.7°; $a = 167.7$ m/s^2 $\angle$ 262.7°; 1.69 m $\angle$ 57°.
8.8 (a) 4.47 m $\angle$ 26.6°; (b) 2.0 s; (c) At this time the particle is at its nearest distance for $t > 2$ s; (d) $v = 2.24$ m/s $\angle$ 116.6°, $a = 3.16$ m/s^2 $\angle$ 161.6°.
8.9 7.65 m $\angle$ 87.7°, $v = 5$ m/s $\angle$ 144.5°, $a = 3.55$ m/s^2 $\angle$ 234.5°.
8.11 (a) $v_m = 31.4$ m/s, $a_m = 1974$ m/s^2; (b) $v = +25.1$ m/s, $a = -1184$ m/s^2; (c) 0.0356 s.
8.12 (a) $\Omega_m = 6.58$ rad/s; (b) $\alpha_m = 82.7$ rad/s^2; (c) $\Omega = 6.2$ rad/s, $\alpha = -27.6$ rad/s^2.
8.13 $T = 12$ s, $A = 0.3$ m.
8.14 $x = 6.4 \sin (4t + 38.7°)$.
8.15 (a) 0.603 m $\angle$ 12.7°; (b) $v = 17.8$ m/s $\angle$ 204.8°; (c) $a = 262$ m/s^2 $\angle$ 153.7°.
8.16 23.1 m/s, 46.2 m.
8.17 74.5 min, 6 km, 54 min.
8.18 (a) 18.75 km/h; (b) 13.4 min.

Chapter 9

9.1 29.62 N; beam balance will indicate 19.62 N.
9.2 $v = \sqrt{[2gs (\sin \theta - \mu \cos \theta)]}$.
9.3 $a = g\,[m_2 - m_1 (\mu \cos \theta + \sin \theta)]/(m_1 + m_2)$.

9.6 $64.3°$.
9.7 (a) $g\sqrt{(4 \sin^2\theta + \cos^2\theta)}$; (b) $\sqrt{(2gR \sin \theta)}$; (c) $3mg \sin \theta$.
9.8 (a) 40.19 m/s; (b) 6.84 s; (c) 175 m.
9.9 $T_{AB} = 4g/3$ N, $T_{BC} = 2g/3\sqrt{3}$ N.
9.10 (a) $\dot{x} = -\Omega a \sin\theta$; $\ddot{x} = -\Omega^2 a \cos\theta$; (b) $N = m(g - \Omega^2 a \cos\theta)$; (c) $\Omega = \sqrt{(g/a)}$.
9.11 (a) $\dot{y} = -4.16$ m/s; (b) $\ddot{x} = -4.13$ m/s^2, $\ddot{y} = -2.37$ m/s^2.
9.13 $\theta = 44.4°$.
9.14 8.68 m/s.
9.15 $t = 0.055$ s, $s = \log_e 4/3$ m.
9.16 $T = 2\pi\sqrt{(R/g)}$.
9.17 (a) 1 m/s; (b) 0.2 m; (c) 0.943 s.
9.18 $\boldsymbol{v} = 10$ m/s $\angle 38.7°$; $\boldsymbol{a} = 18.87 \angle 96.7°$; $P = 1$ N; $N = 1.6$ N; P does not vary.
9.19 $a_1 = -21g/51$, $a_2 = 15g/51$, $T_1 = 72\,mg/51$.

Chapter 10

10.1 6 m/s to the left.
10.2 Body comes instantaneously to rest after 0.64 s, velocity after 5 s is 46.8 m/s to the right.
10.3 (a) $t_2 = 2.94$ s (it comes to rest at this time in its downward motion and instantaneously reverses); (b) $t_3 = 3.88$ s; (c) 24 m/s.
10.4 $v = 6.61$ m/s, 17850 N.
10.5 2 m, 300 N m.
10.6 $m\sqrt{(5gR)}$.
10.7 $0.9R$; it carries on in the same direction with velocity $2\sqrt{(gR/5)}$.
10.8 Taking the direction of the velocity of A before the impact as positive (a) $v_A = -6.57$ m/s, $v_B = +1.99$ m/s; (b) 12.8 N s; (c) 13.7 N m.
10.9 (a) 125 m/s; (b) 41.7 m/s; (c) 55.2 m/s.
10.10 904 m/s, 58.1 N s.
10.11 (a) 10 m/s; (b) 3 m/s; (c) 3.75 m/s, 0.968 m; (d) $v_A = 7.5$ m/s, $v_B = -2.5$ m/s; (e) v_A, 0 to 7.5 m/s, v_B, 10 to -2.5 m/s.
10.16 $(mv^2 \tan^2\phi)/2 + mgR(1 - \cos\phi)$.
10.17 3.
10.18 Taking the velocity of A as positive (a) $v_A = (L_1 + L_2)\sqrt{[Km_2/m_1(m_2 + m_1)]}$, $v_B = -v_A m_1/m_2$; (b) $s_A = s_B = (m_2L_2 - m_1L_1)/(m_1 + m_2)$; (c) $s_A = [m_2(L + L_2) - m_1L_1]/(m_2 + m_1)$, $s_B = [m_2L_2 - m_1(L + L_1)]/(m_2 + m_1)$.
10.21 10 W.

Chapter 11

11.1 ${}_E v_B = 4.35$ m/s $\angle 36.6°$, ${}_E a_B = 10.14$ m/s^2 $\angle 249.5°$.
11.2 $\Omega = 2.38$ rad/s direction indeterminate; $\alpha = 1.66$ rad/s^2 clockwise; 28 m/s^2 $\angle -30.4°$.
11.3 ${}_E v_H = 4.07$ m/s $\angle 0°$; $\Omega_{BDF} = 5.7$ rad/s clockwise.

11.4 (a) ${}_{\mathrm{E}}\boldsymbol{v}_{\mathrm{C}} = 0.68$ m/s $\angle$ 147°; (b) $\Omega_{\mathrm{HJ}} = 1.86$ rad/s clockwise.
11.5 (i) (a) – 5 rad/s, (b) + 7 rad/s; (ii) (a) – 35 rad/s, (b) 1 rad/s; (iii) (a) – 20 rad/s, (b) + 4 rad/s.
11.6 $\Omega_{\mathrm{AC}} = 1.73$ rad/s anticlockwise, ${}_{\mathrm{E}}\boldsymbol{v}_{\mathrm{D}} = 2.75$ m/s $\angle$ 23.7°.
11.7 (a) $\Omega_{\mathrm{BC}} = 5$ rad/s anticlockwise, $\alpha_{\mathrm{BC}} = 43.3$ rad/s^2 clockwise; (b) ${}_{\mathrm{E}}\boldsymbol{a}_{\mathrm{G}} =$ 86.6 m/s^2 $\angle$ 210°.
11.8 (a) ${}_{\mathrm{E}}\boldsymbol{v}_{\mathrm{P}} = 5.7$ m/s $\angle$ 30°; (b) ${}_{\mathrm{E}}\boldsymbol{a}_{\mathrm{P}} = 18$ m/s^2 $\angle$ 210°; (c) ${}_{\mathrm{E}}\boldsymbol{a}_{\mathrm{G}} = 96$ m/s^2 $\angle$ 260°; (d) $\Omega_{\mathrm{CP}} = 2.54$ rad/s clockwise; (e) 28 rad/s^2 clockwise; (f) retarding.
11.9 (a) ${}_{\mathrm{E}}\boldsymbol{v}_{\mathrm{F}} = 3.9$ m/s $\angle$ 174°, ${}_{\mathrm{E}}\boldsymbol{a}_{\mathrm{F}} = 42.5$ m^2 $\angle$ – 8°; (b) $\Omega_{\mathrm{AB}} = 4.58$ rad/s anticlockwise, $\alpha_{\mathrm{AB}} = 28.3$ rad/s^2 anticlockwise.
11.10 (a) ${}_{\mathrm{E}}\boldsymbol{v}_{\mathrm{F}} = 0.96$ m/s $\angle$ 30°; (b) $\Omega_{\mathrm{AB}} = 0.98$ rad/s anticlockwise; (c) $\alpha_{\mathrm{BD}} =$ 24.3 rad/s^2 anticlockwise.
11.11 ${}_{\mathrm{E}}\boldsymbol{a}_{\mathrm{D}} = 10.6$ m/s^2 $\angle$ 0°; sliding acceleration of C relative to the slot is 13.5 m/s^2 $\angle$ 117.3°.
11.12 ${}_{\mathrm{E}}\boldsymbol{v}_{\mathrm{A}} = 2.04$ m/s $\angle$ 0°; ${}_{\mathrm{E}}\boldsymbol{a}_{\mathrm{A}} = 152.4$ m/s^2 $\angle$ 0°.
11.13 ${}_{\mathrm{E}}\boldsymbol{a}_{\mathrm{FG}} = 10.7$ m/s^2 $\angle$ – 90°; $\alpha_{\mathrm{BC}} = 22.8$ rad/s^2 anticlockwise.
11.14 $\Omega_{\mathrm{AB}} = 2.886$ rad/s clockwise; $\alpha_{\mathrm{AB}} = 5.966$ rad/s^2 anticlockwise; ${}_{\mathrm{E}}\boldsymbol{v}_{\mathrm{D}} = 3.82$ m/s $\angle$ 10.9°; ${}_{\mathrm{E}}\boldsymbol{a}_{\mathrm{D}} = 5.31$ m/s^2 $\angle$ 253.6°.
11.15 ${}_{\mathrm{E}}\boldsymbol{v}_{\mathrm{B}} = 2.89$ m/s $\angle$ 0°; ${}_{\mathrm{E}}\boldsymbol{a}_{\mathrm{B}} = 53.9$ m/s^2 $\angle$ 180°.
11.16 $\Omega_{\mathrm{AB}} = 0.88$ rad/s clockwise; $\alpha_{\mathrm{AB}} = 0.052$ rad/s^2 anticlockwise; $\mathrm{d}r/\mathrm{d}t = +3.09$ m/s; $\mathrm{d}^2r/\mathrm{d}t^2 = +7.63$ m/s^2.
11.17 (a) $\Omega_{\mathrm{PQ}} = 17.6$ rad/s clockwise, $\alpha_{\mathrm{PQ}} = 2590$ rad/s^2 clockwise; (b) ${}_{\mathrm{E}}\boldsymbol{v}_{\mathrm{Z}} =$ 9.93 m/s $\angle$ – 45°, ${}_{\mathrm{E}}\boldsymbol{a}_{\mathrm{Z}} = 1004$ m/s^2 $\angle$ – 65.3°; (c) ${}_{\mathrm{E}}\boldsymbol{v}_{\mathrm{Q}} = 12.5$ m/s $\angle$ – 82.5°, ${}_{\mathrm{E}}\boldsymbol{a}_{\mathrm{Q}} = 1680$ m/s^2 $\angle$ 253.7°.
11.18 $\Omega_{\mathrm{AB}} = 2.886$ rad/s clockwise, $\alpha_{\mathrm{AB}} = 4.81$ rad/s^2 anticlockwise; ${}_{\mathrm{E}}\boldsymbol{a}_{\mathrm{G}} = 4.81$ m/s^2 $\angle$ – 90°.

Chapter 12

12.3 $(I_{xx})_{\mathrm{G}} = \rho t R^4(\pi/8 - 8/9\pi)$, $(I_{zz})_{\mathrm{G}} = \rho t R^4(\pi/4 - 8/9\pi)$.
12.4 $I_{yy} = 13000$ kg m^2.
12.5 $I_{xx} = 4050$ kg m^2.
12.6 $I_{zz} = 16611$ kg m^2.
12.7 $I_{z1z1} = I_{z3z3} = 1624\rho$ kg m^2, $I_{z2z2} = 595\rho$ kg m^2.
12.8 $I_{xx} = 47.57$ kg m^2.
12.9 $I_{yy} = 44.25\rho$ kg m^2, $I_{xx} = 56.025\rho$ kg m^2, $I_{y1y1} = 57.75\rho$ kg m^2.
12.10 $(I_{xx})_{\mathrm{G}} = 10.687\rho$ kg m^2, $(I_{zz})_{\mathrm{G}} = 17.062\rho$ kg m^2.
12.11 (a) $I_{\mathrm{axis}} = 8.247$ kg m^2; (b) $I_{\mathrm{dia}} = 6.1$ kg m^2.
12.12 9.675 kg m^2, 85.2%.

Chapter 13

13.1 Body accelerates to the right with $\bar{\boldsymbol{a}} = 100$ m/s^2 $\angle$ 0° and $\alpha = 41.7$ rad/s^2 clockwise.
13.2 $P = mg$, $\bar{a} = g$.
13.3 $P = 1.9mg$.

13.4 (a) $\alpha = 0.000768$ rad/s^2 clockwise; (b) $\Omega = 0.0577$ rad/s (direction indeterminate); (c) $\bar{a}_x = 2.04$ m/s^2 $\angle 0°$, $\bar{a}_y = 2.33$ m/s^2 $\angle -90°$, $T_S = 238000$ N, $T_B = 496000$ N $\angle -70.1°$.

13.6 $\mu = 0.2$.

13.7 (a) $\alpha = -(2g \cos\theta)/L$; (b) $\Omega = \sqrt{[4g(\sin\beta - \sin\theta)/L]}$; (c) $a_x = g\cos\theta(2\sin\beta - 3\sin\theta)$, $a_y = g(2\sin^2\theta - \cos^2\theta - 2\sin\beta\sin\theta)$; (d) $\sqrt{[(4g\sin\beta)/L]}$.

13.8 $a = 12.1$ m/s^2, $P = 1490$ N, $N_E = 85$ N, $N_D = 510$ N.

13.9 $F = 9.26$ N $\angle 0°$.

13.10 (a) 595 N m clockwise; (b) 187.5 N $\angle 120°$.

13.11 $\alpha = (3g\cos\theta)/4$, $\Omega = \sqrt{[(3g\sin\theta)/2]}$, 174 N $\angle 129.3°$.

13.13 (a) 1.93 m/s^2; (b) 1080 N m; (i) 1022 N in the direction of motion; (ii) 57.9 N in the opposite direction to motion.

13.15 $f = 0.249$ Hz, 2.49 rev/s.

13.16 (a) $\alpha = 1.504$ rad/s^2 clockwise; (b) 7.06 N at wall, 11 N at floor.

13.17 2673 N, 16.04 kW.

13.18 24.25 N m anticlockwise; power in = 50 W, power out = 70 W.

13.19 10.65 N m anticlockwise, 30.74 W out.

13.20 $\Omega = 10.93$ rad/s, $\bar{v} = 9.87$ m/s.

13.21 $\Omega_A = +53.6$ rad/s; $\Omega_B = -134$ rad/s.

13.22 $\Omega_A = \Omega_B$, $\Omega_A = \Omega_C + 200$, $\Omega_C = -21.05$ rev/min, $\Omega_A = 179$ rev/min.

13.23 $\Omega = 3I/5ma$.

13.24 $V = 6.3$ m/s, $\alpha = -2.01$ rad/s^2 (that is, opposite to its angular velocity).

13.25 ($k_G{}^2 = 0.0428$ m^2) (a) 5.23 rad/s; (b) 0.284 m/s; (c) 28370 N; (d) 11 890 N.

13.26 $\Omega_{AB} = \Omega_{BC} = 6u/7L$.

13.27 6.85 rad/s.

13.29 (a) 250 N m; (b) 39.8 N m, 18.03 rad/s.

13.30 (a) 140503 N m; (b) 197050 N m.

13.31 $v = 11.083$ m/s, 10908 N m.

13.34 $a = 3\mu g$, $\bar{a} = \mu g$.

13.35 1.19 m.

13.36 Slipping does not take place at P; $\bar{a}_A = 0.477g$, $\bar{a}_B = 0.147g$, $\alpha = 0.11g$ rad/s^2.

13.38 Energy dissipated $= mgs\,\mu\cos\beta(\sin\beta - 3\mu\cos\beta)/(\sin\beta - \mu\cos\beta)$.

13.39 (a) $\Omega_1 = \sqrt{(2M_1 gr)/(I_G + M_1 r^2)}$; (b) $v_2 = \Omega_1 L(I_G + M_1 r^2)/(I_G + M_1 r^2 + M_2 L^2)$; (c) $v_3 = 2v_2$; (d) $s = 4M_1 L^2 r(I_G + M_1 r^2)/\mu(I_G + M_1 r^2 + M_2 L^2)^2$.

13.40 (b) 0.0122 m.

Index